I0066564

Agronomy: Science and Technology of Plants

Volume I

Agronomy: Science and Technology of Plants
Volume I

Edited by **Jamie Hanks**

R CALLISTO
REFERENCE

New York

Published by Callisto Reference,
106 Park Avenue, Suite 200,
New York, NY 10016, USA
www.callistoreference.com

Agronomy: Science and Technology of Plants
Volume I
Edited by Jamie Hanks

© 2015 Callisto Reference

International Standard Book Number: 978-1-63239-061-5 (Hardback)

This book contains information obtained from authentic and highly regarded sources. Copyright for all individual chapters remain with the respective authors as indicated. A wide variety of references are listed. Permission and sources are indicated; for detailed attributions, please refer to the permissions page. Reasonable efforts have been made to publish reliable data and information, but the authors, editors and publisher cannot assume any responsibility for the validity of all materials or the consequences of their use.

The publisher's policy is to use permanent paper from mills that operate a sustainable forestry policy. Furthermore, the publisher ensures that the text paper and cover boards used have met acceptable environmental accreditation standards.

Trademark Notice: Registered trademark of products or corporate names are used only for explanation and identification without intent to infringe.

Printed in the United States of America.

Contents

Preface

Agronomy has seen rapid progress in recent years. The experimental researches in this field have presented some thought provoking results which can revolutionize the way we grow our yield. This field is concerned with improving the agricultural practices by drawing out methods that would increase the productivity by providing favourable conditions to the crops. Agronomic activities like restoring soil fertility, preparation of good seed bed, improving crop varieties, and managing soil resources help in getting a better yield and good quality of produce. Agronomy also takes care of crop rotation, irrigation, weed control, insecticides and pesticides and plant breeding. It also uses biotechnology to achieve its objectives, like introducing desired characteristics to a crop or eliminating undesired traits. It is an approach that will change the face of agriculture in the world.

The book is the collective effort of scientists and researchers. I would like to thank my publisher for giving me this incredible opportunity to broaden the horizon of agronomy.

Editor

Morphological Variability of Wild Cardoon (*Cynara cardunculus* L. var. *sylvestris*) Populations in North of Tunisia

Imen Ben Ammar,[1] Fethia Harzallah-Skhiri,[2] and Bouthaina Al Mohandes Dridi[1]

[1] *High Institute of Agronomy of Chott-Mariem, University of Sousse, 4042 Chott Mariem, Tunisia*
[2] *High Institute of Biotechnology of Monastir, University of Monastir, Tunisia*

Correspondence should be addressed to Imen Ben Ammar; i_benammar@yahoo.fr

Academic Editors: O. Merah and K. L. Sahrawat

In north of Tunisia, wild cardoon (*Cynara cardunculus* L. var. *sylvestris* (Lamk) Fiori) is called "khurshef." It is consumed mainly for its fleshy stems and leafstalks in some traditional dishes. In some regions, heads were used to prepare cheese. North Tunisian germplasm has been currently damaged by severe genetic erosion, pollution, urbanization, and bad farming practices. In order to preserve this species and to assess morphological relationship between accessions, the present study aims to prospect and to characterize individuals in several areas of the north of Tunisia. Six populations were collected and then 20 individuals per population were evaluated using UPOV (International Union for the Protection of New Varieties of Plant) descriptors related to leaves, leafstalks, and heads. Multivariate analyses were used to elucidate relationship among the studied populations. Principal components analysis revealed more diversity within each population. Cluster study reveals large variability among populations. This analysis allows classifying the germplasm of wild cardoon into five groups. Similarities observed between ecotypes despite their distinctiveness of geographic origin suggest a narrow genetic base. These analyses are very useful for the management and the use of wild cardoon in future breeding programs for *Cynara* germplasm.

1. Introduction

Cynara cardunculus L. (Asteraceae), commonly named "cardoon," is widespread in the Mediterranean area [1]. It comprises two botanical varieties: *C. cardunculus* L. var. *altilis* DC (domestic cardoon) and *C. cardunculus* L. var. *sylvestris* (Lamk) Fiori (wild cardoon), considered to be the wild ancestor of globe artichoke [2, 3]. Two gene pools can be distinguished within wild cardoon: the eastern Mediterranean type, mainly distributed in Italy, Greece, and Tunisia, and the western gene pool, diffused in the Iberian Peninsula. It is a nondomesticated robust perennial plant characterized by a rosette of large spiny leaves and branched flowering stems [4]. The main differences between cultivated and wild cardoon are the larger production potential of the former and the distribution of the assimilates between shoot and root, with the wild type investing more carbohydrates in the roots providing more resistance to adverse climatic conditions [5]. The wild cardoon is considered the presumed wild progenitor of the artichoke and the cultivated cardoon [2–4, 6]. These three taxa are fully interfertile [3, 7]. Recent studies have suggested that a high level of differentiation is present in the wild cardoon gene pool and that samples from the western Mediterranean range more closely resemble the cultivated cardoon than the wild samples from the eastern Mediterranean one [4, 8]. Assessment of genetic diversity and determining the relationship between ecotypes allow the management of the germplasm and may increase the efficiency of efforts to improve species. Thus, morphological characterization is the first step in the description and classification of germplasm [9]. This study aims to investigate genetic diversity in several Tunisian populations of wild cardoon collected from north Tunisia as inferred by variations in their morphology as a necessary step for the best conservation of wild cardoon gene pool and for future improvement of crop characters. Moreover, it is likely that a source of resistance to *Verticillium* has been found in wild material.

TABLE 1: List of the six populations of wild cardoon included in this study.

Government	Locality	Code
Bizerte	Ain Berda 1	AB1
	Ain Berda 2	AB2
	Beni Amor	BA
Beja	Dhahirat	DH
	Daouar Mahjouba	DM
Siliana	Ain Dissa	SN

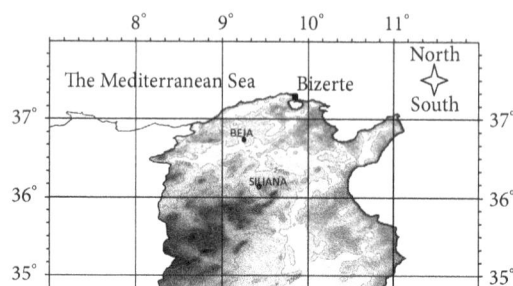

FIGURE 1: Localities of the six populations.

2. Materials and Methods

2.1. Plant Material and Study Areas. This study was conducted in 6 localities (Ain Berda 1, Ain Berda 2, Beni Amor, Daouar Mahjouba, Dhahirat, and Ain Dissa) from 3 governorates (Bizerte, Beja, and Siliana) located in the north of Tunisia (Figure 1). Six populations of wild cardoon (AB1, AB2, BA, DH, DM, and SN) (Table 1) were assessed. Twenty samples per population were collected. For each one, samples were coded from 1 to 20.

2.2. Morphological Traits. At flowering stage, simple random sampling method was followed for collecting the individuals. Forty-three morphological characters were recorded. The morphological traits (Table 2) were evaluated based on the UPOV (International Union for the Protection of New Varieties of Plant) descriptor list for artichoke (since there are no descriptors edited for cardoon). Several specific botanical characters for cardoon were chosen as descriptors. Twenty-five are qualitative traits and eighteen are quantitative traits. The quantitative traits were plant height, the number of leaves per plant, the number of lateral shoots, the number of heads, the length of leafstalk, the width of leafstalk, the depth of leafstalk, the thickness of leafstalk, the length and width of leaf, and the length of spines. The qualitative traits were depth of leaf lobation (determined visually following this scale: from very shallow, shallow, medium, deep, and very deep); leaf and leafstalk color (visually measured); anthocyanin at leafstalk base (visually measured following this scale: no anthocyanin, anthocyanin present in less than 50% of the leafstalk, anthocyanin present in 50% of the leafstalk, and anthocyanin present in more than 50% of the leafstalk); and leafstalk texture (visually measured following this scale: hollow leafstalk, 50% of the leafstalk hollow, and hard and

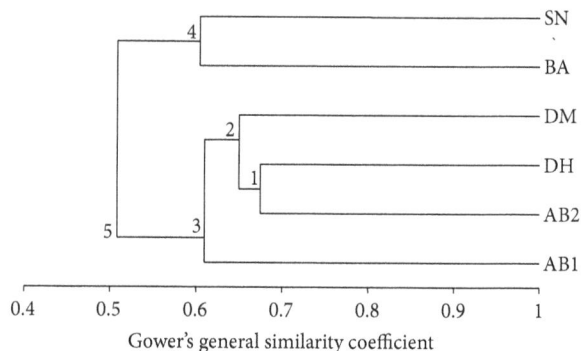

FIGURE 2: UPGMA dendrogram (based on Gower's coefficient of similarity) revealing morphological diversity among the six populations collected from different regions of north Tunisia.

full leafstalk). For some traits, a discrete classification was adopted. This was particularly important for germplasm characterization. Furthermore, all the characteristics concerning colour are subjective and are often affected by climatic and soil conditions.

2.3. Statistical Analysis. For all data obtained from morphological study, cluster analysis was made using the MVSP (Multivariate Statistical Package for Windows 3.1) and was carried out by UPGMA method. Two trees illustrating the genetic distance among populations (Figure 2) and within ecotypes (Figure 3) were constructed based on Gower's general similarity coefficient [10] for morphological data. Quantitative and qualitative data on morphological traits were subjected to a principal coordinate analysis (PCA) using MVSP statistical software. The first two principal coordinates were used to produce a two-dimensional scatter plot to understand how each axis influenced the variation among groups and whose morphological characteristics have stronger discriminating power.

3. Results

3.1. Cluster Analysis. Morphological characterization revealed high polymorphism among six populations. The dendrogram based on Gower's general similarity coefficient clustered populations into five groups (Figure 2). The first cluster gathered the two populations: Dhahirat (DH) and Ain Berda 2 (AB2) at 67% of similarity. The population Ain Berda 1 (AB1) was the most divergent from other populations (Ain Berda 2 (AB2), Dhahirat (DH), and Daouar Mahjouba (DM)) in cluster 3 ($d = 0.61$). Cluster 4, constituted with populations Siliana Nord (SN) and Beni Amor (BA), presents with cluster 3 (AB1, AB2, DH, and DM) the most level of divergence ($d = 0.50$). Figure 3 presented the dendrogram based Gower's general similarity coefficient which illustrates diversity within accessions. It shows discrimination of ecotypes evaluated on morphological traits into two main groups at 70% of similarity. The first group GI is divided into two subgroups (a) and (b). The subgroup (a) includes 6 subsamples of accessions DM (Daouar Mahjouba) and 1 of DH (Dhahirat). The subgroup

TABLE 2: Morphological traits recorded in the wild cardoon ecotypes.

Parameters number	Code	Parameters and descriptive value
Character of plant		
1	H	Height of plant (cm)
2	Nlf/p	Number of leafs/plant
3	Nlsh	Number of lateral shoots on main stem
4	Nh	Number of heads
5	Hp	Heads position (1: terminal/2: lateral)
6	H	Habit (1: erect/2: semierect/3: horizontal)
Characters of leafstalk		
7	Lls	Length of leafstalk (cm)
8	Wls	Width of leafstalk (cm)
9	Dls	Depth of leafstalk (cm)
10	Thls	Thickness of leafstalk (cm)
11	Tex	Texture of leafstalk (1: hollow/2: medium/3: strong)
12	Ca	Colour of anthocyanin at leafstalk base (1: absent/2: sparse/3: medium/4: strong)
13	Cls	Colour of leafstalk (1: whitish; 2: greenish white; 3: whitish green; 4: light green; 5: green; 6: green ash; 7: grayish green)
14	Hls	Hairs on leafstalk (1: absent/9: present)
15	Lsls	Length of spines at leafstalk (cm)
Characters of leaf		
16	Llf	Length of leaf (cm)
17	Clf	Leaf colour (1: yellowish green; 2: light green; 3: greenish; 4: dark green; 5: green ash; 6: grayish green)
18	Icv	Intensity of green colour (1: light/2: medium/3: dark)
19	HGC	Hue of green colour (1: absent/2: yellowish/3: grayish)
20	Igh	Intensity of grey hue (1: weak/2: medium/3: strong)
21	Hlf	Hairs on the upper side of the leaf (1: absent/9: present)
22	Lslf	Length of spines at leaf (cm)
Characters of lobations		
23	Shlb	Shape of lobes (1: acute/2: nearly right angle/3: obtuse)
24	Nlb	Number of lobes (3: few/5: medium/7: many)
25	Llb	Length of the longest lobe (cm)
26	Wlb	Width of the longest lobe (cm)
27	Nlbs	Number of the secondary lobes (1: none or very few/3: few/5: medium/7: many/9: very many)
28	Slbs	Shape of the secondary lobes (1: acuminate/2: acute/3: rounded)
29	Lslb	Length of spines at lobes (cm)
Characters of head		
30	HD	Diameter of head (cm)
31	HS	Size of head (1: small/2: medium/3: large)
32	Sc	Shape of central head (1: circular/2: broad elliptic/3: ovate/4: triangular/5: transverse broad elliptic)
33	Hc	Colour of central head flower (1: violet; 2: purplish blue; 3: white)
34	Ta	Time of appearance (1: early/2: medium/3: late)
35	Drec	Diameter of receptacle (cm)
36	Trec	Thickness of receptacle (cm)
37	Srec	Shape of receptacle (1: flat/slightly depressed/strongly depressed)
38	D	Density of inner bracts (1: sparse/2: medium/3: dense)
Characters of outer bract		
39	Sob	Shape of outer bract (1: broader than long/2: as broader as long/3: longer than broad)
40	Saob	Shape of apex of outer bract (1: acute/2: flat/3: emarginate)
41	Cob	Colour of outer bract (1: green/2: green striped with violet/3: violet striped with green/4: mainly violet/5: entirely violet)
42	Tob	Thickness at base of outer bract (3: thin/5: medium/7: thick)
43	Lsob	Length of spines of outer bract (cm)

TABLE 3: Means, maximal and minimal values of the recorded quantitative traits of subgroups obtained after classification.

Traits*		a**	b**	c**	d1**	d2**	d3**	d4**	d5**	d6**
H	Mean	84.67 ± 5.2	85.39 ± 32.6	88.86 ± 25.4	67.51 ± 11.2	52.17 ± 10.4	71.06 ± 17	77.57 ± 20	58.35 ± 15.1	43.92 ± 22.3
	Max.	93	162	118	52.5	74.2	98.5	114	87.9	92.2
	Min.	77	36	71	20	34	35	36	35.2	23
Nlf/p	Mean	20 ± 6.4	25.83 ± 5.3	21.66 ± 7.4	21.4 ± 5.2	16.88 ± 4.2	16.03 ± 4.6	18.73 ± 5.6	16.1 ± 3.6	20.62 ± 9.2
	Max.	32	35	30	31	26	25	33	24	36
	Min.	14	18	16	11	9	9	10	12	9
Nlsh	Mean	5.71 ± 2.4	5 ± 1.3	4 ± 0	2.72 ± 0.6	2.27 ± 1.4	2.65 ± 1.3	2.63 ± 0.7	2.9 ± 1.9	2 ± 0.5
	Max.	11	7	4	4	6	7	4	8	3
	Min.	4	3	4	2	1	1	2	2	1
Nh	Mean	11.14 ± 5	14.94 ± 4.9	6.33 ± 4.7	9.83 ± 2.7	3.55 ± 1.6	4.84 ± 2.8	5.68 ± 3.3	5.4 ± 3.4	3.5 ± 1.8
	Max.	21	25	10	12	7	13	15	15	7
	Min.	7	7	1	2	2	2	2	3	2
Llf	Mean	41.01 ± 9	40.49 ± 4	35.72 ± 13	45.77 ± 5.4	33.79 ± 10.8	34.99 ± 7.6	33.03 ± 4.4	40.38 ± 4.5	27.42 ± 5.1
	Max.	53.10	49.97	45.67	39	49.7	52.3	41.43	47.8	38.67
	Min.	28.79	32.93	21.33	19.7	19.6	21.2	25.47	34.3	20.7
Lslf	Mean	0.81 ± 0.3	0.77 ± 0.2	0.58 ± 0.6	0.83 ± 0.1	0.49 ± 0.2	0.66 ± 0.2	0.65 ± 0.1	0.76 ± 0.4	0.44 ± 0.1
	Max.	1.4	1.24	1.26	0.74	0.96	1.28	0.9	1.4	0.7
	Min.	0.5	0.4	0	0.3	0	0.4	0.4	0	0.3
Lls	Mean	65.7 ± 14.0	63.68 ± 14	58.03 ± 15.1	59.71 ± 5.3	42.28 ± 8.5	58.33 ± 14.3	51.42 ± 8.6	47.38 ± 8.8	31.62 ± 12
	Max.	89	99.4	69	43	54.7	77.2	75.2	60	60.2
	Min.	46.7	33.6	40.8	25.9	28.8	16.2	36.1	31.2	22
Wls	Mean	1.83 ± 0.4	2.04 ± 0.5	1.46 ± 0.5	2.53 ± 0.4	1.43 ± 0.3	1.26 ± 0.3	1.37 ± 0.4	1.34 ± 0.4	1.25 ± 0.2
	Max.	2.2	3.35	1.75	2.2	2.1	1.9	2.24	2.04	1.42
	Min.	0.9	1.6	0.9	1	0.9	0.66	0.71	0.63	0.96
Dls	Mean	1.08 ± 0.5	1.75 ± 0.3	1.41 ± 0.7	1.75 ± 0.2	0.98 ± 0.2	1.08 ± 0.2	1.29 ± 0.4	0.89 ± 0.2	0.92 ± 0.2
	Max.	1.6	2.51	2.2	1.35	1.5	1.75	1.9	1.18	1.2
	Min.	0.52	1.17	0.94	0.7	0.58	0.78	0.5	0.6	0.6
Thls	Mean	0.62 ± 0.19	0.80 ± 0.17	0.63 ± 0.19	1.11 ± 0.14	0.56 ± 0.13	0.49 ± 0.12	0.57 ± 0.18	0.57 ± 0.2	0.56 ± 0.18
	Max.	0.83	1.24	0.85	0.95	0.77	0.8	0.91	0.91	0.8
	Min.	0.4	0.52	0.46	0.44	0.3	0.3	0.3	0.26	0.3
Lsls	Mean	1.48 ± 0.2	1.65 ± 0.2	0.77 ± 0.2	2.17 ± 0.1	1.33 ± 0.1	1.07 ± 0.1	1.33 ± 0.2	1.38 ± 0.2	1.24 ± 0.2
	Max.	2.2	2.25	1.5	1.4	2.09	1.48	2.1	3.2	1.9
	Min.	1.15	0.87	0	0.74	0.5	0.76	0.8	0.24	0.6
Llb	Mean	13 ± 2.7	12.16 ± 2.3	13.2 ± 4	13.73 ± 1.5	12.29 ± 3.3	11.20 ± 2.1	10.64 ± 1.8	15.58 ± 2.4	10.08 ± 1.9
	Max.	15	16.3	16.7	10.2	18	15	14	19.7	13
	Min.	7.8	8.2	8.9	5.6	6.7	7.8	7.1	11.4	6.6
Wlb	Mean	1.95 ± 0.8	2.31 ± 0.9	3.16 ± 1.2	2.3 ± 0.5	2.13 ± 0.6	2.27 ± 0.5	2.25 ± 0.4	2.93 ± 0.5	2.1 ± 0.4
	Max.	3.1	4.3	4.3	2.1	3.2	3.7	3.2	3.6	2.6
	Min.	0.7	0.8	2	0.7	0.7	1.4	1.7	2	1.4
Lslb	Mean	1.10 ± 0.2	1.20 ± 0.3	0.81 ± 0.7	0.93 ± 0.2	0.69 ± 0.4	0.96 ± 0.3	1.02 ± 0.2	0.76 ± 0.4	0.95 ± 0.3
	Max.	1.4	1.9	1.4	0.8	1.73	1.7	1.5	1.24	1.44
	Min.	0.8	0.6	0	0.24	0	0.45	0.65	0	0.4
Nlbs	Mean	4.14 ± 1.6	4.22 ± 1.7	1.66 ± 1.2	6.83 ± 1.3	4.22 ± 1.6	4 ± 1.7	4.57 ± 1.3	2.6 ± 1.6	2.5 ± 1.4
	Max.	7	7	3	7	7	7	7	5	5
	Min.	3	1	1	3	1	1	3	1	1
HD	Mean	6.45 ± 0.4	5.47 ± 0.5	4.14 ± 0.4	7.73 ± 0.9	3.05 ± 0.7	3.98 ± 0.8	4.84 ± 0.6	3.04 ± 0.4	2.95 ± 0.5
	Max.	6.8	7	4.5	5	4.14	5.5	6.04	3.66	3.67
	Min.	5.74	5	3.71	1.73	1.94	2.62	3.85	2.6	2

TABLE 3: Continued.

Traits*		a**	b**	c**	d1**	d2**	d3**	d4**	d5**	d6**
Drec	Mean	6.03 ± 0.6	4.36 ± 1.1	3.11 ± 0.4	6.58 ± 0.3	2.30 ± 0.7	3.41 ± 0.9	4 ± 0.6	2.13 ± 0.5	2.08 ± 0.5
	Max.	6.9	5.81	3.5	3.9	4.16	5.76	5.37	2.84	2.8
	Min.	5.36	0.37	2.75	2.8	1.03	2.25	2.82	1.41	1.53
Trec	Mean	2.16 ± 0.3	1.58 ± 0.2	1.27 ± 0.2	2.37 ± 0.1	0.96 ± 0.2	1.16 ± 0.3	1.44 ± 0.3	1.01 ± 0.2	1.05 ± 0.2
	Max.	2.7	1.94	1.4	1.55	1.24	1.9	2.25	1.37	1.4
	Min.	1.75	1.2	1.04	1.1	0.6	0.7	1	0.8	0.85
Lsob	Mean	0.85 ± 0.2	1.96 ± 0.3	0.69 ± 0.3	1.89 ± 0.3	0.75 ± 0.3	0.91 ± 0.3	0.82 ± 0.2	0.67 ± 0.3	1.07 ± 0.4
	Max.	1.27	1.81	0.94	1.62	1.15	1.61	1.17	1	1.9
	Min.	0.55	0.75	0.35	0.53	0.13	0.49	0.6	0	0.61

*See Table 2; **see Figure 3.

(b) gathered 13 subsamples of AB1 (Ain Berda 1), 3 of AB2 (Ain Berda 2), and 2 of DH (Dhahirat). The major group GII is subdivided into 7 subgroups (c), (d1), (d2), (d3), (d4), (d5), and (d6). Subgroup (c) comprises only 3 subsamples. The major subgroup is (d3) that assembles 26 subsamples of DH (Dhahirat), DM (Daouar Mahjouba), AB1 (Ain Berda 1), and AB2 (Ain Berda 2). Group (c) presents the highest groups (88.86 cm); however, the highest ecotypes belong to group (b) with 162 cm. Table 3 shows the main quantitative data of several groups obtained by classification with the dendrogram presented in Figure 3. Group (b) presented the group of the highest number of leaves/plant (39) with also the maximal value (35 leaves/plant) and a good potential to produce lateral shoots (5 lateral shoots). Group (a) included the cluster of ecotypes with high number of lateral shoots (5.71) with also the maximal value (11 lateral shoots for DM13). On basis of the length of spines on different organs, it can be noticed that groups representing the group of ecotypes with the lowest length of spine, respectively, are (d2) on leaf (0.44 cm), (c) on leafstalk (0.77 cm), (d2) on lobation (0.69 cm), and (d1) on outer bracts (0.67 cm). Figure 2 illustrates also the genetic distance among the studied ecotypes of wild cardoon gathered into several groups. It is noticed that the nearest distance (d = 0.90) was exhibited among the ecotypes AB14 and AB210 representing the highest similarity, while the farthest genetic distance (0.69) was exhibited between AB11 and (AB18, AB15, AB113, and AB12) subgroup on group (b) from the same accession and the same area. The cluster showed also a high diversity within DH14, DH5, and SN5 (0.70), knowing that DH and SN are geographically very distant but clustered in the same subgroup.

3.2. Principal Component Analysis. Principal component analysis (PCA) was performed taking into account all parameters. The eigenvalue obtained by PCA indicates that the first two PCA axes described 28.89% of the total of variance (Table 4). The first axis accounts for factor 19.75% of the variability. It is correlated with the following traits: H, Nlf/p (number of leaves per plant), Nlsh (number of lateral shoots), Nh (number of heads), Lslf, Lls, Ils, Dls (depth of leafstalk), Thls (thickness of leafstalk), Lsls (length of spines of leafstalk), Lslb (length of spines of lobation), HD (diameter of head), HS (size of head), DO (degree of opening), Drec (diameter of receptacle), Trec (thickness of receptacle), Tob (thickness of outer bract), and Lsob (length of spines on outer bracts) (Figure 4).

The second axis represents 9.14% of the variance; it represents Hp (head position), Llf (length of leaf), Cls (colour of leafstalk), Llb (length of lobation), Wlb (width of lobation), Ta (time of appearance), and Srec (shape of receptacle) (Figure 4).

4. Discussion and Conclusion

The main objective of this study was to analyze morphological variation among and within several accessions of wild cardoon from the north of Tunisia. On one hand, morphological analysis based on different characters allowed clustering the six populations into five groups. On the other hand, 120 ecotypes were assessed and split into two groups demonstrated by the topology of the dendrogram based on Gower's general similarity coefficient. This study allows grouping ecotypes of several accessions into one cluster in spite of differences between the origins and the environmental conditions. There is no association between clusters with geographic location of populations; this reflects a genetic basis of the plant form in wild cardoon. In agreement with [4, 7], this discrimination may be based on the hypothesis that wild cardoon gene pool is differentiated according to its geographical distribution. In addition, on the basis of the geographical isolation, Lanerti et al. [11] identified two distinct gene pools, the Sardinian and Sicilian populations, which are clearly differentiated. Moreover, the plant width is an indicator of plant's growth habit. Indeed, most of Tunisian ecotypes have a semierect and horizontal habit. This descriptor is important in terms of crop management because it can help in terms of defining the area of each plant, harvesting, and feasibility of farming practices in the breeding program. The influence of environmental conditions is strongly correlated with quantitative traits but not with the qualitative ones, particularly the shoot number and leafstalk texture [12]. This is not in agreement with Lanerti et al. [11] study that proved that the pigmentation

TABLE 4: Definition of the first two components of PCA made on the basis of morphological traits of wild cardoon accessions.

Principal components	PCA1	PCA2
% variance	19.75	9.14
Cumulative %	19.75	28.89
Traits*	Eigenvalue	
H	0.213	−0.116
Nf	0.071	0.236
Nc	0.177	−0.073
Npl	0.195	0.044
Pc	−0.010	−0.014
Pt	−0.156	0.173
Long f	0.068	−0.260
Léf	0.125	−0.165
Cf	−0.171	0.064
Icv	−0.185	−0.184
Tcv	0.124	0.174
Icg	0.080	0.223
Plf	−0.095	0.189
Longp	0.175	−0.217
Larp	0.191	0.075
Pp	0.207	0.066
Ep	0.057	0.153
Lép	0.097	0.013
Text	−0.101	0.259
Ca	−0.041	0.197
Cp	−0.171	0.128
Plp	0.034	0.293
Fl	0.000	0.000
Nl	0.011	−0.135
Longl	−0.043	−0.254
Largl	−0.011	−0.236
Lél	0.218	−0.121
Nls	0.212	−0.035
Fls	−0.217	−0.053
D	0.229	0.016
Tc	0.226	−0.021
Cc	0.196	0.081
Ea	−0.221	−0.029
Dr	0.208	−0.035
Er	0.213	0.045
Fr	−0.121	−0.249
Ds	0.234	−0.063
Fb	0.136	−0.010
Fa	0.092	0.086
Cb	0.001	0.199
Ep	0.188	0.021
Léb	0.098	0.232

*See Table 2.

intensity (explained by the anthocyanin color at the base of leafstalk in our study) is known to be very sensitive to temperature. Our results suggest that man's traits selection is significant to understand role in the variation within wild cardoon accessions. For example, thorn length is a main character. So it seems interesting to select ecotypes with short thorn in order to use them in breeding programs. Indeed, we found some ecotypes in groups (c), (d2), and (d1) without spines on leaf and on lobation. Other ecotypes, one in group (c) and one in group (d5), are without spines, respectively, on leafstalk and on outer bract. The results that have shown the good potential of some ecotypes to produce biomass (number of leafs and lateral shoots) may be explained by the particular adaptability of the *Cynara* genus into the Mediterranean environment. This agrees with the study of Foti et al. [13]. The genetic diversity may be related to natural hybridization and fluctuations in environmental conditions. Plants might respond to their environments through developmental plasticity in many aspects of their phenotypes [14]. Some groups, such as (a) and (b), showed the highest number of heads (flowers). This could be attributed to the natural habitats, which may exert some stress from the surrounding environments. Consequently, these ecotypes tend to produce more flowers as an attempt to maintain their existence and to ensure the reproduction under unpredictable conditions. Sultan [15] proved that plants increased their reproductive output when they grow in rich resource conditions. Archontoulis et al. [16] proved that, under water stress, the light and nitrogen distributions are more complicated because water stress affects not only appearance and elongation of leaves and uptake and partitioning of nitrogen, but also morphological aspects of leaf positioning, leaf angle, and azimuth angle.

Moreover, in 2000, Sultan [17] reported that plants might respond to environmental conditions not only by altering their offspring through changes in the quantity and quality of seed provisioning. Notable differences in the obtained values of leaves number (from 9 to 36 leaves) and leaf dimensions (from 19.9 cm to 53.10 cm) were found. It may be concluded that variability in leaf traits seems to be an environmentally affected character. The above finding is in agreement with the study of Balaguer et al. [18] who reported that the variability in leaf size affects the ability of plants to capture light and therefore compete with neighbors. *Cynara* is a rather complex crop in terms of morphology and growth. Moreover, studies concerning temperature and photoperiod effects on the development rate of *Cynara* are lacking in the literature, however, irrespective of the rate of development that varies among regions and varieties. Under Mediterranean climate, the biomass yield of this crop is very variable and depends on the variation in air temperature [19].

The morphological characterization is the first step in the description and the classification of germplasm [9]. According to Greene et al. [20], plant breeders can use genetic similarity information to complement phenotypic information in the development of breeding populations. Morphological markers have been traditionally used to highlight differences among cultivars. However, this type of characterization does not always reflect the real genetic variation, because the phenotype is determined by genetic

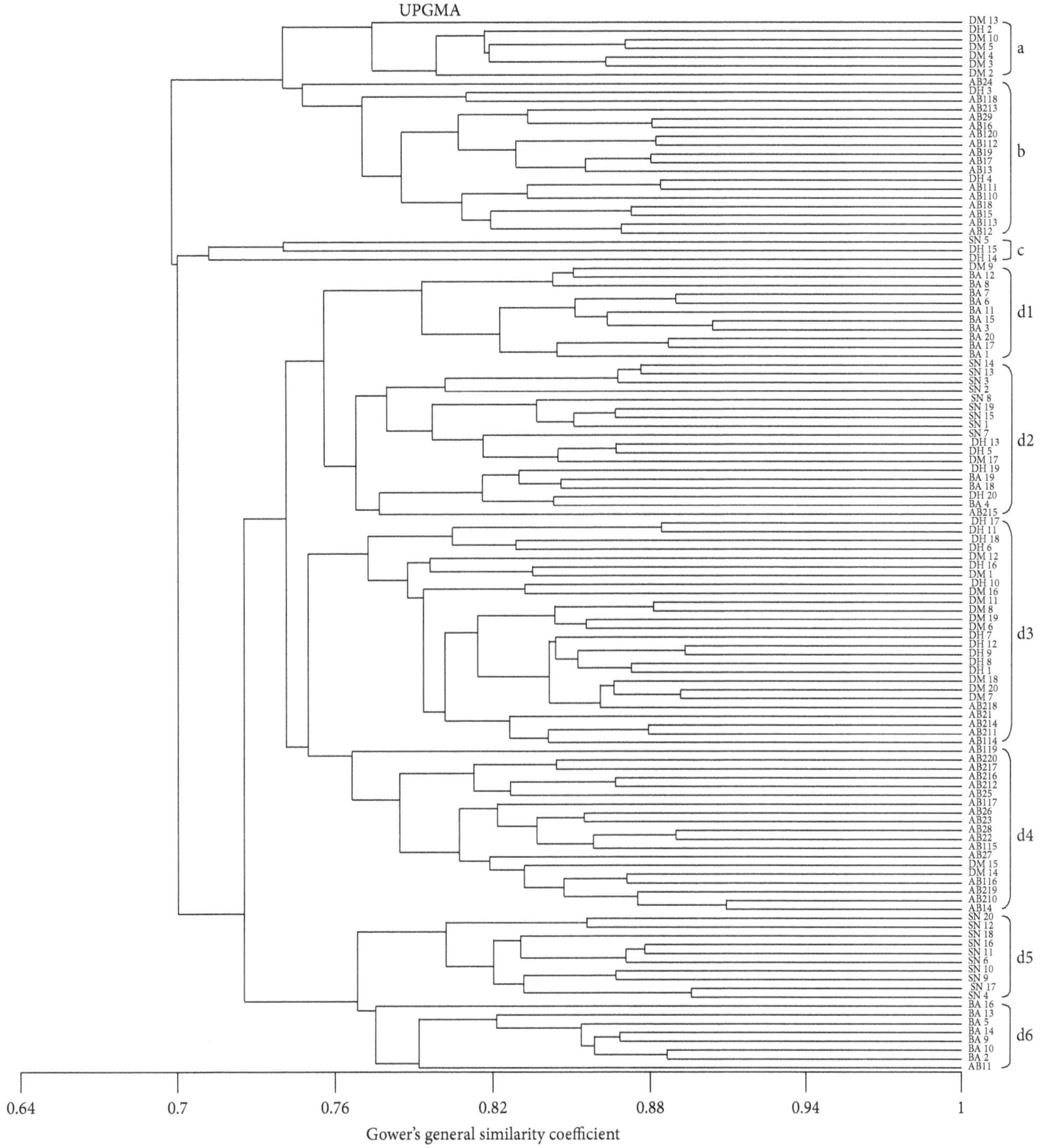

FIGURE 3: UPGMA dendrogram (based on Gower's coefficient of similarity) of the 120 ecotypes presenting morphological variability within the six populations.

information of the individual and also the results of its interaction with the environment; thus, in many cases a trait may be the expression of phenotypic plasticity [21].

The cluster analysis based on morphological traits showed an important diversity in the germplasm of north Tunisian wild cardoon. The traits used in this study can provide reliable information on the variability in wild cardoon populations when we reduce the number of descriptors without decreasing the discrimination between populations. Similarities observed among several ecotypes and some populations, despite their distinctiveness of geographic origin, suggest a narrow genetic base. Cluster analysis, obtained with characterization, will be utilized to choose ecotypes from within clusters to be then considered for inclusion in the core collection. Thus, it is very interesting to manage these genetic resources by establishment, for example, of *ex*

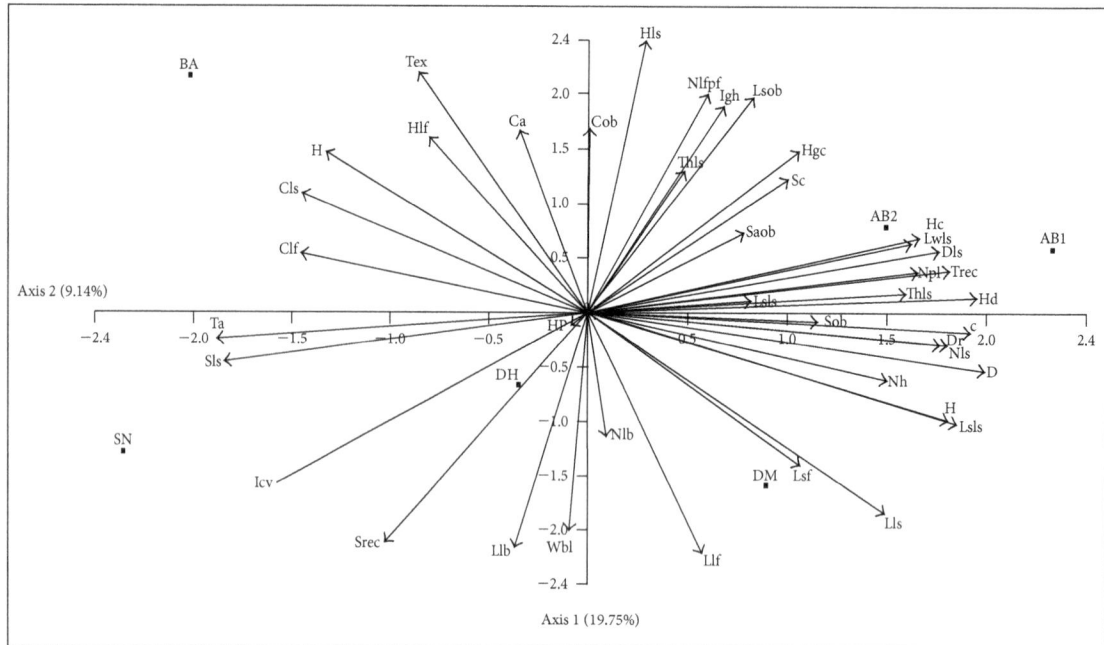

FIGURE 4: PCA scatter plot of first two principal components detecting the relationship among and within 120 ecotypes of six populations of Tunisian wild cardoon based on morphological traits.

situ collection. Therefore, it is necessary to accomplish this work with estimating the genetic diversity of the accessions understudied by using the molecular and biochemical characterization.

Conflict of Interests

The authors declare that there is no conflict of interests regarding the publication of this paper.

References

[1] G. Pottier-Alapetite, *Flora of Tunisia*, Imprimerie Officielle de la République Tunisienne, 1979.

[2] C. Foury, "Ressources génétiques et diversification de l'artichaut (*Cynara scolymus* L.)," *Acta Horticulturae*, vol. 242, pp. 155–166, 1989.

[3] A. Rottenberg and D. Zohary, "The wild ancestry of the cultivated artichoke," *Genetic Resources and Crop Evolution*, vol. 43, no. 1, pp. 53–58, 1996.

[4] A. Wiklund, "The genus *Cynara* L. (Asteraceae-Cardueae)," *Botanical Journal of the Linnean Society*, vol. 109, no. 1, pp. 75–123, 1992.

[5] S. A. Raccuia and M. G. Melilli, "Genetic variation for assimilate accumulation and translocation in *Cynara* spp," *Acta Horticulturae*, vol. 660, pp. 241–248, 2004.

[6] A. Rottenberg, D. Zohary, and E. Nevo, "Isozyme relationships between cultivated artichoke and the wild relatives," *Genetic Resources and Crop Evolution*, vol. 43, no. 1, pp. 59–62, 1996.

[7] J. Basnizki and D. Zohary, "Breeding of seed planted artichoke," in *Plant Breeding Reviews*, J. Janick, Ed., vol. 12, pp. 253–269, 1994.

[8] G. Sonnante, A. V. Carluccio, R. Vilatersana, and D. Pignone, "On the origin of artichoke and cardoon from the *Cynara* gene pool as revealed by rDNA sequence variation," *Genetic Resources and Crop Evolution*, vol. 54, no. 3, pp. 483–495, 2007.

[9] J. S. C. Smith and O. S. Smith, "The description and assessment of distances between inbred lines of maize: the utility of morphological, biochemical and genetic descriptors and a scheme for the testing of distinctiveness between inbred lines," *Maydica*, vol. 34, pp. 151–161, 1989.

[10] J. C. Gower, "A genral coefficient of similarity and some of its properties," *Biometrics*, vol. 27, no. 4, pp. 857–871, 1971.

[11] S. Lanerti, L. Ledda, M. G. Mameli, I. di Leo, and E. Portis, "Molecular and morphological variation among and within populations of *Cynara scolymus* L. cv. 'Spinoso sardo'," *Acta Horticulturae*, vol. 681, pp. 333–340, 2005.

[12] I. Lahoz, J. A. Fernández, D. Migliaro, J. I. Macua, and C. Egea-Gilabert, "Using molecular markers, nutritional traits and field performance data to characterize cultivated cardoon germplasm resources," *Scientia Horticulturae*, vol. 127, no. 3, pp. 188–197, 2011.

[13] S. Foti, G. Mauromicale, S. A. Raccuia, B. Fallico, F. Fanella, and E. Maccarone, "Possible alternative utilization of *Cynara* spp. I. Biomass, grain yield and chemical composition of grain," *Industrial Crops and Products*, vol. 10, no. 3, pp. 219–228, 1999.

[14] M. Fernández-López, S. Goormachtig, M. Gao, W. D'Haeze, M. van Montagu, and M. Holsters, "Ethylene-mediated phenotypic plasticity in root nodule development on Sesbania rostrata," *Proceedings of the National Academy of Sciences of the United States of America*, vol. 95, no. 21, pp. 12724–12728, 1998.

[15] S. E. Sultan, "Phenotypic plasticity for fitness components in Polygonum species of contrasting ecological breadth," *Ecology*, vol. 82, no. 2, pp. 328–343, 2001.

[16] S. V. Archontoulis, J. Vos, X. Yin, L. Bastiaans, N. G. Danalatos, and P. C. Struik, "Temporal dynamics of light and nitrogen vertical distributions in canopies of sunflower, kenaf and cynara," *Field Crops Research*, vol. 122, no. 3, pp. 186–198, 2011.

[17] S. E. Sultan, "Phenotypic plasticity for plant development, function and life history," *Trends in Plant Science*, vol. 5, no. 12, pp. 537–542, 2000.

[18] L. Balaguer, E. Martínez-Ferri, F. Valladares et al., "Population divergence in the plasticity of the response of Quercus coccifera to the light environment," *Functional Ecology*, vol. 15, no. 1, pp. 124–135, 2001.

[19] S. V. Archontoulis, P. C. Struik, J. Vos, and N. G. Danalatos, "Phenological growth stages of *Cynara cardunculus*: codification and description according to the BBCH scale," *Annals of Applied Biology*, vol. 156, no. 2, pp. 253–270, 2010.

[20] S. L. Greene, M. Gritsenko, and G. Vandemark, "Relating morphologic and RAPD marker variation to collection site environment in wild populations of red clover (*Trifolium pratense* L.)," *Genetic Resources and Crop Evolution*, vol. 51, no. 6, pp. 643–653, 2004.

[21] C. Jana, X. Diaz, P. S. Laarin, C. Contreras, and V. Alfaro, "Morphological and molecular characterization of accessions of Argentina type globe artichoke variety collected at Coquimbo Region, Chile," *Acta Horticulturae*, vol. 983, pp. 197–202, 2013.

Characterization of the Wine Grape Thermohydrological Conditions in the Tropical Brazilian Growing Region: Long-Term and Future Assessments

Antônio Heriberto de Castro Teixeira,[1] Jorge Tonietto,[2] Giuliano Elias Pereira,[2] and Fernando Braz Tangerino Hernandez[3]

[1] *Embrapa, 13070-115 Campinas, SP, Brazil*
[2] *Embrapa, 95700-000 Bento Gonçalves, RS, Brazil*
[3] *São Paulo State University, 15385-000 Ilha Solteira, SP, Brazil*

Correspondence should be addressed to Jorge Tonietto; heriberto.teixeira@embrapa.br

Academic Editors: J. Hatfield, N. Hulugalle, and L. Mateos

Over the last years, Brazil has appeared among the new tropical wine producing countries. The joined effect of rising air temperature and decreasing precipitation makes it important to quantify the trend of the thermohydrological conditions of the commercial vineyards. The aims of the current research were to classify and delimit these conditions for the winemaking processes under different time scenarios in the Brazilian Northeastern region. Bioclimatic indicators were used together with long-term weather data and projections of the IPCC emission scenarios under simulated pruning dates. The results showed that decreasing of precipitation should be good for wine production when irrigation water is available, but rising air temperature will affect the wine quality and stability mainly for pruning done from November to March. The best pruning periods are around May for any time scenario considered. In general, more care should be taken for pruning happening in other periods of the year, regarding the effect of increasing thermal conditions on wine quality. The classification and delimitation done, joined with other ecological characteristics, are important for a rational planning of the commercial wine production expansion, mainly in situations of climate and land use changes together with rising water competition.

1. Introduction

Long-term observations and models are showing pieces of evidence of alterations in the climate system happening in several places of the world, which can be attributed to human activities. These effects are mainly consequences of changes in the atmospheric composition and hence global average air temperature is projected to rise together with regional variations in precipitation patterns [1, 2].

Grapevine phenology, wine quality, and yield are very dependent on climate at regional, local, and microclimatic scales [3]. Regional climate has been the focus of climate change impact assessments. At the local level, the considerations of grape site selection, cultural practices, and water management are increasing, being very important issues for potential adaptations to climate changes [4].

Air temperature values lower than 10°C will limit the vine shoot growth, inducing the plants to a dormancy period in temperate climates [5]. The optimum thermal range is considered between 25 and 30°C [6]. According to Huglin and Schneider [7], under situations of air temperature higher than 25°C, net photosynthesis decreases. Above 30°C, berry size and weight are reduced, with the metabolic processes dropping near 45°C [8]. Rising air temperature contributes to high sugar concentration in grapes, resulting in larger alcohol content in wines and lower acidity, increasing pH. As consequences, the presence, intensity and quality of the aroma, colour, and wine stability may be affected negatively [9, 10].

Webb et al. [11] analysing air temperature ranges concluded that, by 2030 in Australia, there may be additional areas for growing some grape varieties, but, overall, reductions in

Characterization of the Wine Grape Thermohydrological Conditions in the Tropical Brazilian Growing Region: Long-Term and Future Assessments

11

suitable areas are predicted by 2050 (between 27%, mid-warming, and 44%, high warming). However, in the USA, White et al. [12] showed that predictions based on average air temperature ranges alone might underestimate the climate changes impacts on viticulture.

According to Ganichot [13], the potential alcohol levels of grapes increased by 2% (from 9.7 to 22.7 v/v%), while total acidity decreased from 6 to 4 g L^{-1} H$_2$SO$_4$ and pH increased from 3.0 to 3.3, across all grape cultivars between 1980 and 2001 in southern France. Jones [14] reviewed reports on increased alcohol levels in wines from Alsace, Australia, and Napa founding that 50% of their increases are attributed to climate changes.

Besides the direct effects of rising air temperature on vine physiology and grape composition, there are important secondary ones. Increased soil and water salinity is a phenomenon associated with several semiarid and arid regions relying on irrigation [15, 16]. These joint effects can promote high concentrations of Na, K, and Cl in wines [17]. Rising air temperatures (30°C) may increase suspended solid concentrations, but Brix levels larger than 24°C Brix are likely not due to photosynthesis and sugar transport from leaves and woods. These high levels can be attributed to an increase of evapotranspiration (ET), as the warm air close to the canopies is one of the energy sources for the water fluxes [18].

Warming conditions can directly affect the vineyard water requirements, which, together with precipitation reductions, will lead to high levels of both aridity and irrigation water demand [19]. According to Webb et al. [20], the air temperature rise during the harvest period may reduce berry quality through higher ET rates. A prediction of increasing vineyard water consumption up 30% by 2070, despite the changes in precipitation patterns, was observed in Australia [21].

Jackson and Cherry [22] reported that high rainfall amounts reduce the ripening capacity of grapes. On the other hand, a certain level of water stress during this stage is favourable for the organoleptic wine quality [7]. Dry weather makes irrigation technologies an important asset for controlling water deficiencies and excesses and the need of vineyard water requirements quantification [19, 21].

According to Webb [23] the vineyard-growing season precipitation amounts are predicted to decrease or increase in Australia, depending upon the region, while the aridity is expected to rise in all areas by 2030 and 2050. Generally, these predictions translate the need of a reduction in irrigation water supply through increasing water productivity, which may be considered as the ratio of the actual yield to the water consumption [24].

A range of emission scenarios have been developed in the Intergovernmental Panel on Climate Change (IPCC) Special Reports on Emission Scenarios that reflect ways in which the world might develop [25].

A large number of climate models have classified winemaking regions by using different methodologies (e.g., Bonnardot et al. [26]; Tonietto and Carbonneau [27]). For thermohydrological vineyard delimitation, aiming at grape and wine production, one can apply bioclimatic indices based on the thermohydrological requirements. The Multicriteria

Climatic Classification (MCC) System proposed by Tonietto and Carbonneau [27] has been used under temperate climate conditions in Europe [28, 29] and in South America [30, 31]. However, the method has worked well considering a single, six-month growing season per year under these conditions.

Over the last years, the Brazilian tropical region has appeared among the new wine producing areas. With proper irrigation and cultural management practices, the farmers can produce grapes and carry out winemaking at any time of the year, allowing a potential average of between two to three vineyard-growing cycles per year, in accordance with and depending on each variety [19]. The rise of the air temperature with a consequent increase in aridity in the Brazilian Northeast region will affect the wine quality and water requirements. The joined effect of rising water consumption and decreasing precipitation, together with rapid land use change, makes it important to quantify the vineyard thermohydrological trends on a large scale to subsidize winemaking adaptations and water productivity improvements in the near future.

The objective of this research was to combine bioclimatic indicators, together with long-term weather data and projections of the IPCC A2 and B2 emission scenarios for the years 2020, 2050, and 2080, to classify and delimit the thermohydrological conditions for the winemaking processes under different pruning dates and time scenarios in the Brazilian Northeastern region. The modelling aims to subsidize the rational expansion of wine grape crop and improvements on water productivity, while minimizing undesired climate change effects on wine quality and environmental damages. This information is very important in situations of rising water competition by irrigated agriculture, natural vegetation, and nonagricultural sectors in the present and future conditions.

2. Materials and Methods

Figure 1 shows details of the Brazilian geographic regions, with states of the Northeast region and the location of the rain gauges and the conventional agrometeorological stations used.

The available monthly total precipitation data were from SUDENE (Superintendence of Development of the Northeast) referred to 1455 locations of rain gauges, while the monthly mean air temperature data were from INMET (National Meteorological Institute) recorded in 75 conventional stations.

These weather data are long-term values for the period from 1961 to 1990, taken as the baseline conditions in this study and used for comparisons with the IPCC A2 and B2 emission scenarios for the projected years 2020, 2050, and 2080. In the stations with only precipitation data available, the monthly air temperature (T_{month}) values were estimated from the geographic coordinates [32].

As the air temperature data is easier to be obtained for the whole Brazilian Northeast than the other meteorological parameters involved in the water vapour transfer, the Thornthwaite (TH) method, which needs only the monthly averages (T_{month}) as an input, was first applied to retrieve the monthly reference evapotranspiration (ETo). Available or

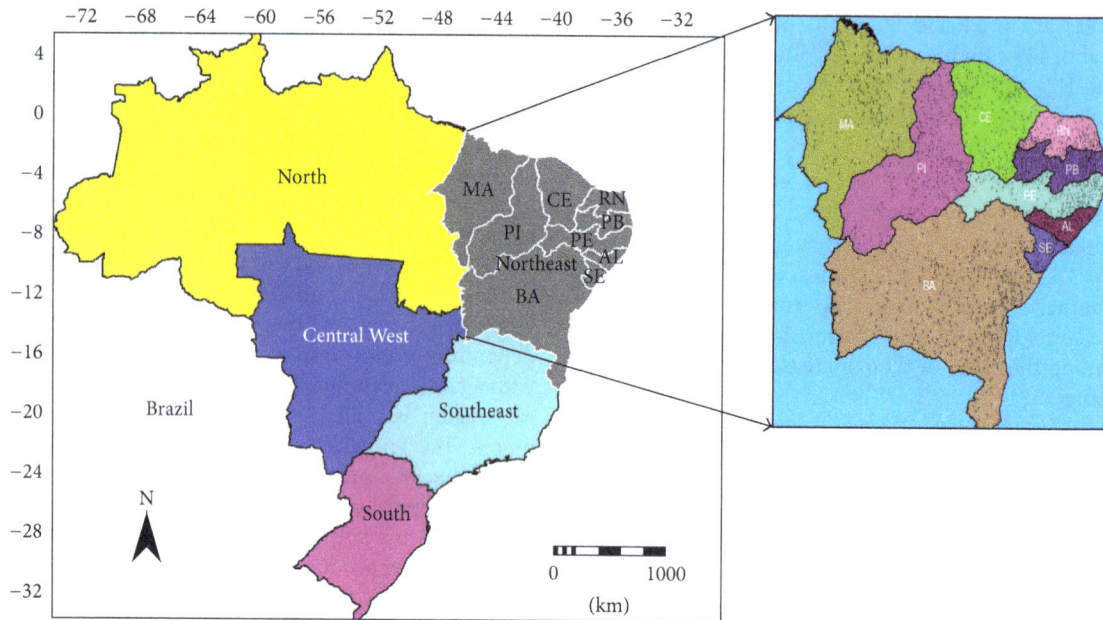

FIGURE 1: Brazilian regions and the Northeastern states of Maranhão (MA), Piauí (PI), Ceará (CE), Rio Grande do Norte (RN), Paraíba (PB), Pernambuco (PE), Alagoas (AL), Sergipe (SE), and Bahia (BA), together with the locations of the rain gauges and conventional agrometeorological stations.

estimated weather data from the conventional stations were used together with astronomical relations [33].

Simultaneous measurements of potential evapotranspiration (ETp) in wine grape cv. *Syrah* under optimal soil moisture conditions, together with ETo by the Penman-Monteith (PM) method [34], allowed the acquirement of crop coefficient (K_c) along a complete growing season [35]. Daily data of solar radiation, air temperature and relative humidity, and wind speed from an automatic agrometeorological station were used for the calculation of ETo.

In the current research, a model based on the relation of K_c and the accumulated degree days (DD_{ac}) was used to retrieve ETp, considering the cv. *Syrah* as a reference [19], after applying regression equations for calibrating ETo_{TH} into ETo_{PM} throughout regression equations. Seven automatic agrometeorological stations in the semiarid conditions of the Brazilian Northeast were used for this calibration:

$$K_c = a GD_{ac}^2 + b GD_{ac} + c, \tag{1}$$

where $a = -2\times10^{-7}$, $b = 4\times10^{-4}$, and $c = 0.54$ are the regression coefficients valid for the Brazilian semiarid conditions, with $R^2 = 0.80$.

For the calibration of $ET0_{TH}$ into $ET0_{PM}$, before using the model relating K_c and DD_{ac}, the regression equations for the first and for the second semesters were applied separately, both presenting R^2 around 0.70. Since the intention is to apply bioclimatic indicators for classifying wine grape areas and not to have accurate ET measurements, this calibration process seems to be plausible considering the absence of all meteorological parameters necessary for PM equation in the whole Brazilian Northeast region.

The ETp for one growing season (GS) was considered as the vineyard water requirements for an average GS of four

months (VWR_{GS}). Taking five K_c modelled values from DD_{ac}, and a base air temperature (T_b) of 10°C, zero being the initial value for DD_{ac} and considering the accumulated ones for the first, second, third, and fourth months, the average crop coefficient ($K_{c_{GS}}$) was multiplied by ETo_{GS} to give VWR_{GS}:

$$VWR_{GS} = K_{c_{GS}} ETo_{GS}. \tag{2}$$

VWR_{GS} together with the mean total precipitation for a growing season (P_{GS}) allowed the development and application of a vineyard water indicator (VWI_{GS}) for the moisture delimitations in the Brazilian Northeast, varying the pruning dates:

$$VWI_{GS} = \frac{P_{GS}}{VWR_{GS}}. \tag{3}$$

The indicator represented by (3) enables the characterization of the water component of the climate, taking into account the input and output of water from and to the vineyards, indicating the potential moisture availability in their root zones.

VWI_{GS} values around 1.00 imply the feasibility for rainfed wine grape crop, while those much higher indicate unsuitable conditions independently of irrigation water availability, due to problems of moisture excess. Low VWI_{GS} values mean natural water deficiencies and the degree of irrigation needs according to the pruning dates.

The future scenarios were considered by using precipitation and air temperature data for the projected years cantered on 2020, 2050, and 2080 from IPCC reports [1, 2] simulated by the model HadCM3 from the Hadley centre [36], which was chosen because it was the one that best agreed with

Characterization of the Wine Grape Thermohydrological Conditions in the Tropical Brazilian Growing Region: Long-Term and Future Assessments

13

observed historical weather data among others tested in South America [37].

The baseline and projected weather data were interpolated in a Geographic Information System (GIS) environment, by the "moving average" method, and the thermohydrological models were applied to the grids of P_{GS} and T_{GS}.

Basic criteria for dividing the Northeastern region into four classes (C1, C2, C3, and C4) of vineyard thermohydrological conditions were used. First, the VWI_{GS} values were considered the most important factor, as the excessive moisture is not suitable for the wine grape crop, independently if irrigated or not.

When VWI_{GS} values are lower or equal to 2.00 (classes C1 and C2), the moisture conditions are the best, because of less plant diseases, root respiration problems, and direct damage to the fruits promoted by excess of precipitation. These conditions will favour better quality of must and wine, where, according to data from the Brazilian Geographical and Statistical Institute (IBGE), the commercial wine grapes are concentrated in the Brazilian Northeastern region.

As warmer conditions affect negatively the wine quality, the most suitable areas in relation to moisture conditions were classified according to a T_{GS} threshold of 24°C. C1 areas were considered those with T_{GS} below 24°C and C2 areas are those above this value. Although the latter class has suitable natural moisture conditions for irrigated wine crop, as long as the air temperature increases above this threshold value, the wine quality should be affected, contributing to high levels of alcohol, low acidity, and large pH values, becoming unbalanced with instability for the phenolic and aromatic composition.

In situations of VWI_{GS} being higher than 2.00, the growing wine grape areas and yield are reduced, according to data from IBGE. A class with VWI_{GS} higher than 2.00 and lower or equal to 4.00 was then considered as an intermediary one; insofar the natural moisture increases, the yield and wine quality should be affected, independently of the thermal conditions.

Areas and pruning dates inside the worst class, those having VWI_{GS} higher than 4.00, should present the biggest problems in grape yield due to high levels of natural humidity that compromise the grape sanity, enological potential, and consequently the wine quality. There is no commercial wine grape crop reported by the IBGE under these conditions in the Brazilian Northeastern states.

3. Results

3.1. Long-Term Vineyard Thermohydrological Conditions. Figure 2 shows the maps of the T_{GS} long-term values, for, respectively, the coldest and the hottest GS, with pruning dates in July and October (Figure 2(a)), respectively, and those of P_{GS}, for the wettest and the driest conditions, with pruning dates in January and June (Figure 2(b)), respectively.

The lowest T_{GS} values occur at the winter solstice time in the Southern hemisphere, while the highest ones are found when the Sun is around the zenith position over the central part of the Brazilian Northeastern region. For pruning in July, many pixels present T_{GS} values lower than 22.0°C, while one

can see a larger area with those higher than 24.0°C, for pruning in October (Figure 2(a)).

Analysing the P_{GS} maps (Figure 2(b)), during the wettest GS, Maranhão (MA) and the Northwestern side of Piauí (PI) state are the most problematic areas, because of the Amazon's climatic influences, with total averaged values higher than 750 mm. When the pruning occurs during the driest period, almost the whole Northeast Brazil shows P_{GS} long-term values below 400 mm, even lower than 150 mm in the central and Southwestern areas of the region.

Figure 3 shows the spatial variation of the VWR_{GS} values for the reference cv. *Syrah*, under different pruning dates and an average GS of four months, based on the long-term data from 1961 to 1990 (baseline conditions), in the Brazilian Northeast region.

The pruning dates with maximum VWR_{GS} values are in September, when the average is 410 mm GS^{-1}, while the lowest ones are found when they are in March, around 300 mm GS^{-1}. According to the standard deviations (SD), the smallest and the largest spatial variations occur, respectively, for pruning done from March to May (SD = 18 mm GS^{-1}) and between August and October (SD = 23 mm GS^{-1}).

Table 1 presents the variation of VWR_{GS} for the reference cv. *Syrah*, among the Brazilian Northeastern states, considering the baseline conditions (1961–1990) and different pruning dates for a four-month average GS.

Highlights are for Piauí (PI), Ceará (CE), and Rio Grande do Norte (RN) as the states with the largest vineyard water demands. The highest average values are found in Ceará (CE), while the lowest ones are for Bahia (BA) state. The extremes of VWR_{GS} values represent a daily range from 2.5 to 3.3 mm d^{-1}.

For the delimitation of the thermohydrological conditions of the reference wine grape cv. *Syrah*, considering different pruning dates and an average four months GS, the bioclimatic indicator VWI_{GS} was first considered for the baseline conditions (Figure 4).

Pruning dates between June and August present the largest area of natural climate dryness conditions. Under these circumstances the VWI_{GS} values are below 0.50 and the SD between 0.25 and 0.40. On the other hand, VWI_{GS} larger than 2.00 and SD above 0.70 occur in larger areas for pruning dates from December to April, mainly in the Northwestern side of Maranhão (MA) state.

Table 2 presents the average VWI_{GS} values per state, for the wine grape cv. *Syrah*, considering different pruning dates along the year and an average growing season of four months, in the Brazilian Northeast, by using weather data for the baseline conditions (1961–1990).

VWI_{GS} values around 2.00 or higher for the rainy period are found in Maranhão (MA), Piauí (PI), and Ceará (CE) states. The lowest ones occur during the driest pruning periods, in Ceará (CE) and Rio Grande do Norte (RN), with averages even bellow 0.10.

Figure 5 shows the delimitation of four classes, according to the thermohydrological conditions for the baseline conditions, considering different pruning dates along the year, the reference wine grape crop cv. *Syrah*, and an average GS of four months.

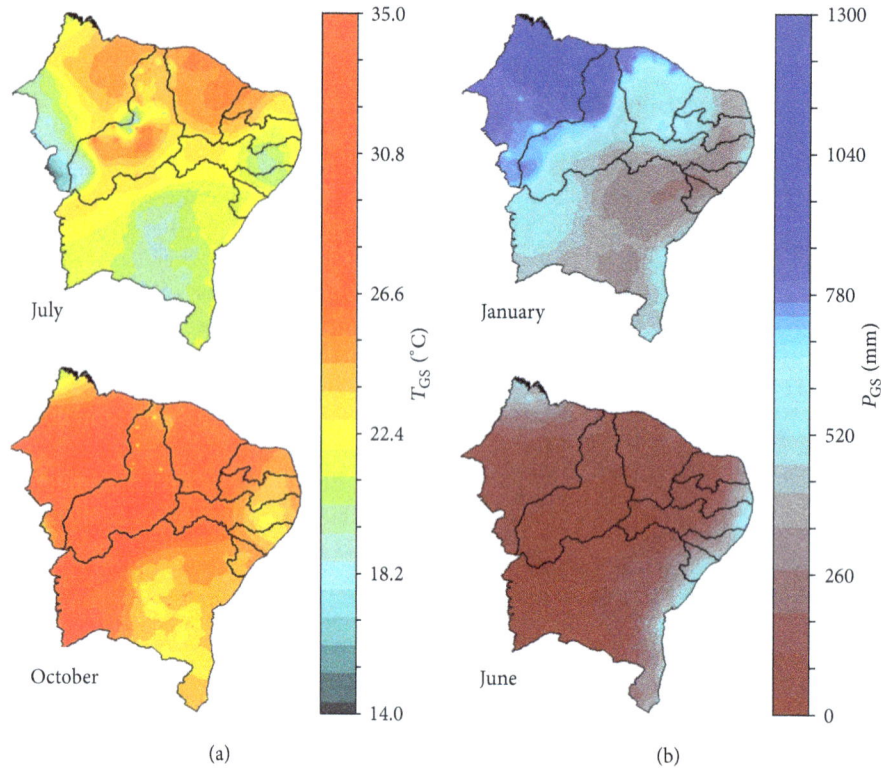

FIGURE 2: Maps of the long-term weather values (1961–1990) for an average wine grape growing season of four months in the Brazilian Northeastern region: (a) mean air temperature for the coldest and the hottest conditions (T_{GS}), with pruning dates, respectively, in July and October and (b) total precipitation, for the wettest and the driest conditions (P_{GS}), with pruning dates, respectively, in January and June.

TABLE 1: Mean values of the vineyard grape water requirements, cv. *Syrah*, for different pruning dates, considering an average growing season of four months (VWR$_{GS}$) and the baseline conditions (1961–1990) in the Brazilian Northeastern states of Maranhão (MA), Piauí (PI), Ceará (CE), Rio Grande do Norte (RN), Paraíba (PB), Pernambuco (PE), Alagoas (AL), Sergipe (SE), and Bahia (BA).

Pruning date	VWR$_{GS\_MA}$ (mm)	VWR$_{GS\_PI}$ (mm)	VWR$_{GS\_CE}$ (mm)	VWR$_{GS\_RN}$ (mm)	VWR$_{GS\_PB}$ (mm)	VWR$_{GS\_PE}$ (mm)	VWR$_{GS\_AL}$ (mm)	VWR$_{GS\_SE}$ (mm)	VWR$_{GS\_BA}$ (mm)
January	278.6	312.9	327.0	336.2	319.2	319.2	318.3	331.4	311.2
February	271.2	304.0	313.2	323.8	306.1	304.9	306.7	321.5	301.4
March	269.5	298.0	309.0	316.0	297.6	294.3	295.2	308.3	288.6
April	287.7	308.8	324.9	325.1	307.8	302.0	300.9	311.5	295.0
May	319.0	329.0	348.2	341.4	324.3	316.6	312.3	320.4	312.0
June	352.5	358.3	375.0	363.1	348.0	341.0	332.5	336.4	336.5
July	392.8	401.8	405.2	390.9	378.7	374.6	363.0	366.0	374.8
August	406.5	420.5	415.0	400.1	389.4	388.4	374.3	376.5	385.9
September	408.7	428.9	422.5	409.1	389.4	400.4	386.6	389.1	399.8
October	382.6	408.5	409.5	399.8	389.1	390.3	378.0	383.1	376.6
November	344.0	371.0	383.5	379.8	367.0	367.2	359.0	366.7	351.1
December	312.2	341.5	357.8	361.4	346.1	345.8	341.9	353.0	333.7
Year	335.4	356.9	365.9	362.2	346.9	345.4	339.1	347.0	338.9

The largest sizes for the C1 and C2 classes occur, respectively, when the pruning dates are from April to June (62% of the region) and from September to November (78% of the region). The classes C3 and C4 occur mostly when the pruning is done from December to February (28% of the region) and from January to March (5% of the region), respectively. The highest concentration is for C2 class when the pruning dates are in October, representing more than 80% of the Brazilian Northeast, while pruning periods in May present more than 65% of areas classified as C1.

Characterization of the Wine Grape Thermohydrological Conditions in the Tropical Brazilian Growing Region: Long-Term and Future Assessments

15

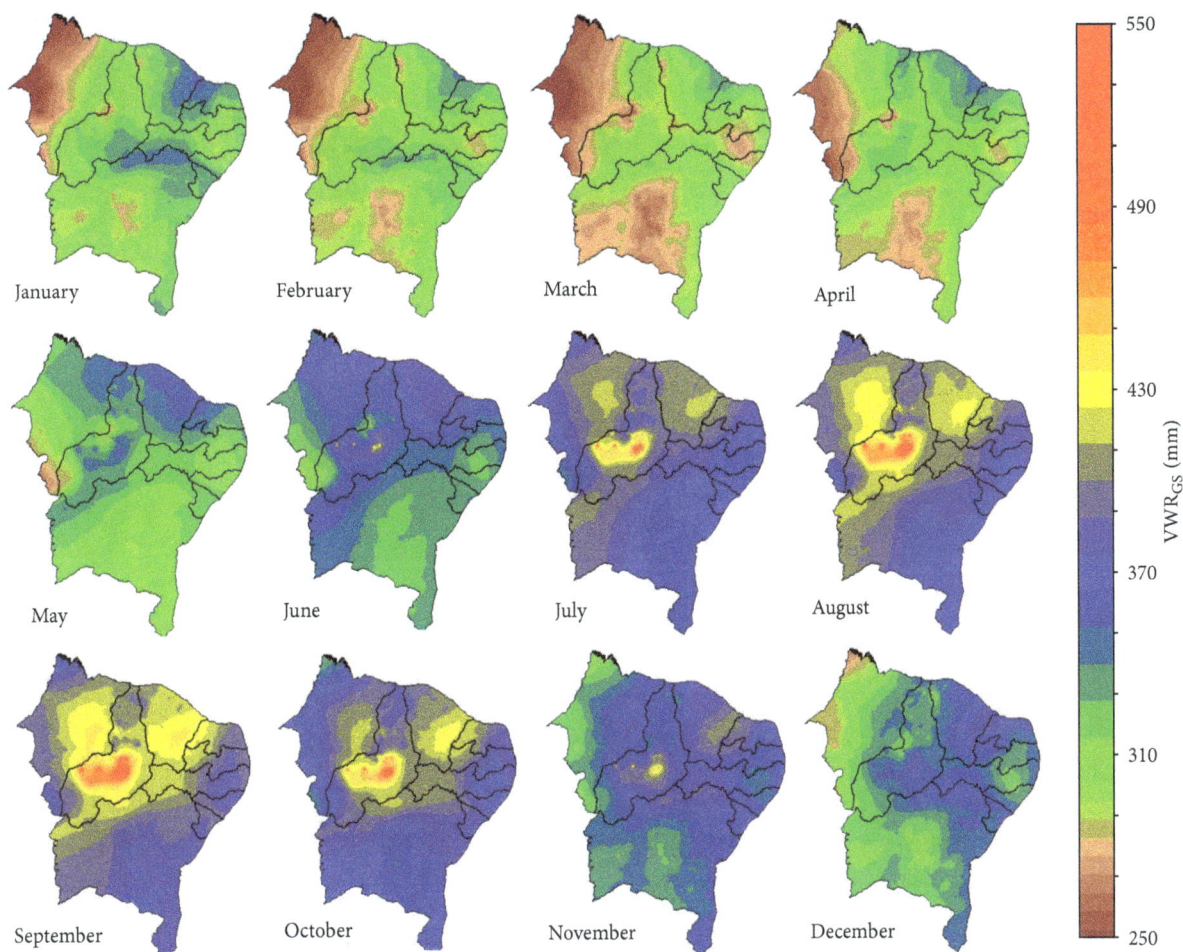

FIGURE 3: Maps of the wine grape water requirements, cv. *Syrah*, for different pruning dates, an average growing season of four months (VWR$_{GS}$), and the baseline conditions (1961–1990), in the Brazilian Northeastern region.

TABLE 2: Mean values of the vineyard water index for the wine grape, cv. *Syrah*, considering an average growing season (VWI$_{GS}$) of four months and the baseline conditions (1961–1990), in the Brazilian Northeastern states of Maranhão (MA), Piauí (PI), Ceará (CE), Rio Grande do Norte (RN), Paraíba (PB), Pernambuco (PE), Alagoas (AL), Sergipe (SE), and Bahia (BA).

Pruning date	VWI$_{GS\_MA}$	VWI$_{GS\_PI}$	VWI$_{GS\_CE}$	VWI$_{GS\_RN}$	VWI$_{GS\_PB}$	VWI$_{GS\_PE}$	VWI$_{GS\_AL}$	VWI$_{GS\_SE}$	VWI$_{GS\_BA}$
January	3.52	2.18	1.92	1.39	1.46	1.28	0.99	0.94	1.32
February	3.36	1.89	2.07	1.59	1.61	1.35	1.30	1.35	1.11
March	2.67	1.41	1.79	1.51	1.55	1.30	1.55	1.67	0.90
April	2.48	1.36	1.70	1.47	1.50	1.27	1.52	1.65	0.88
May	0.77	0.24	0.49	0.64	0.75	0.69	1.37	1.62	0.41
June	0.39	0.11	0.19	0.37	0.49	0.49	1.03	1.23	0.34
July	0.34	0.17	0.09	0.20	0.30	0.33	0.68	0.85	0.38
August	0.45	0.36	0.07	0.10	0.18	0.23	0.42	0.58	0.59
September	0.71	0.61	0.12	0.10	0.18	0.27	0.33	0.44	0.84
October	1.24	1.00	0.32	0.20	0.29	0.40	0.34	0.41	1.11
November	1.93	1.44	0.69	0.45	0.55	0.63	0.46	0.49	1.32
December	2.76	1.89	1.30	0.90	1.00	0.98	0.69	0.63	1.36
Year	1.72	1.06	0.90	0.74	0.84	0.77	0.89	0.99	0.88

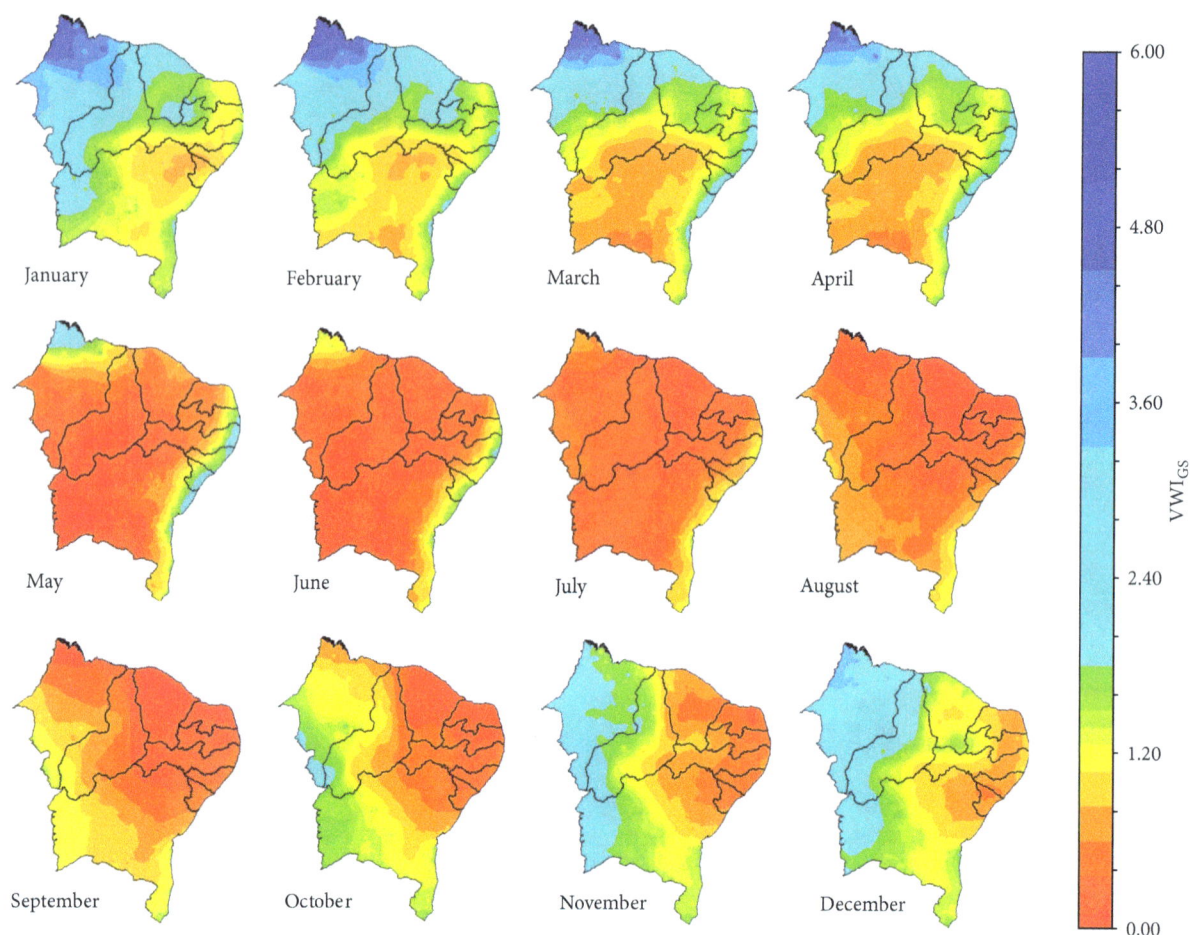

FIGURE 4: Maps of the vineyard water indicator for the wine grape, cv. *Syrah*, for different pruning dates, an average growing season of four months (VWI$_{GS}$), and the baseline conditions (1961–1990), in the Brazilian Northeastern region.

3.2. Projected Vineyard Thermohydrological Conditions.
Table 3 presents the averaged values of the differences (Δ) for the wine grape thermohydrological indicators, between the baseline conditions (1961–1990) and the projected years (2020, 2050, and 2080), considering the IPCC A2 and B2 emission scenarios for the entire Brazilian Northeastern region.

A generalized T_{GS} increase is evident for any pruning date; however, the lowest ΔT_{GS} values are predicted with pruning occurring from June to August. On the other hand, the highest ones are from November to March, being even larger than $5.0°C\,GS^{-1}$, considering the projected year of 2080 and the IPCC A2 emission scenario. These T_{GS} increases will make the vineyard water consumption rise, as evidenced by the predicted ΔVWR_{GS} averages, higher than $180\,mm\,GS^{-1}$ for pruning dates in December.

In parallel to the T_{GS} increases, decreases in P_{GS} values, according to the pruning dates, are also generally evident. The most negative ΔVWR_{GS} values are during the rainy period, when the P_{GS} values are expected to have an average reduction of more than $250\,mm\,GS^{-1}$ by the year 2080, for pruning dates between February and March.

The simultaneous increases in VWR_{GS} and reductions in P_{GS} will cause a progressive negative trend in VWI_{GS}. The

average annual values are expected to have a reduction of more than -0.60 by the projected year of 2080 from both IPCC scenarios. The exceptions are for pruning dates from November to December for the projected year of 2020 and the IPCC A2 emission scenario, when the values remain stable, as a consequence of the expected increases in P_{GS}.

The spatial variations of the P_{GS} differences between the baseline and the projected years (ΔP_{GS}) in percentages, for three different wine grape pruning dates along the year, taking as reference the cv. *Syrah*, in the Brazilian Northeastern region, are shown in Figure 6. Positive values mean increases, while the negative ones represent decreases.

For both IPCC emission scenarios strong P_{GS} reductions are predicted in large areas, mainly when the pruning dates are around May. However, in some places, more notable for pruning centred in January in the southern side of Bahia (BA) state, there are signs of rainfall increases, especially for the projected years of 2020 and 2050.

The spatial variations of the VWR_{GS} differences between the baseline and the projected years (ΔVWR_{GS}), for three different wine grape pruning date conditions along the year, taking the cv. *Syrah* as reference, in the Brazilian Northeastern region, are presented in Figure 7, also in percentages.

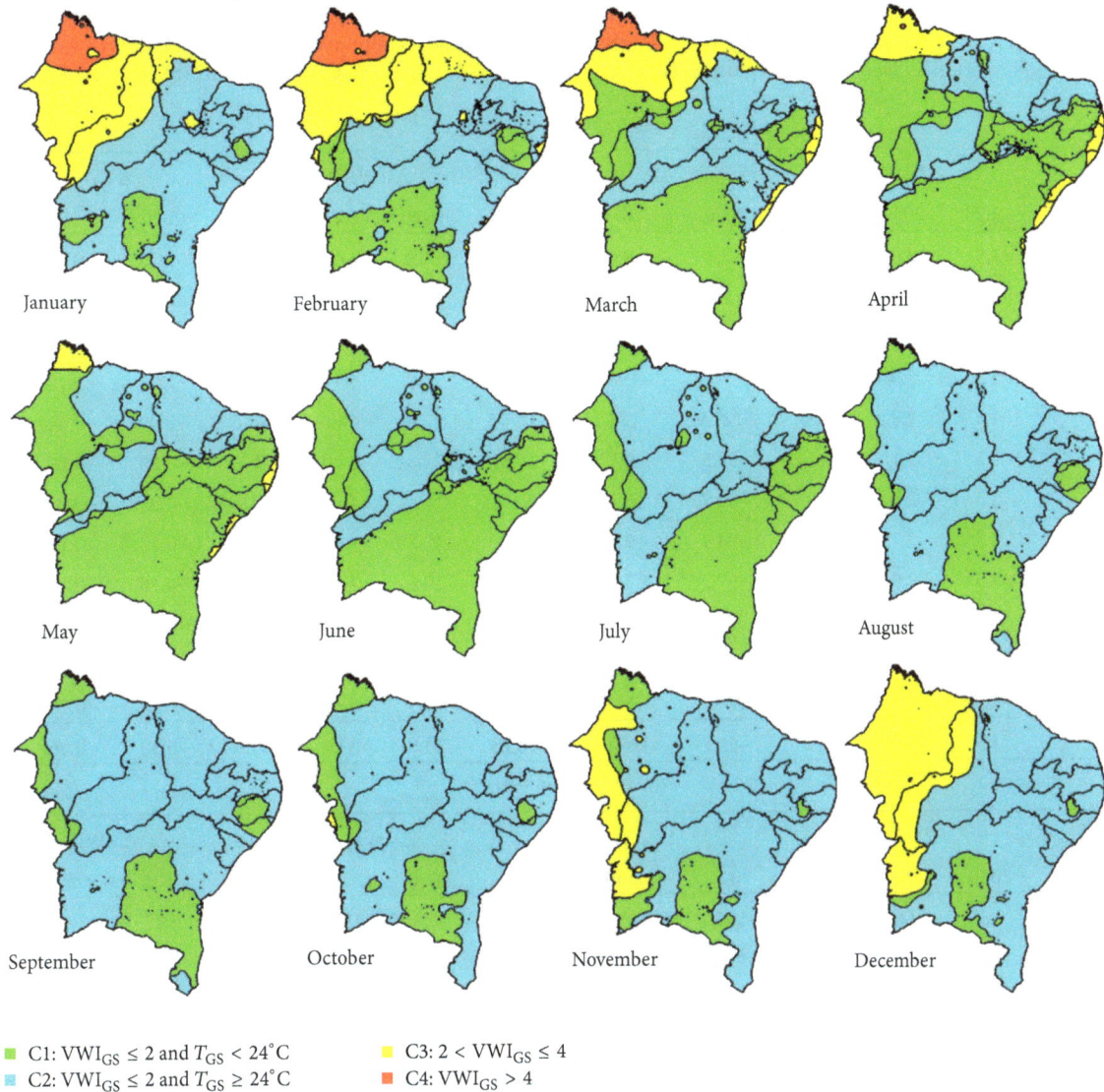

FIGURE 5: Class maps of the thermohydrological conditions for the wine grape, cv. *Syrah*, under different pruning dates, an average growing season of four months, and the baseline conditions (1961–1990), in the Brazilian Northeastern region.

Although all areas have predictions for rising vineyard water demands, lower ΔVWR_{GS} values are verified for the 2020 projected year, with the minor ranges for pruning around September, mainly in the southern side of Bahia (BA) state. Considering the projected year of 2080 strong increases are predicted, especially during the wettest period represented by the pruning centred in January for IPCC A2 emission scenario, with highlights for Ceará (CE) and Piauí (PI) and the western part of Rio Grande do Norte (RN), Paraíba (PB), and Pernambuco (PE) states.

The spatial variation for the differences between the baseline and the projected years of the vineyard water indicator (ΔVWI_{GS}), for three different wine grape pruning date conditions along the year, taking as reference the cv. *Syrah*, in the Brazilian Northeastern region, is shown in Figure 8.

In general, there are decreases in VWI_{GS}, with the highest ones occurring for pruning dates around January as a result of

the predictions of reducing P_{GS} and increasing VWR_{GS}. The most negative ΔVWI_{GS} values are in the Northwestern side of Maranhão (MA) state. For the Southwestern side of Bahia (BA) state, more increases in VWI_{GS} are predicted.

Analyses were done on how the initial four thermohydrological classes for wine grape could be changed along the IPCC projected years, considering three different pruning date conditions along the year, for the reference cv. *Syrah*, in the Brazilian Northeastern region (Figure 9).

By the projected year of 2080, almost all the Brazilian Northeastern states are predicted to present larger C2 areas. However, for pruning dates around May, there are still predictions of the C1 class, mainly in the southern side of Bahia state.

C3 areas will progressively be reduced, being restricted for pruning done around January, while the worst class C4 arising in the Northwestern part of Maranhão state, for pruning dates

TABLE 3: Average differences for the wine grape growing season values of air temperature (ΔT_{GS}), precipitation (ΔP_{GS}), vineyard water requirements (ΔVWR_{GS}), and vineyard water index (ΔVWI_{GS}) between the baseline conditions (1961–1990) and the projected years of 2020, 2050, and 2080, in the Brazilian Northeastern region, considering the reference cv. *Syrah*: IPCC A2 (a) and B2 (b) emission scenarios.

(a) A2 emission scenario

Pruning date	ΔT_{GS} (°C)			ΔP_{GS} (mm)			ΔVWR_{GS} (mm)			ΔVWI_{GS} (—)		
	2020	2050	2080	2020	2050	2080	2020	2050	2080	2020	2050	2080
January	1.5	3.0	5.6	−7.4	−88.0	−217.1	31.8	68.9	170.9	−0.20	−0.57	−1.14
February	1.6	3.1	5.5	−110.4	−165.9	−269.3	32.3	69.3	156.7	−0.51	−0.79	−1.22
March	1.6	3.1	5.3	−161.8	−206.5	−265.8	27.6	60.6	132.9	−0.64	−0.84	−1.10
April	1.5	3.0	5.0	−177.1	−198.9	−233.9	25.9	53.1	108.9	−1.08	−1.17	−1.30
May	1.4	2.8	4.6	−121.9	−137.1	−148.6	22.7	44.6	86.1	−0.39	−0.44	−0.48
June	1.3	2.6	4.1	−89.5	−97.3	−102.7	22.8	42.0	71.7	−0.26	−0.29	−0.30
July	1.3	2.6	4.0	−84.7	−89.6	−94.8	19.8	40.4	72.4	−0.23	−0.25	−0.26
August	1.3	2.7	4.2	−84.9	−91.7	−114.0	19.7	43.1	89.5	−0.22	−0.23	−0.31
September	1.5	2.9	4.5	−57.2	−76.5	−124.3	19.3	46.3	112.5	−0.16	−0.18	−0.36
October	1.5	2.9	5.0	13.4	−38.5	−118.0	19.1	48.5	119.3	−0.01	−0.18	−0.44
November	1.5	3.0	5.4	59.9	−5.6	−129.3	28.6	62.4	155.8	0.07	−0.28	−0.62
December	1.5	3.0	5.6	75.4	−11.1	−142.1	29.3	65.9	186.1	0.09	−0.28	−0.85
Average	1.5	2.9	4.9	−62.2	−100.6	−163.3	24.9	53.8	121.9	−0.30	−0.46	−0.70

(b) B2 emission scenario

Pruning date	ΔT_{GS} (°C)			ΔP_{GS} (mm)			ΔVWR_{GS} (mm)			ΔVWI_{GS} (—)		
	2020	2050	2080	2020	2050	2080	2020	2050	2080	2020	2050	2080
January	1.3	2.6	4.0	−106.1	−69.8	−203.9	27.8	58.2	107.9	−0.47	−0.48	−0.99
February	1.4	2.8	3.6	−133.8	−162.8	−252.1	29.2	60.8	86.6	−0.57	−0.75	−1.07
March	1.3	2.6	3.4	−172.0	−207.2	−267.0	23.7	50.5	68.0	−0.65	−0.82	−1.03
April	1.2	2.5	3.4	−179.4	−198.9	−222.8	21.8	43.0	65.3	−1.08	−1.16	−1.25
May	1.2	2.3	3.2	−124.7	−135.7	−143.9	19.5	36.7	54.2	−0.39	−0.43	−0.46
June	1.1	2.2	2.9	−91.1	−97.8	−99.4	20.0	35.2	49.8	−0.27	−0.29	−0.29
July	1.3	2.1	2.8	−85.1	−93.4	−93.3	16.5	31.9	46.9	−0.23	−0.25	−0.25
August	1.1	2.2	3.0	−82.6	−90.3	−105.0	16.4	34.5	53.8	−0.22	−0.24	−0.28
September	1.3	2.5	3.0	−56.3	−72.2	−110.6	16.3	37.8	50.4	−0.16	−0.21	−0.31
October	1.2	2.5	3.6	−62.2	−17.1	−107.0	14.9	54.2	78.2	−0.19	−0.11	−0.37
November	1.3	2.7	4.0	−30.0	22.9	−98.1	25.7	37.4	100.9	−0.16	−0.10	−0.47
December	1.4	2.8	3.7	−25.1	11.5	−130.7	26.4	57.7	93.5	−0.18	−0.20	−0.64
Average	1.3	2.5	3.4	−95.7	−92.6	−152.8	21.5	44.8	71.3	−0.38	−0.42	−0.62

during the rainy conditions in January, will practically disappear along the years (see Figures 7 and 9).

4. Discussion

According to the long-term data, there are no thermal limitations for wine grape crop in the Brazilian Northeast, with pruning dates in the middle of the year. On one hand, several situations are inside the optimum average range of air temperature from 25 to 30°C [6]. On the other hand, the air temperature values are not below the threshold of 10°C [5], which could introduce a dormancy stage.

When the pruning is around October, many areas with thermal conditions outside the upper limit (over 30.0°C) often occur and could bring some limitations for wine quality. These latter situations will contribute to high sugar content in grapes but wines with increasing levels of alcohol, low acidity, and large pH values. These effects together will promote

Characterization of the Wine Grape Thermohydrological Conditions in the Tropical Brazilian Growing Region: Long-Term and Future Assessments

19

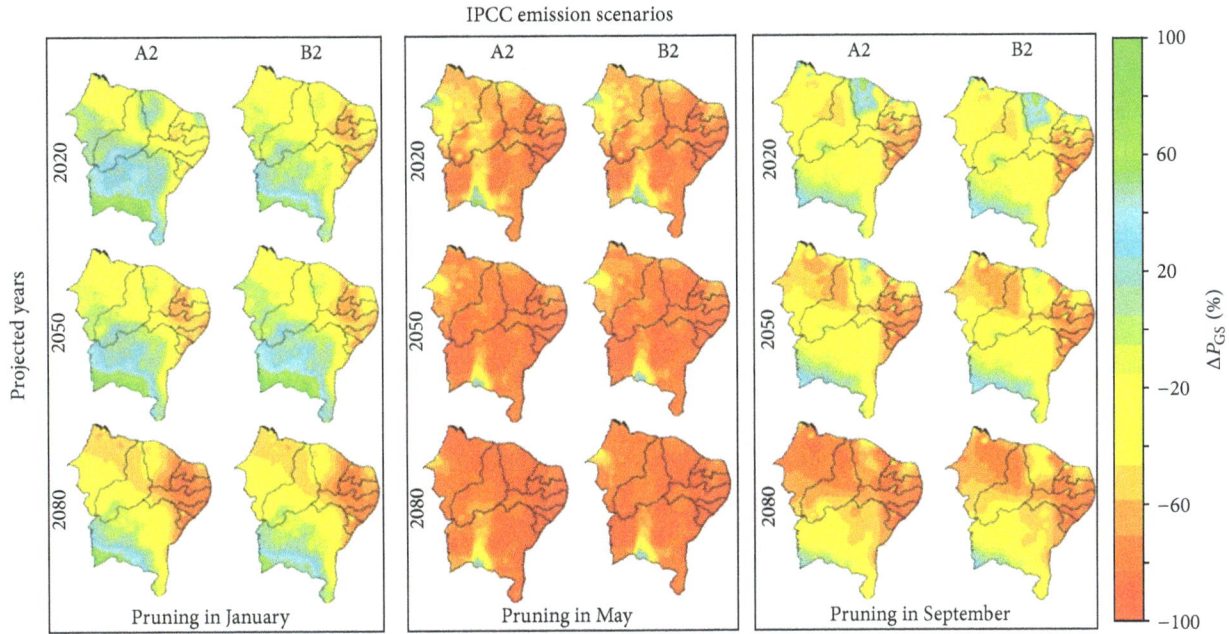

FIGURE 6: Spatial variation of the growing season precipitation differences in percentage (ΔP_{GS}), between the baseline (1961–1990) and the projected years of 2020, 2050, and 2080, considering three different pruning dates along the year, an average growing season of four months for the cv. *Syrah*, and the IPCC A2 and B2 emission scenarios in the Brazilian Northeastern region.

a wine unbalance with instability for the phenolic and aromatic composition [8–10]. In addition, secondary effects related to declines in soil structure and increases of soil salinity could occur [15, 16], favouring high concentrations of Na, K, and Cl in the tropical wines [17].

Normally, the highest vineyard water requirements occur when the pruning dates are centred in September and the lowest when they are around March. Among the Brazilian Northeastern states, Piauí (PI), Ceará (CE), and Rio Grande do Norte (RN) are highlighted, showing the largest evapotranspiration rates, which may correspond to good yield if water is available. However, some care must be taken in case of water scarcity, having ample room for wine grape water productivity improvements in situations of lower atmospheric demands [24].

Comparing the long-term regional daily averaged vineyard water requirements of 2.5 to 3.3 mm day^{-1} from the current study with water use from five experiments under different conditions [38–42], the results from the upscaling techniques used in the current study seem to be plausible. However, as the growing seasons in tropical climates are shorter, the total evapotranspiration rates are lower than those from temperate conditions.

After computing the ongoing and outgoing components of the climatic water balance, it was observed that the lowest natural vineyard moisture conditions occur for pruning dates from May to August. These conditions will avoid plant diseases, root respiration problems, and direct damage to the berries promoted by excess of precipitation, favouring better quality of both must and wine [19]. The highest baseline moisture levels for pruning done from January to April should promote the worst wine commercial production. Excessive

rainfalls reduce the ripening capacity of grapes, and the impossibility of water stress application is unfavourable for the organoleptic wine quality [7]. However, the effects of moisture excess are minimized along the IPCC projected years.

Considering the long-term weather data of 1961–1990, the pruning periods with the best thermohydrological conditions for wine grape crop (class C1) are from May to June. Highlights are for Bahia (BA), Pernambuco (PE), Paraíba (PB), Alagoas (AL), Sergipe (SE), and the west side of Maranhão (MA) with the largest areas inside this class. The worst thermohydrological class (C4) only occurs in the Northwestern part of Maranhão (MA) state, when the pruning is between January and March. In this last class, the highest levels of natural humidity compromise the grape sanity and the enological potential [19]. It is important to note that, even if C3 and C4 areas not being suitable during the wettest period of the year in the Brazilian Northeast, they will present favourable conditions (C1 class) when the pruning dates are from June to October.

Taking into account the projected years, there are general air temperature increases during the growing seasons with increments of C2 and decreases in C1 areas, which may contribute to more problems for elaborating the Brazilian tropical wines, because of increments of alcohol content and decreases of acidity, increasing pH [9, 10].

The most important effect from climate changes for wine production in the Brazilian Northeast region should be to take more care of the increasing thermal conditions regarding the wine quality. Examples of attention to this problem are the addition of tartaric acid to address the imbalance of acidity and the reverse osmosis procedures to dealcoholise wine [21].

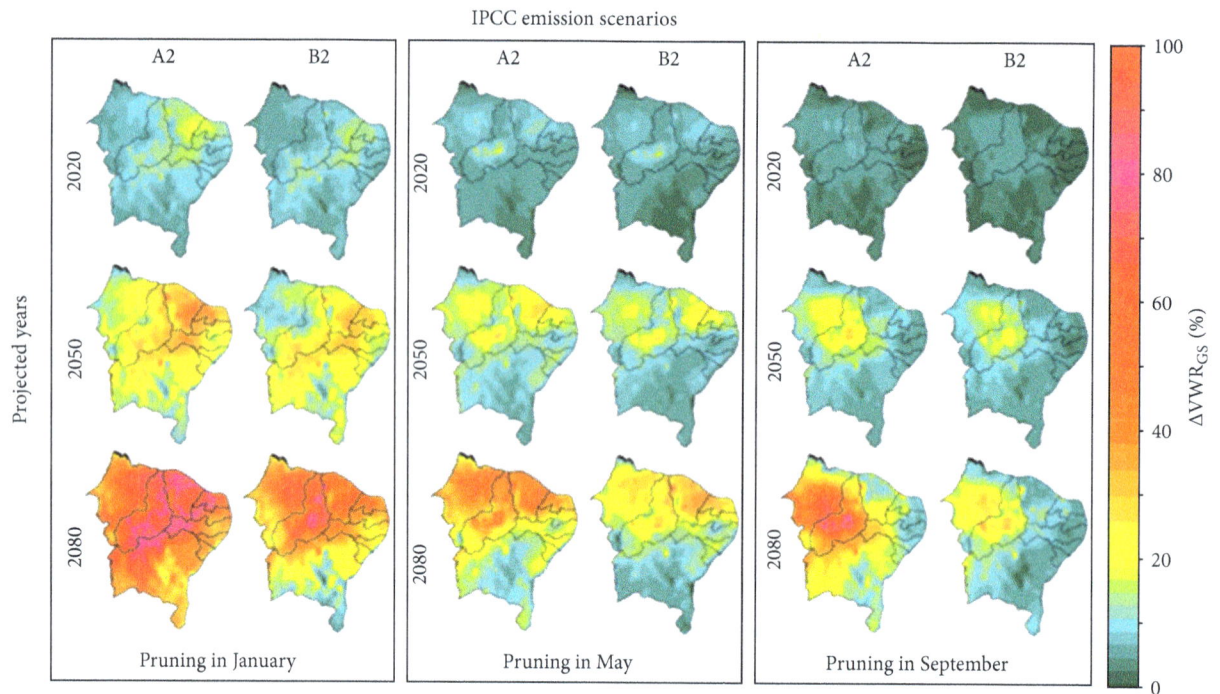

FIGURE 7: Spatial variation of the growing season vineyard water requirement differences in percentage (ΔVWR_{GS}) between the baseline (1961–1990) and the projected years of 2020, 2050, and 2080, considering three different pruning date conditions along the year, an average growing season of four months for the cv. *Syrah*, and the IPCC A2 and B2 emission scenarios in the Brazilian Northeastern region.

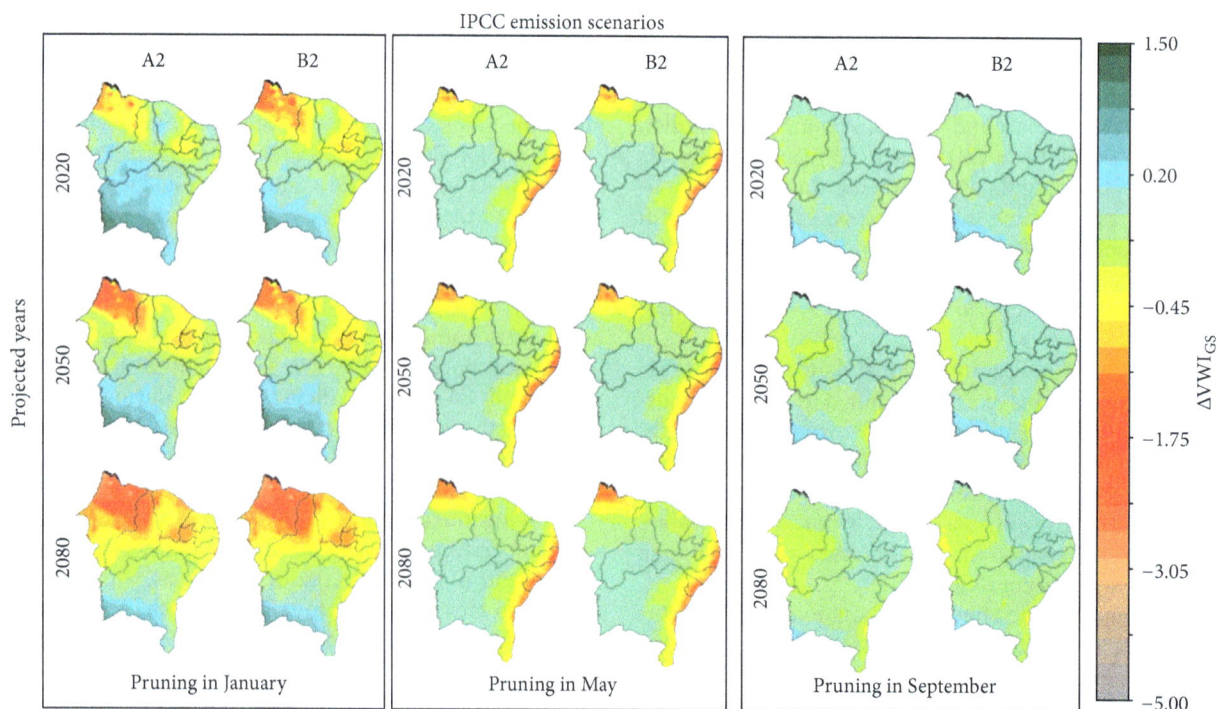

FIGURE 8: Spatial variation of the wine grape growing season vineyard water index differences (ΔVWI_{GS}), between the baseline (1961–1990) and the projected years of 2020, 2050, and 2080, considering three different pruning date conditions along the year, an average growing season of four months for the cv. *Syrah*, and the IPCC A2 and B2 emission scenarios in the Brazilian Northeastern region.

Characterization of the Wine Grape Thermohydrological Conditions in the Tropical Brazilian Growing Region: Long-Term and Future Assessments

21

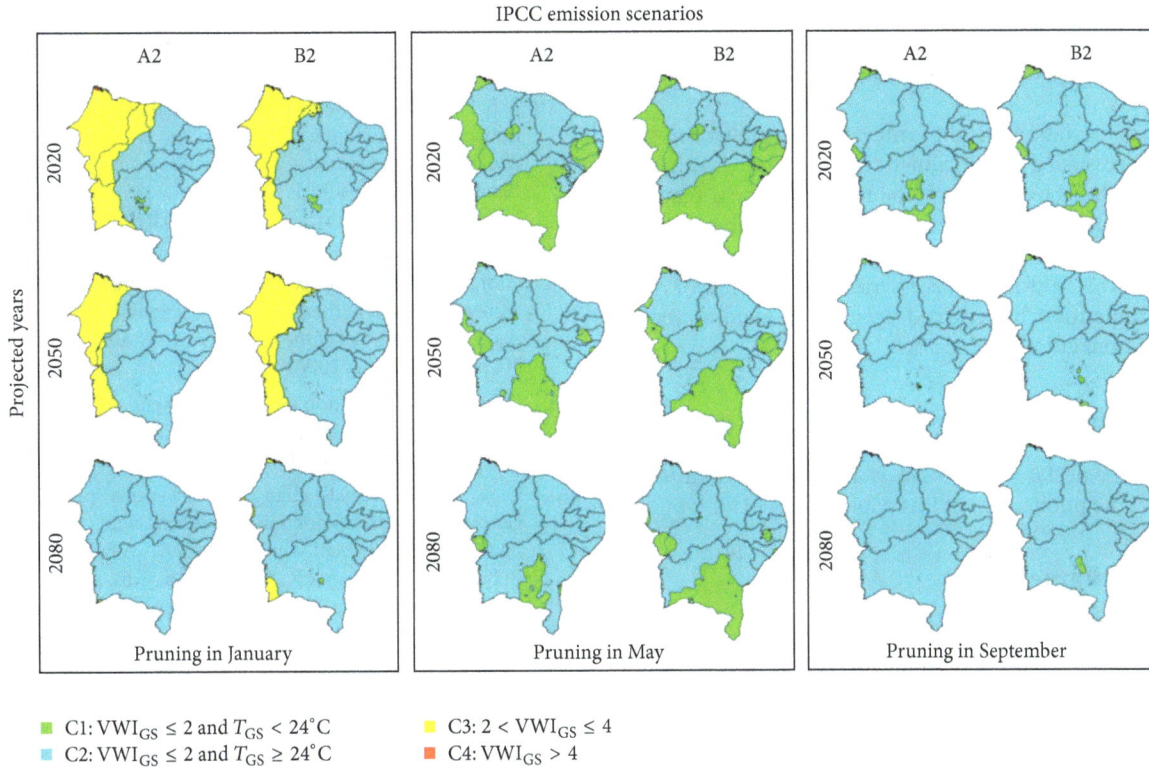

FIGURE 9: Class maps for the thermohydrological conditions of wine grape, cv. *Syrah*, under three different pruning dates and with an average growing season of four months considering the projected years of 2020, 2050, and 2050 for the IPCC A2 and B2 emission scenarios in the Brazilian Northeastern region.

The differentiating effects of changing precipitation patterns among areas in the current research are in agreement with Webb [23], who predicted decreasing or increasing rainfall during the growing seasons, depending upon the vineyard areas in Australia, by 2030 and 2050. On the other hand, increments in water requirements along the years agreed with Anderson et al. [21], who reported vineyard evapotranspiration rising up to 30% by 2070 in that country.

In the Brazilian Northeast, the increases of wine grape water consumption during the rainy period are important, because the subsequent months strong reductions in precipitation are also predicted, which means higher irrigation requirements during the climatologically driest periods. The simultaneous increases in atmospheric demand and decreases in rainfall amounts will call more attention for improvements of vineyard water productivity [24].

Accounting for the most suitable areas for wine grape crop in the Brazilian Northeast, their reductions are more drastic than those reported by Webb et al. [11]. According to these authors, considering a high emission scenario, decline in suitable areas by 2050 around 44% is predicted in Australia. However, the current results are similar in magnitude to those reported by White et al. [12] in USA, who concluded that, by the end of the 21st century, suitable grape production areas could reduce by 81%.

It is important to keep the current analyses of the projected vineyard thermohydrological conditions with caution,

as uncertainties arise inherent to the future social-economic status of the population and technological changes, as these aspects are the basis for the IPCC scenarios simulations.

To make the developed models in the Brazilian semiarid conditions applicable to other wine grape varieties and climatic areas, probably one needs to adjust the regression coefficients of the equation relating the crop coefficient and the accumulated degree days.

5. Conclusions

Bioclimatic indicators, based on the relations between the crop coefficient, the reference evapotranspiration, and the accumulated degree days, together with precipitation and air temperature data, allowed the large-scale classification of the thermohydrological conditions for wine grape production in the Brazilian Northeast (region), considering long-term and projected time scenarios.

The growing season values of the vineyard water requirements together with those of the total of precipitation were used for the development and application of vineyard water indicator. This indicator joined with growing season air temperature data was used to classify four different thermohydrological conditions in the Brazilian Northeaster region.

The modelling has the advantage of incorporating the thermal effects during the vineyard crop stages, extrapolating the water variables to different spatial-temporal scales.

To make them applicable to other climatic areas, probably, the only thing needed is to adjust the regression coefficients of the equation relating the crop coefficient and the accumulated degree days.

The delimitation of the vineyard thermohydrological conditions by using the bioclimatic indices, joined with other ecological characteristics, is suitable for a rational planning of commercial wine production expansion in the Brazilian Northeast considering different pruning dates and time scenarios. This information is essential in situations of rising water competition by irrigated agriculture, natural vegetation, and nonagricultural sectors.

Conflict of Interests

The authors declare that there is no conflict of interests regarding the publication of this paper.

Acknowledgments

The research herein was supported by FACEPE and CNPq, which are acknowledged for financial support to current projects on water productivity at different spatial-temporal scales in Brazil.

References

[1] International Panel on Climate Change, "Climate change 2001: the physical scientific basis," Working Group 1, IPCC Third Assessment Report, Cambridge University Press, Cambridge, UK, 2011.

[2] International Panel on Climate Change, "Climate change 2007: the physical scientific basis. Summary for policymakers," Contribution of Working Groups I to the Forth Assessment Report of the Intergovernmental Panel on Climate Change, Cambridge University Press, Cambridge, UK, 2007.

[3] G. V. Jones, M. A. White, O. R. Cooper, and K. Storchmann, "Climate change and global wine quality," *Climatic Change*, vol. 73, no. 3, pp. 319–343, 2005.

[4] J. Gladstones, "Climate and Australian viticulture," in *Viticulture, Volume 1: Resources*, P. Dry and B. G. Coomb, Eds., pp. 90–118, Winetitles, Adelaide, Australia, 2004.

[5] A. J. Winkler, J. A. Cook, W. M. Kliewer, and L. A. Lider, *General Viticulture*, University of California Press, Berkeley, Calif, USA, 1974.

[6] A. Costacurta and G. Roselli, "Critères climatiques et édaphiques pour l'établissement des vignobles," *Le Bulletin de l'Organisation Internationale de la Vigne et du Vin*, vol. 53, pp. 783–786, 1980.

[7] P. Huglin and C. Schneider, *Biologie et écologie de la vigne*, Lavoisier, Paris, France, 1998.

[8] C. R. Hale and M. S. Buttrose, "Effect of temperature on ontogeny of berries of Vitis vinifera L. cv. Cabernet Sauvignon," *Journal of the American Society for Horticultural Science*, vol. 99, pp. 390–394, 1974.

[9] J. M. Tarara, J. Lee, S. E. Spayd, and C. F. Scagel, "Berry temperature and solar radiation alter acylation, proportion, and concentration of anthocyanin in Merlot grapes," *American Journal of Enology and Viticulture*, vol. 59, no. 3, pp. 235–247, 2008.

[10] R. M. de Orduña, "Climate change associated effects on grape and wine quality and production," *Food Research International*, vol. 43, no. 7, pp. 1844–1845, 2010.

[11] L. B. Webb, P. H. Whetton, and E. W. R. Barlow, "Climate change impacts on Australian viticulture," in *Proceedings of the 13th Australian Wine Industry Technical Conference (AWITC '08)*, R. Blair, P. Williams, and P. Sakkie, Eds., pp. 99–105, Australian Wine Industry Technical Conference, Urrbrae, Australia, 2008.

[12] M. A. White, N. S. Diffenbaugh, G. V. Jones, J. S. Pal, and F. Giorgi, "Extreme heat reduces and shifts United States premium wine production in the 21st century," *Proceedings of the National Academy of Sciences of the United States of America*, vol. 103, no. 30, pp. 11217–11222, 2006.

[13] B. Ganichot, "Evolution de la date des vendanges dans les Côtes du Rhône méridionales," in *Actes des 6émes rencontres rhodaniennes*, pp. 38–41, Institut Rhodanien, Orange, France, 2002.

[14] C. V. Jones, "Climate change: observations, projections, and general implications for viticulture and wine production," in *Economics Department Working Paper No. 7*, E. Essick, P. Griffin, B. Keefer, S. Miller, and K. Storckmann, Eds., pp. 1–7, Whitman College, Washington, DC, United States, 2007.

[15] L. R. Clark, R. W. Fitzpatrick, R. S. Murray, and M. G. McCarthy, "Vineyard soil degradation following irrigation with saline groundwater for twenty years," in *Proceedings of the 17th World Congress of Soil Science*, no. 33, pp. 1115 CD ROM, International Union of Soil Science, Bangkok, Thailand, 2002.

[16] A. L. Richards, J. L. Hutson, and M. G. McCarthy, "Monitoring and modeling transient rootzone salinity in drip irrigated viticulture," in *Proceedings of the 13th Australian Wine Industry Technical Conference (AWITC '08)*, R. Blair, P. Williams, and P. Sakkie, Eds., pp. 212–217, Australian Wine Industry Technical Conference, Urbrae, Australia, 2008.

[17] R. R. Walker, D. H. Blackmore, P. R. Clingeleffer et al., "Salinity effects on vines and wines," *Le Bulletin de l'Organisation Internationale de la Vigne et du Vin*, vol. 76, pp. 200–227, 2003.

[18] M. Keller, "Managing grapevines to optimise fruit development in a challenging environment: a climate change primer for viticulturists," *Australian Journal of Grape and Wine Research*, vol. 16, no. 1, pp. 56–69, 2010.

[19] A. H. C. Teixeira, *Water Productivity Assessments from Field to Large Scale: A Case Study in the Brazilian Semi-Arid Region*, LAP Lambert Academic Publishing, Saarbrücken, Germany, 2009.

[20] L. B. Webb, P. H. Whetton, and E. W. R. Barlow, "Modelled impact of future climate change on the phenology of winegrapes in Australia," *Australian Journal of Grape and Wine Research*, vol. 13, no. 3, pp. 165–175, 2007.

[21] K. Anderson, C. Findlay, S. Fuentes, and S. Tyerman, *Garmaute Climate Change Review: Viticulture, Wine and Climate Change*, University of Adelaide, Adelaide, Australia, 2008.

[22] D. I. Jackson and N. J. Cherry, "Prediction of a district's grape-ripening capacity, using a latitude-temperature index (LTI)," *American Journal of Enology and Viticulture*, vol. 1, pp. 19–28, 1988.

[23] L. B. Webb, *The impact of projected greenhouse gas-induced climate change on the Australian wine industry [Ph.D. thesis]*, School of Agriculture and Food Systems, University of Melbourne, Victoria, Australia, 2006.

[24] A. H. C. Teixeira and L. H. Bassoi, "Crop water productivity in semi-arid regions: from field to large scales," *Annals of Arid Zone*, vol. 48, pp. 1–13, 2009.

[25] N. Nakicenovic, J. Alcamo, G. Davis et al., *IPCC Special Report on Emissions Scenarios*, Cambridge University Press, Cambridge, UK, 2000.

[26] V. Bonnardot, O. Planchon, V. A. Carey, and S. Cautenet, "Diurnal wind, relative humidity and temperature variation in the Stellenbosch-Groot Drakenstein wine producing area," *South African Journal for Enology and Viticulture*, vol. 23, pp. 62–71, 2002.

[27] J. Tonietto and A. Carbonneau, "A multicriteria climatic classification system for grape-growing regions worldwide," *Agricultural and Forest Meteorology*, vol. 124, no. 1-2, pp. 81–97, 2004.

[28] S. Hormazábal, G. Lyon, and A. Carbonneau, "Variabilité et limite du macroclimat viticole méditerranéen des Départements de l'Aude, de l'Hérault et du Gard, dans le Midi de la France," *Progrès Agricole et Viticole*, vol. 119, pp. 102–110, 2002.

[29] D. Blanco-Ward, J. M. García Queijeiro, and G. V. Jones, "Spatial climate variability and viticulture in the Miño River Valley of Spain," *Vitis*, vol. 46, no. 2, pp. 63–70, 2007.

[30] M. Ferrer, R. Pedocchi, M. Michelazzo, G. Gonzalez-Neves, and A. Carbonneau, "Delimitación y descripción de regiones vitícolas del Uruguay en base al método de clasificación climática multicriterio utilizando índices bioclimáticos adaptados a las condiciones del cultivo," *Agrociencia*, vol. 11, pp. 47–56, 2007.

[31] C. Montes, J. F. Perez-Quezada, A. Peña-Neira, and J. Tonietto, "Climatic potential for viticulture in Central Chile," *Australian Journal of Grape and Wine Research*, vol. 18, no. 1, pp. 20–28, 2012.

[32] E. P. Cavalcanti and V. P. R. Silva, "Programa computacional para a estimativa da temperatura do ar para a região Nordeste do Brasil," *Revista Brasileira de Engenharia Agrícola e Ambiental*, vol. 10, pp. 140–147, 2006.

[33] C. W. Thornthwate, "An approach toward a rational classification of climate," *Geographical Review*, vol. 38, pp. 55–94, 1948.

[34] R. G. Allen, L. S. Pereira, D. Raes, and M. Smith, "Crop evapotranspiration: guidelines for computing crop water requirements," FAO Irrigation and Drainage Paper 56, FAO, Rome, Italy, 1998.

[35] A. H. D. C. Teixeira, W. G. M. Bastiaanssen, and L. H. Bassoi, "Crop water parameters of irrigated wine and table grapes to support water productivity analysis in the São Francisco river basin, Brazil," *Agricultural Water Management*, vol. 94, pp. 31–42, 2007.

[36] International Panel on Climate Change, "IPCC SRES climate scenarios (the IPCC Data Distribution Centre)," 2006, http://www.ipcc-data.org/sres/gcm_data.html.

[37] J. A. Marengo, I. F. A. Calvalcanti, P. Satyamurty et al., "Assessment of regional seasonal rainfall predictability using the CPTEC/COLA atmospheric GCM," *Climate Dynamics*, vol. 21, no. 5-6, pp. 459–475, 2003.

[38] L. E. Williams, C. J. Phene, D. W. Grimes, and T. J. Trout, "Water use of mature Thompson Seedless grapevines in California," *Irrigation Science*, vol. 22, no. 1, pp. 11–18, 2003.

[39] I. A. M. Yunusa, R. R. Walker, and P. Lu, "Evapotranspiration components from energy balance, sapflow and microlysimetry techniques for an irrigated vineyard in inland Australia," *Agricultural and Forest Meteorology*, vol. 127, no. 1-2, pp. 93–107, 2004.

[40] G. Rana, N. Katerji, M. Introna, and A. Hammami, "Microclimate and plant water relationship of the "overhead" table grape vineyard managed with three different covering techniques," *Scientia Horticulturae*, vol. 102, no. 1, pp. 105–120, 2004.

[41] L. E. Williams and J. E. Ayars, "Grapevine water use and the crop coefficient are linear functions of the shaded area measured beneath the canopy," *Agricultural and Forest Meteorology*, vol. 132, no. 3-4, pp. 201–211, 2005.

[42] S. Ortega-Farias, M. Carrasco, A. Olioso, C. Acevedo, and C. Poblete, "Latent heat flux over Cabernet Sauvignon vineyard using the Shuttleworth and Wallace model," *Irrigation Science*, vol. 25, no. 2, pp. 161–170, 2007.

Effect of Soil Moisture Deficit Stress on Biomass Accumulation of Four Coffee (*Coffea arabica*) Varieties in Zimbabwe

Abel Chemura,[1] Caleb Mahoya,[2] Pardon Chidoko,[2] and Dumisani Kutywayo[3]

[1] *School of Agricultural Sciences and Technology, Chinhoyi University of Technology, Private Bag 7724, Chinhoyi, Zimbabwe*
[2] *DR&SS Coffee Research Institute, Chipinge, Zimbabwe*
[3] *DR&SS Head Office, Agricultural Research Centre, Harare, Zimbabwe*

Correspondence should be addressed to Abel Chemura; achemura@gmail.com

Academic Editors: A. D. Arencibia and B. Kindiger

A study was conducted to evaluate four common coffee (*Coffea arabica*) varieties in Zimbabwe for drought tolerance and ability to recover. The plants were subjected to drought stress for 21 and 28 days with evaluation of recovery done 14 days after interruptive irrigation. Coffee varieties were not significantly different in initial fresh and dry biomass before stressing ($P > 0.05$). CR95 had significantly accumulated more ($P < 0.05$) dry root mass (0.8 g) than the rest of the varieties after 21 days of drought stress. SL28 and CR95 had an 8.3% increase in dry biomass while Cat128 did not gain any dry biomass after 21 days of drought stress. CR95 had significantly more ($P < 0.05$) total dry biomass after 21 days and 28 days of drought stress while SL28 was consistently the least in both periods. Cat129 had the highest recovery gains in dry root, dry shoot, and total dry biomass after 21 days and 28 days of drought stress. Initial root biomass was negatively correlated with changes in total fresh and dry biomass of young coffee ($r > 0.60$) after both 21 and 28 days of drought stress, indicating that root biomass may be the most important factor determining drought tolerance in coffee varieties.

1. Introduction

Coffee (*Coffea arabica*) is produced in many developing countries contributing significantly to poverty alleviation and national economic development. In addition to the importance of coffee in many African national economies in terms of GDP and export earnings, it is directly linked to poverty alleviation as the majority of producers are smallholder farmers, and many rely only on coffee for socioeconomic development [1, 2]. The majority of the coffee produced in Southern Africa is Arabica coffee (*Coffea arabica* L.) which requires well-distributed rainfalls totaling over 1000 mm per year and temperatures between 24 and 26°C [3–5]. In Zimbabwe, coffee is produced in the eastern highlands districts of Chipinge, Chimanimani, Mutare, and Mutasa where natural climatic conditions approximate requirements and in the northern parts of the country in Guruve, Harare, and Mhangura districts under managed conditions [6, 7].

Coffee is a unique and legal source of income for many smallholder farmers and, as such, production has expanded from traditional areas to marginal areas where meeting crop water requirement is a serious challenge. In addition, in traditional production areas rainfall patterns have become unpredictable and unreliable exposing the coffee plants to frequent and often severe droughts [8, 9]. Changing weather patterns due to climate change and variability are projected to reduce the suitability of areas under coffee production in many producing zones, increase the risk of coffee pests and diseases, and increase coffee production costs [7, 9–13].

Soil moisture stress due to drought limits production through inhibiting growth and causing plant wilting and dieback. These in turn reduce yields and quality and expose the plant to opportunistic pathogens and pests [14–16]. Coffee production faces heightened risk of droughts because coffee is a perennial crop that is in the field throughout the year

[10, 17]. The major abiotic stresses that are projected to be increased by a changing climate are droughts, heat, salinization, water logging, pest, and disease epidemics [18]. The impact of droughts is more pronounced on the smallholder coffee sector that often relies on rainfed systems and lack both alternatives and coping options. Studies have indicated that coffee performs better under irrigation than under rainfed conditions [6, 16, 19]. Irrigation is therefore becoming a necessity for successful coffee production in many areas. However, even in areas where a reliable source of good quality water supply can be secured, irrigation comes with increased costs of installation, water pumping, and maintenance which reduces profitability of coffee production and makes it out of reach for many resource constrained smallholder farmers.

Among the most promising adaptation options is genetic improvement for development of crop varieties that can maintain productivity under less than optimal climatic conditions [10, 20]. According to DaMatta et al. [21], mechanisms involved in plant drought tolerance are drought avoidance, dehydration tolerance, and dehydration postponement. These are achieved through physiological adjustments, morphological change, changing resource allocation, and altering biochemical components of the plant system in response to moisture deficit and other environmental stresses [14, 22, 23]. The physiological responses of the coffee plant include leaf folding, leaf and branch dieback, and leaf shape change [24]. Biomass allocation between the roots, the stems, and leaves of the plant could be changed as a plant adjusts to moisture stress, optimizing the available moisture resources to the most important systems for survival [24, 25].

Development of deep roots to increase the soil water catchment and biophysical control of water loss through reducing leaf area and closure of stoma can be used by the plant to maintain a positive plant water status [21]. For example, the rates and direction of root growth of plants are principally determined by soil water gradients [4]. In addition, osmotic adjustment such as increasing net solute concentration significantly contributes to turgor maintenance thereby necessitating metabolic processes such as photosynthesis, growth, and in some plants reproduction and seed dispersal [26]. Maintaining turgor especially in the leaves can also be attained by increases in tissue elasticity. Increasing tissue elasticity results in attainment of a low water potential without the plant experiencing the detrimental effects of water deficit [8, 21, 26]. These characteristics vary between plant species and varieties [27]. The ability of a crop or crop variety to achieve these and other related water use and water retention characteristics enables it to be regarded as drought tolerant. Drought tolerant varieties are suitable for a wide range of agroecological environments and reduce irrigation requirements and associated costs.

To complement genetic improvement efforts, it is necessary to take stock of current germplasm in terms of drought tolerance in order to inform future breeding and selection programs. In Zimbabwe, there are many formally and informally introduced coffee genetic materials that have not been adequately evaluated. This limits the potential use of these materials in breeding programs and in crop production. There has been very limited deliberate research

to assess the drought tolerance of coffee varieties under production, many of which have been primarily developed focusing on disease tolerance. In addition, an understanding of the yield differences between these varieties under different environments and management practices could be obtained from their physiological response to soil moisture deficit stress. The objectives of this study were to evaluate the morphological response of different coffee varieties under moisture deficit stress and determine the initial physiological characteristics that are related to resilience.

2. Materials and Methods

2.1. Study Site. The trial was carried out in the greenhouse at Coffee Research Institute (CoRI) between November and December 2012. CoRI is located at 20˚12S and 32˚37E at an altitude of 1100 m.a.s.l [28]. The average annual rainfall is 1180 mm of which 80% falls in five months from November to March. The mean maximum temperature is 20˚C and minimum is 14˚C. Most of the soils in this area are leached and strongly weathered and in the orthoferralitic group derived from Umkondo quartzites and sandstone [29].

2.2. Experimental Materials and Design. Four varieties were assessed for their response to moisture deficit stress. These were SL28, Costa Rica 95 (CR95), Catimor 128, and Catimor 129. Seedlings of these varieties were grown in the nursery in recommended growth medium for raising coffee seedlings in black polythene pots. Watering was done on a weekly interval and all other routine nursery management activities were based on nursery recommendations from the Coffee Handbook [4]. Eight-month-old seedlings were subjected to two moisture stress regimes, one for 21 days and the other for 28 days from November 14, 2012, under greenhouse conditions to protect them from rain. After moisture stress period, the seedlings were irrigated and assessed for their recovery ability with biomass measurements after 2 weeks. For each variety and each assessment (initial, 21 days, 28 days, 21 days, 14 days recovery and 28 days and 14 days recovery) three coffee plants were used. The trial was laid out in a Complete Randomized Design.

2.3. Measurements and Data Analysis. Destructive sampling was done to weigh the roots, shoots, and total biomass. The roots were carefully excavated and cleaned with tap water before being weighed. The fresh and oven-dried (70˚C for 8 hrs) mass of roots shoots and total biomass were measured using a digital balance. Data were analyzed using analysis of variance (ANOVA) and Pearson correlation was used to determine the correlation between initial biomass and changes in biomass after soil moisture deficit stress in R Statistical Software [30].

3. Results

3.1. Root Biomass. Varieties did not show significant differences ($P > 0.05$) in root biomass before stressing (Table 1). There were significant differences in dry root biomass after 21

TABLE 1: Root biomass of coffee seedlings before, and after stressing and after recovery.

Variety	Initial		21 days		28 days		21 + 14 days		28 + 14 days	
	Fresh	Dry	Fresh	Dry	Fresh	Dry	Fresh	Dry	Fresh	Dry
SL28	1.8	0.4	1.8	0.4^a	1.4^a	0.4^a	1.6^a	0.4	2.8	0.5
Cat129	2.3	0.4	2.3	0.5^{ab}	1.8^a	0.5^a	1.9^{ab}	0.5	3.6	0.7
Cat128	2.0	0.5	1.8	0.5^{ab}	1.7^a	0.6^a	2.1^{ab}	0.6	2.2	0.5
CR95	3.4	0.7	3.4	0.8^b	5.0^b	0.9^b	3.6^b	0.9	5.1	0.9
P	0.167	0.099	0.338	0.025	0.007	0.004	0.042	0.119	0.171	0.442

Means with the same letter within a column are not significantly different according to Tukey's test ($P = 0.05$).

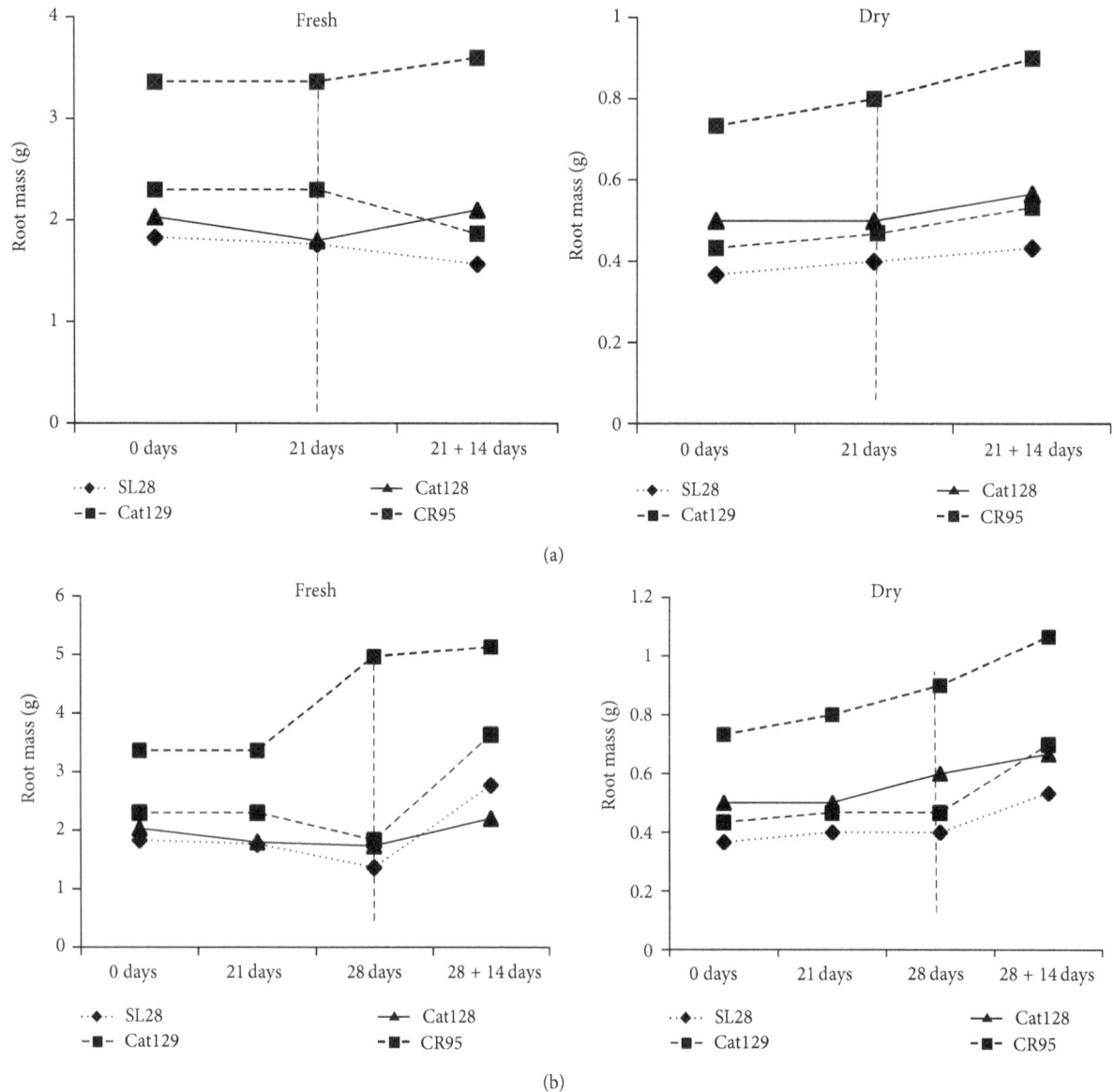

(a)

(b)

FIGURE 1: Changes in fresh and dry root biomass after stressing and recovery (a broken line indicates the time of interruptive irrigation).

days of moisture stress ($P < 0.05$). CR95 had the heaviest root mass (0.8 g) while SL28 had the least mass (0.4 g). After 28 days of soil moisture stress, there were significant differences in fresh and dry root biomass between varieties ($P < 0.05$). CR95 had higher fresh and dry biomass (5 g and 0.9 g, resp.)

than the rest of the varieties which were not significantly different from each other (Table 1).

Fresh root biomass was significantly different ($P < 0.05$) after 21 days of stressing and 14 days of recovery while dry biomass was not significantly different after the same period

TABLE 2: Percentage changes in root biomass after 21 and 28 days of stressing and recovery.

Roots variety	21 days		28 days		21 + 14 days		28 + 14 days	
	Fresh	Dry	Fresh	Dry	Fresh	Dry	Fresh	Dry
SL28	−3.8	8.3	−34.1	8.3	−11.3	8.3	102.4	33.3
Cat129	0.0	7.1	−25.5	7.1	−18.8	14.3	98.2	50.0
Cat128	−13.0	0.0	−17.3	16.7	16.7	13.3	26.9	11.1
CR95	0.0	8.3	32.2	18.5	6.9	12.5	3.4	18.5

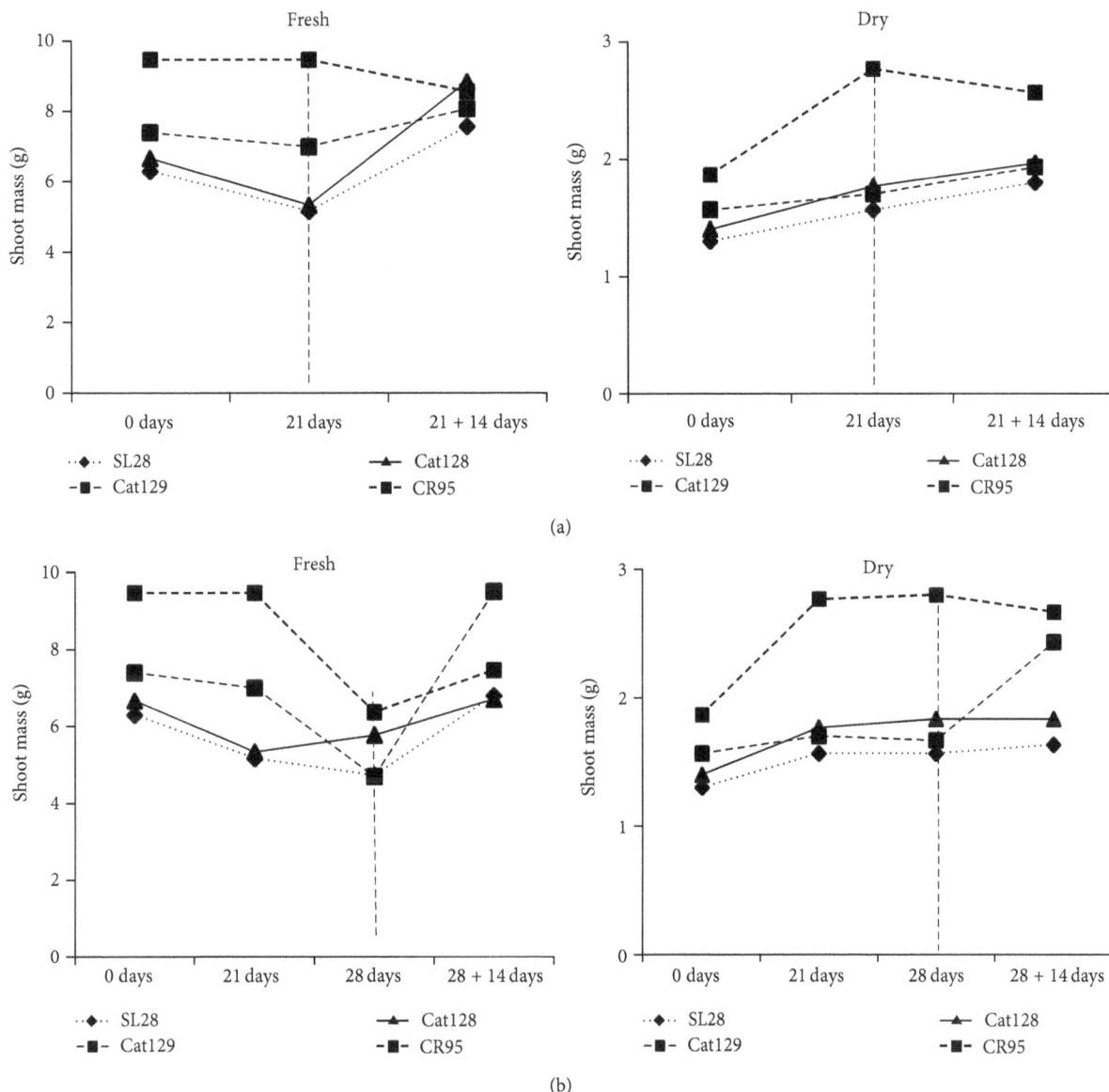

(a)

(b)

FIGURE 2: Changes in fresh and dry shoot biomass after stressing and recovery (dashed line shows time of interruptive irrigation application).

($P > 0.05$). No significant differences ($P > 0.05$) were realized between varieties after 28 days of stressing and 14 days of recovery. The highest change in fresh biomass after 21 days of moisture stress was for Cat128 which lost 13% followed by SL28 which lost 3.8 of fresh weight (Table 2).

There were no changes in fresh weight in Cat129 and CR95. In terms of dry weight, Cat128 did not show any change after 21 days of soil moisture stress while SL28 and CR95 gained 8.3% of dry biomass during the same period. After 28 days of soil moisture stress, only CR95 had positive change in fresh root biomass (32.2%). Although CR95 had significantly heavier dry root biomass than the rest of the varieties, its percentage gain in dry root biomass after 21 days of soil moisture stress was the same as that of SL28 (Table 2).

TABLE 3: Shoot biomass of coffee seedlings before, and after stressing and after recovery.

Shoots variety	Initial		21 days		28 days		21 + 14 days		28 + 14 days	
	Fresh	Dry	Fresh	Dry	Fresh	Dry	Fresh	Dry	Fresh	Dry
SL28	6.3	1.3	5.2	1.6	4.7	1.6[a]	7.6	1.8	6.8	1.6
Cat129	7.4	1.6	7.0	1.7	4.7	1.7[a]	8.1	1.9	9.5	2.4
Cat128	6.7	1.4	5.3	1.8	5.8	1.8[a]	8.8	2.0	6.7	1.5
CR95	9.5	1.9	9.5	3.4	6.4	2.8[b]	8.6	2.6	7.5	2.4
P	0.465	0.398	0.255	0.094	0.351	0.009	0.836	0.344	0.764	0.311

Means with the same letter within a column are not significantly different according to Tukey's test ($P = 0.05$).

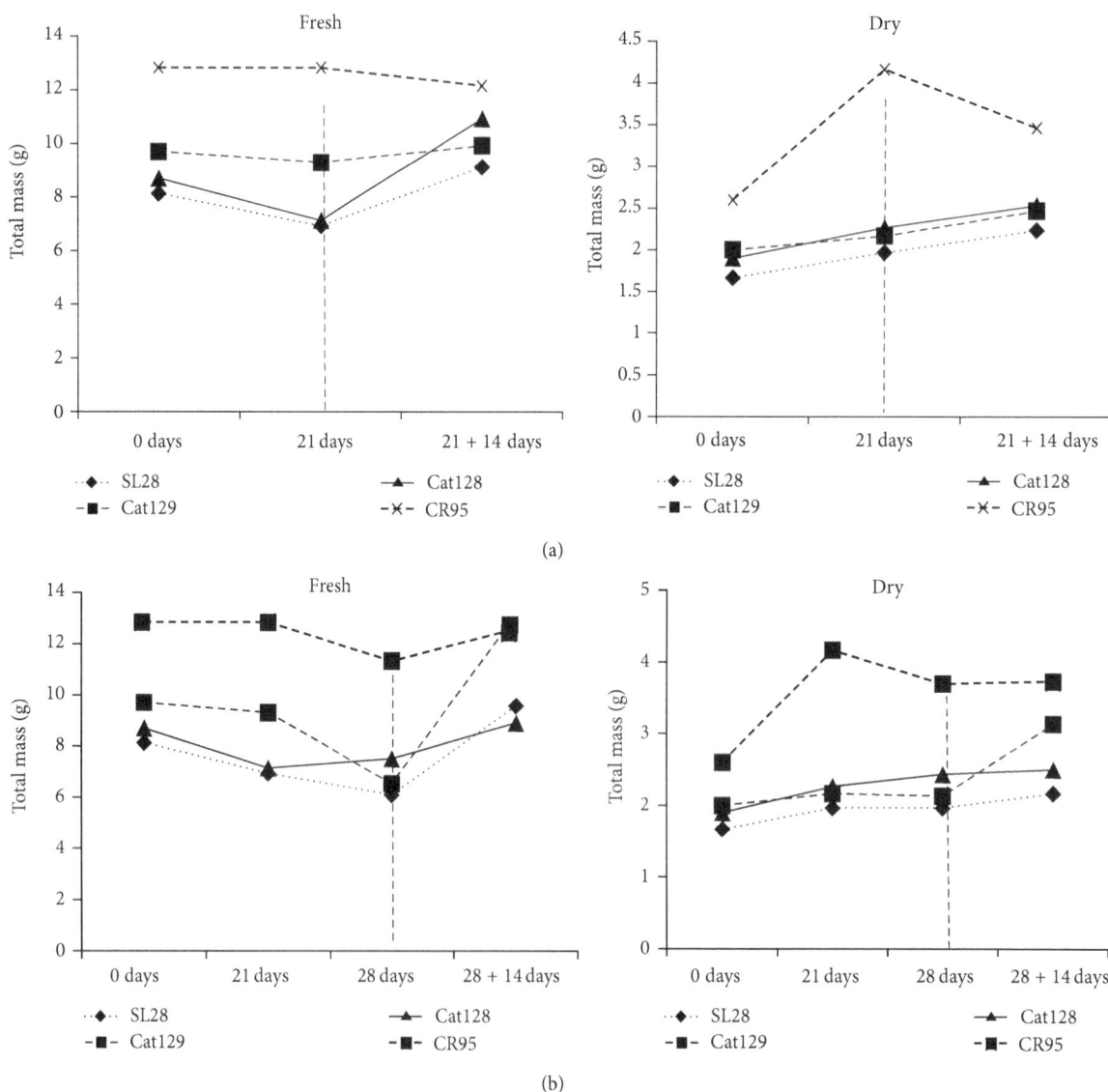

(a)

(b)

FIGURE 3: Changes in total fresh and dry biomass after stressing and recovery (dashed line shows time of interruptive irrigation application).

TABLE 4: Percentage changes in shoot biomass after 21 and 28 days of stressing and recovery.

Shoots variety	21 days		28 days		21 + 14 days		28 + 14 days	
	Fresh	Dry	Fresh	Dry	Fresh	Dry	Fresh	Dry
SL28	−21.9	17.0	−33.1	17.0	46.5	14.9	43.7	4.3
Cat129	−5.7	7.8	−57.4	6.0	15.2	13.7	102.1	46.0
Cat128	−25.0	20.8	−15.6	23.6	65.6	11.3	16.2	0.0
CR95	0.0	32.5	−48.7	33.3	−9.5	−7.2	17.3	−4.8

TABLE 5: Total plant biomass before stressing, after stressing, and after recovery.

Total	Initial		21 days		28 days		21 + 14 days		28 + 14 days	
variety	Fresh	Dry	Fresh	Dry	Fresh	Dry	Fresh	Dry	Fresh	Dry
SL28	8.1	1.7	6.9	2.0^a	6.1^a	2.0^a	9.1	2.2	9.6	2.2
Cat129	9.7	2.0	9.3	2.2^{ab}	6.5^{ab}	2.1^a	9.9	2.5	13.1	3.1
Cat128	8.7	1.9	7.1	2.3^{ab}	7.5^{ab}	2.4^a	10.9	2.5	8.9	2.0
CR95	12.8	2.6	12.8	4.2^b	11.3^b	3.7^b	12.2	3.5	12.6	3.3
P	0.327	0.23	0.246	0.038	0.037	<0.001	0.379	0.123	0.661	0.36

Means with the same letter within a column are not significantly different according to Tukey's test ($P = 0.05$).

TABLE 6: Percentage changes in total biomass after 21 and 28 days of stressing and recovery.

Total	21 days		28 days		21 + 14 days		28 + 14 days	
variety	Fresh	Dry	Fresh	Dry	Fresh	Dry	Fresh	Dry
SL28	−17.3	15.3	−33.3	15.3	31.7	13.6	56.8	10.2
Cat129	−4.3	7.7	−48.5	6.3	6.8	13.8	101.0	46.9
Cat128	−22.0	16.2	−16.0	21.9	53.3	11.8	18.7	2.7
CR95	0.0	37.6	−13.2	29.7	−5.2	−16.8	11.2	0.9

The greatest loss in fresh root biomass was for SL28 which lost 34.1%. CR95 gained 18.5% of dry root biomass after 28 days of soil moisture stress. Corresponding to the losses in fresh root biomass, SL28 and Cat129 had the least increase in dry biomass after 28 days of stressing, gaining 8.3% and 7.1% dry root biomass. These varieties (SL28 and Cat129) continued to lose fresh biomass during the 14 days recovery period after 21 days moisture stress. Cat128 gained the highest percentage of fresh root biomass while Cat129 gained the highest amount of dry biomass during this recovery period.

When the coffee plants were irrigated for 14 days after a 28-day stressing period, SL28 and Cat129 had the highest percentage increases in fresh root biomass (102.4% and 98.2%, resp.). The least gain in fresh root biomass was for CR95 which only increased fresh biomass by 3.4%. The highest percentage increase in dry root biomass was for Cat129 whose dry root biomass increased by 50% (Figure 1). In terms of fresh root biomass, Cat128 had the most gains after 21 days of soil moisture stress while SL28 and Cat129 had the highest gain in fresh root biomass after 28 days of soil moisture stress.

3.2. *Shoot Biomass.* Fresh and dry shoot biomass did not significantly differ between varieties before and after 21 days of soil moisture stress ($P > 0.05$). CR95 had superior ($P < 0.05$) shoot dry biomass than other varieties after 28 days of soil moisture stress (Table 3). There were no significant differences in shoot biomass between varieties after both recovery periods ($P > 0.05$).

SL28, Cat129, and Cat128 lost fresh shoot biomass after 21 days of soil moisture stress. There was no change in fresh shoot biomass for CR95. CR95 had the highest percentage change in dry shoot biomass, gaining 32.5% during the 21 days soil moisture stressing period (Table 4). The least percentage change in shoot biomass was for Cat128 which gained 7.8% of dry biomass after 21 days stressing period. After 28 days of

soil moisture stress, all varieties had lost fresh shoot biomass with the greatest loss shown by Cat129 (57.4%) followed CR95 (48.7%).

Corresponding to the greatest fresh shoot biomass losses, Cat129 had the least gains in dry shoot biomass after 28 days of soil moisture stress (6%) compared to CR95 which gained 33.3% of dry biomass during the same period. Interestingly, unlike other varieties, CR95 lost both fresh and dry biomass during recovery from 21 days of soil moisture stress while Cat128 and SL28 had big percentage gains in fresh shoot biomass (65.6% and 46.5%, resp.) during the same period. During recovery from 28 days of soil moisture stress, Cat129 had the highest percentage gains in fresh and dry shoot biomass while CR95 had an increase of 17.3% in fresh biomass but suffered a loss of 7.3 in dry shoot biomass (Figure 2).

3.3. *Total Biomass.* Initial fresh and dry biomass was not significantly different between varieties ($P > 0.05$). CR95 had significantly ($P < 0.05$) more total dry biomass after 21 days of soil moisture stress. Significantly more biomass ($P < 0.05$) was also realized from CR95 after 28 days of soil moisture stress with SL28 consistently being the least in both (Table 5). However, there were no significant differences ($P > 0.05$) in fresh and dry biomass between SL28, Cat129, and Cat128. Varieties did not show any significant differences ($P > 0.05$) in fresh and dry biomass 2 weeks after application of irrigation after both 21 days and 28 days of soil moisture stress (Table 5).

All varieties except CR95 had a reduction in total fresh biomass after 21 days of soil moisture stress. CR95 increased in total dry biomass by 37.6% and 29.7% after 21 days and 28 days of soil moisture stress respectively. Contrary, Cat129 gained only 7.7% and 6.3% of total biomass after 21 days and 28 days of soil moisture stress (Table 6). There were no changes in total fresh biomass of CR95 after 21 days of soil moisture stress, but after 28 days, the variety had lost 13.2% of fresh biomass.

CR95 continued to lose both fresh (−16.8%) and dry (−5.2%) biomass into the 2 weeks recovery after 21 days of soil moisture stress. During the same period, other varieties had increases in both fresh and dry biomass. Cat128 gained the most total fresh biomass (53.3%) while SL28 and Cat129 had the highest gains in total dry biomass (13.6% and 13.8%, resp.).

The greatest recovery potential after 28 days of soil moisture stress was shown by Cat129 which gained 101.0% of total fresh biomass and 46.9% of total dry biomass (Figure 3). CR95 was the least gaining only 11.2% of its total fresh biomass and 0.9% of its total dry biomass in the same period. Cat129 had the highest gains in dry root, shoot, and total biomass after 21 days and 28 days of soil moisture stress followed by SL28.

3.4. Correlation between Initial Biomass and Change after Stress. There was a strong positive relationship between a variety's initial fresh root biomass and changes in fresh root biomass ($r = 0.855$), dry root biomass ($r = 0.788$), and fresh and dry total biomass ($r = 0.83$ and $r = 0.912$, resp.) after 21 days of soil moisture deficit stress (Table 7).

After 21 days of soil moisture deficit stress, the strongest correlations were obtained for the relationship between a variety's initial shoot biomass and change in fresh shoot biomass ($r = 0.974$), initial fresh biomass and change in dry shoot biomass ($r = 0.968$), and initial total fresh biomass and change in total fresh biomass ($r = 0.960$). Initial dry root biomass was correlated strongly with changes in dry total biomass ($r = 0.952$) while initial dry root biomass was correlated with changes in dry shoot biomass after 21 days of moisture stress ($r = 0.936$).

The strongest correlation was obtained for the positive relationship between initial dry root biomass and changes in dry root biomass after 28 days of soil moisture deficit stress ($r = 964$), followed by the negative relationship between initial dry root biomass and changes in dry shoot biomass ($r = -0.904$).

The results indicated that initial root biomass was negatively correlated with changes in total fresh and dry biomass of young coffee ($r > 0.60$, Table 7) after both 21 and 28 days of soil moisture deficit stress, compared to initial shoot biomass which had weak correlation to both fresh and dry total biomass ($r < 0.5$).

4. Discussion

The absence of significance differences in initial root shoot and total biomass indicates that no coffee variety was significantly superior in biomass before soil moisture deficit stress. This confirms that under optimal conditions there are no morphological differences and yield differences between some of the coffee varieties tested in varietal trials [31]. It has however been established that significant morphological and physiological differences can occur between varieties due

to different soil fertility and environmental factors [32, 33] meaning that when conditions are similar, there may not be any differences in common coffee varieties such as the ones used in this study.

Exposing the varieties to soil moisture stress for 21 and 28 days significantly affected fresh and dry biomass indicating that the growth performance of coffee varieties is significantly affected by soil moisture deficit stress. This confirms that drought stress has significant effect on morphological and other growth characteristics of coffee plants [23–25]. For most of the varieties there was a reduction in fresh biomass and a slow buildup of dry biomass during period of soil moisture deficit stress.

Reduction in fresh biomass could largely be due to the loss of water which considerably contributes to the fresh biomass of the coffee plants. This water is important in maintaining tissue elasticity in plants and its loss is evident in the morphological characteristics of plants such as wilting and leaf folding [24, 26]. Whether the large losses in fresh shoot biomass in all varieties after 28 days of drought stress are a consequence of or a defense mechanism against drought stress is not certain. However, it is known that, in the early stages of soil moisture deficit, a plant would seek to maintain a constant rate of carbon assimilation which is achieved by reduction of leaf area, shedding other leaves, and/or other mechanisms that limit transpiration losses [21, 23].

CR95 had positive changes in root biomass after 21 and 28 days of soil moisture stress indicating that the root system is an integral part of its tolerance to soil moisture deficit stress. This is because its fresh and dry shoot biomass decreased due to soil moisture stress but its root biomass maintained a positive development. This supports the fact that the root system of a variety or a crop species is an important aspect in drought tolerance [4, 21, 24]. This is largely because a deeper and wide reaching root system in drought tolerant varieties is able to source more water for a longer time for the plant's functioning. The findings that heavier root biomass in CR95 contributes significantly to its drought tolerance corroborate the conclusion by Ramos and Carvalho [34] that drought tolerance in coffee is associated with more root biomass. In as much as this is the case, breeding for root biomass (and associated features such as root length) in coffee presents many challenges given the large environmental and local influences on root systems and very complex inheritance of root characteristics in coffee [4, 23].

Unlike the initial biomass in which varieties were not significantly different, soil moisture stress resulted in significant differences between varieties in dry root biomass and total dry biomass after 21 days, and fresh root biomass, dry root biomass, dry shoot biomass, total fresh, and dry biomass after 28 days of soil moisture stress. In all these cases, CR95 showed that it was the most resilient variety because it had significantly more biomass and positive change in biomass after the soil moisture stress period.

Coffee plants demonstrated differences in responding to moisture availability after drought stress because of the differences in incremental biomass accumulations after both 21 and 28 days of soil moisture deficit stress. Thus, some

TABLE 7: Pearson correlation coefficients (*r*) between initial biomass of some coffee physiological variables and changes in mass of varieties after stress.

Variable's initial biomass	Variable responding after stress	21 days		28 days	
		Fresh	Dry	Fresh	Dry
Fresh root	Roots	0.855	0.788	0.936	0.81
Dry root		0.657	0.597	0.968	0.964
Fresh shoots	Shoots	0.974	0.784	−0.666	0.736
Dry shoots		0.968	0.772	−0.666	0.726
Fresh total	Total	0.960	0.827	0.440	0.766
Dry total		0.914	0.872	0.550	0.835
Fresh root	Shoots	−0.818	−0.889	0.394	−0.844
Dry roots		−0.790	−0.936	0.300	−0.904
Fresh root	Total	−0.83	−0.912	−0.625	−0.827
Dry root		−0.797	−0.952	−0.718	−0.889
Fresh shoots	Total	0.265	0.293	0.290	0.385
Dry shoots		0.330	0.364	0.338	0.450

varieties grow faster than others after a stress period [25]. This is partly explained by observations that photosynthate allocation is altered during period of drought stress and during recovery period the normal growth functioning is restored enabling the plant to return to its normal state [21, 23]. What then makes differences is the quickness in which crop varieties trigger this mechanism depending on their genetic and physiological thresholds. However, some studies did not confirm this process in some varieties [24]. In terms of recovery from soil moisture stress, varieties were not consistent between recovering from 21 days and 28 days of soil moisture deficit stress signifying that the one week between these two periods is important during plant distress.

The general observation from this study is that variety CR95 is more drought tolerant than the rest of the varieties which were not significantly different from each other. This tolerance is both manifested and explained by superior root, shoot, and total fresh and dry biomass. These findings confirm the field observations that varieties with dense crowns (and thus more biomass) have lower boundary layer conductance which makes them better able to delay dehydration than those with open crowns [23]. Resultantly, densely crowned varieties are able to maintain productivity under water limiting environments and also require less irrigation for successful cropping.

It was observed that only initial fresh and dry root biomass was negatively correlated with change in biomass after physiological stress, while the rest of the factors had a positive correlation. This indicates that a coffee variety which has high root biomass is likely to withstand soil moisture deficit stress better than those with less root biomass or more shoot biomass. The fact that the other variables had a positive correlation to changes in biomass after stress shows that fresh and dry shoot biomass do not contribute to long term resilience of a variety as they are quickly lost when the plant is under stress. This is especially so for fresh shoot mass which is dominated by tissue water content [16].

Although characteristics that result in drought stress tolerance such as root and shoot biomass are positively

correlated with productivity in coffee [34], selection for drought resistance through a variety's behavioral traits and assuming that this automatically equates to improved yield by that variety under stress may be a very shallow approach bound to result in negative consequences [35]. This is because potential yield in coffee may be negatively correlated with some of the drought-adaptive traits and thus adaptability may reduce economic yield. Regardless to this fact, as pointed out by DaMatta [23] drought prone coffee farms are associated with low input systems and as such varieties that have better survival and yield stability under drought stress are of much greater value than those with greater yield potential under optimal conditions but cannot survive or perform under limiting environments.

5. Conclusions

The coffee variety CR95 demonstrated that it is drought tolerant compared to SL28, Catl29, and Catl28 owing to its superior biomass accumulation. However, once CR95 has reached critical levels of wilting, it requires more time to recover. Catl29 exhibited better ability to recover from periods of soil moisture stress. It was therefore concluded that physiological and molecular characteristics responsible for delaying wilting in varieties are different from those responsible for recovery. Field screening may be required to confirm the performance of coffee varieties under field conditions and the relationship between the physiological factors such as girth, branching, and flowering to tolerance. In addition more work looking at biochemical response, photosynthetic ability, evapotranspiration rates, and other attributes of these coffee varieties under drought stress will provide important information. An understanding of drought tolerance in coffee at a molecular level is also required. This study provides important information into selection of varieties for drought tolerance for use in the coffee sector. However, a holistic approach to variety selection that incorporates drought, disease, pest, and frost tolerance may provide a stronger basis for variety selection for broad recommendations to farmers and crop breeders.

Conflict of Interests

The authors declare that there is no conflict of interests regarding the publication of this paper.

Acknowledgments

The authors want to acknowledge the contributions of staff in the Agronomy Section at Coffee Research Institute under the supervision of Mr. E Njeni for assisting in the management of the trial.

References

[1] ADBG, "Coffee production in Africa and the Global Market situation," *Commodity Market Brief 1*, 2010.

[2] P. Baker, J. Bentley, C. Charveriat, H. Dugne, T. Leftoy, and H. Munyua, "The coffee smallholder," in *Coffee Futures: A Source Book of Some Critical Issues Confronting the Coffee Industries*, P. Baker, Ed., p. 111, CABI-FEDERACAFE-USDA-ICO, Chinchiná, Colombia, 2001.

[3] J. Coste, *Coffee: The Plant and the Product*, Longman, New York, NY, USA, 1992.

[4] W. J. C. Logan and J. Biscoe, *Coffee Handbook*, Coffee Growers Association, Harare, Zimbabwe, 1987.

[5] T. S. Murphy, N. A. Phiri, K. Sreedharan, D. Kutywayo, and C. Chanika, "Integrated stem borer management in smallholder coffee farms in India, Malawi and Zimbabwe," Final Technical Report, 2008.

[6] A. Chemura, C. Mahoya, D. Kutywayo, and P. Chidoko, "The growth response of coffee plants to organic manure, inorganic fertilizers and integrated soil fertility management under different irrigation levels," in *Proceedings of the RCZ International Research Symposium*, vol. 1, HICC, Research Council of Zimbabwe, Harare, Zimbabwe, February 2013.

[7] D. Kutywayo, A. Chemura, W. Kusena, P. Chidoko, and C. Mahoya, "The impact of climate change on the potential distribution of agricultural pests: the case of the coffee white stem borer (*Monochamus leuconotus* P.) in Zimbabwe," *PloS ONE*, vol. 8, no. 8, Article ID e73432, 2013.

[8] M. Maestri, F. M. Da Matta, A. J. Regazzi, and R. S. Barros, "Water relations of coffee leaves (*Coffea arabica* and *C. canephora*) in response to drought," *Journal of Horticultural Science*, vol. 68, pp. 741–746, 1993.

[9] J. Haggar and K. Schepp, *Coffee and Climate Change*, University of Greenwich, London, UK, 2011.

[10] P. Baker and J. Haggar, "Global warming: effects on global coffee," in *Proceedings of the Specialty Coffee Association of America Conference Handout (SCAA '07)*, p. 14, Long Beach, Calif, USA, 2007.

[11] International Coffee Organization, "Climate change and coffee," in *Proceedings of the 103rd Session of International Coffee Organizatoin (ICO '09)*, p. 14, International Coffee Council, London, UK, 2009.

[12] P. Laderach, A. Jarvis, and J. Ramirez, "The impact of climate change in coffee-growing regions: the case of 10 municipalities in Nicaragua," CafeDirect/GTZ, GTZ, 2006.

[13] G. Schroth, P. Laderach, J. Dempewolf et al., "Towards a climate change adaptation strategy for coffee communities and ecosystems in the Sierra Madre de Chiapas, Mexico," *Mitigation and Adaptation Strategies for Global Change*, vol. 14, no. 7, pp. 605–625, 2009.

[14] C. Gay, F. Estrada, C. Conde, H. Eakin, and L. Villers, "Potential impacts of climate change on agriculture: a case of study of coffee production in Veracruz, Mexico," *Climatic Change*, vol. 79, no. 3-4, pp. 259–288, 2006.

[15] R. Ghini, W. Bettiol, and E. Hamada, "Diseases in tropical and plantation crops as affected by climate changes: current knowledge and perspectives," *Plant Pathology*, vol. 60, no. 1, pp. 122–132, 2011.

[16] S. G. Tesfaye, M. R. Ismail, H. Kausar, M. Marziah, and M. F. Ramlan, "Plant water relations, crop yield and quality of arabica coffee (*Coffea arabica*) as affected by supplemental deficitirrigation," *International Journal of Agriculture & Biology*, vol. 15, pp. 665–672, 2013.

[17] B. M. Gichimu, *Arabica Coffee Breeding: Challenges Posed by Climate Change*, Kampala, Uganda, 2013.

[18] M. P. Reynolds, D. Hays, and S. Chapman, "Breeding for adaptation to heat and drought stress," in *Climate Change and Crop Production*, M. P. Reynolds, Ed., pp. 71–92, CABI, Oxfordshire, UK, 2010.

[19] D. Kutywayo, V. Chingwara, C. Mahoya, A. Chemura, and J. Masaka, "The effect of different levels of irrigation water and nitrogen fertilizer on vegetative growth components and yield of coffee in Zimbabwe," *Journal of Science and Technology MSU*, vol. 2, pp. 45–54, 2010.

[20] A. Arendse and T. A. Crane, "Impacts of climate change on smallholder farmers in Africa and their adaptation strategies: what are the roles for research?" in *Proceedings of the International Symposium and Consultation: Centro Internacional de Agricultura Tropical (CIAT '10)*, p. 29, Pan-Africa Bean Research Alliance (PABRA), Arusha, Tanzania, 2010.

[21] F. M. DaMatta, A. R. M. Chaves, H. A. Pinheiro, C. Ducatti, and M. E. Loureiro, "Drought tolerance of two field-grown clones of *Coffea canephora*," *Plant Science*, vol. 164, no. 1, pp. 111–117, 2003.

[22] F. R. C. F. César, S. N. Matsumoto, A. E. S. Viana, M. A. F. Santos, and J. A. Bonfim, "Leaf morphophysiology of coffee plants under different levels of light restriction," *Coffee Science*, vol. 5, no. 3, pp. 262–271, 2010.

[23] F. M. DaMatta, "Exploring drought tolerance in coffee: a physiological approach with some insights for plant breeding," *Brazilian Journal of Plant Physiology*, vol. 16, no. 1, pp. 1–6, 2004.

[24] P. C. Dias, W. L. Araujo, G. A. B. K. Moraes, R. S. Barros, and F. M. DaMatta, "Morphological and physiological responses of two coffee progenies to soil water availability," *Journal of Plant Physiology*, vol. 164, no. 12, pp. 1639–1647, 2007.

[25] M. Worku and T. Astatkie, "Dry matter partitioning and physiological responses of *Coffea arabica* varieties to soil moisture deficit stress at the seedling stage in Southwest Ethiopia," *African Journal of Agricultural Research*, vol. 5, no. 15, pp. 2066–2072, 2010.

[26] N. C. Turner, "Further progress in crop water relations," *Advances in Agronomy*, vol. 58, pp. 293–338, 1997.

[27] S. Haffani, M. Mezni, and W. Chaïbi, "Effect of drought on growth and chlorophyll content in three Vetch species," *IOSR Journal of Agriculture and Veterinary Science*, vol. 2, pp. 50–56, 2013.

[28] P. Chidoko, C. Mahoya, A. Chemura, and D. Kutywayo, "Evaluation of the efficacy of *Lantana camara*, *Albizia versicolor*

and *Allium sativum* for the control of coffee leaf rust under laboratory conditions," in *Proceedings of the 1st International Conference on Pesticidal Plants (ICPP '13)*, J. O. Ogendo, C. W. Lukhoba, P. K. Bett, and A. K. Machocho, Eds., ADAAPT Network, Nairobi, Kenya, 2013.

[29] A. Chemura, R. Madhlazi, and C. Mahoya, "Recycled coffee wastes as potential replacements of inorganic fertilizers for coffee production," in *Proceedings of the 22nd International Conference on Coffee Science (ASIC '08)*, pp. 1197–1201, Campinas, Brazil, September 2008.

[30] R. Core Team, *R: A Language and Environment for Statistical Computing*, R Foundation for Statistical Computing, Vienna, Austria, 2013.

[31] DR&SS, "Annual Summary Report Division of Crops Research," Tech. Rep., Department of Research & Specialist Services, Harare, Zimbabwe, 2011.

[32] A. Chemura, C. Mahoya, and D. Kutywayo, "Effect of organic nursery media on germination and initial growth of coffee seedlings," in *Proceedings of the 23rd Colloquium of the Association for Science and Information on Coffee (ASIC '10)*, p. 11, Bali, Indonesia, October 2010.

[33] P. Chidoko, *An Assessment of Genetic Diversity in Zimbabwean Coffee Varieties Using Morphological Markers [MSc]*, University of Zimbabwe, Harare, Zimbabwe, 2012.

[34] R. L. S. Ramos and A. Carvalho, "Shoot and root evaluations on seedlings from *Coffea genotypes*," *Bragantia*, vol. 56, no. 1, pp. 59–68, 1997.

[35] A. Blum, "Drought resistance, water-use efficiency, and yield potential—are they compatible, dissonant, or mutually exclusive?" *Australian Journal of Agricultural Research*, vol. 56, no. 11, pp. 1159–1168, 2005.

4

Varietal Trials and Physiological Components Determining Yield Differences among Cowpea Varieties in Semiarid Zone of Nigeria

Nkeki Kamai, Nuhu Adamu Gworgwor, and Joshua Wasinaninda Wabekwa

Department of Crop Production, University of Maiduguri, Maiduguri 600001, Nigeria

Correspondence should be addressed to Nkeki Kamai; nkekikamai@yahoo.com

Academic Editors: O. Ferrarese-Filho and Y. Ito

Field trials were conducted at the Teaching and Research Farm, Faculty of Agriculture, University of Maiduguri, Maiduguri ($11°47.840'$N; $13°12.021'$E; elevation 319 m asl), in Borno State in semiarid zone of Nigeria during the 2010 and 2011 rainy seasons. The objectives of the study were to evaluate the agronomic performances of some improved cowpea varieties and to identify the physiological traits associated with high grain yield in the semiarid zone of Nigeria. The trial consisted of eight treatments, which included two local varieties, namely, *Kannanado White* and *Borno Brown* and six improved varieties, namely, IT90K-277-2, IT97K-568-18, IT89KD-288, IT97K-499-35, IT98K-131-2, and IT89KD-391. The treatments were laid out in a randomized complete block design replicated three times. The gross plot size was 5.0 m × 4.0 m (20 m^2) while the net plot size was 3.6 m × 3.0 m (10.8 m^2). The results showed that the improved varieties, namely, IT90K-277-2, IT97K-499-35, IT98K-131-2, and IT89KD-288, had significantly higher grain yield per hectare and matured earlier to escape drought in this agroecological zone. The local varieties also had significantly heavier grains, took more days to reach first and 50% flowering, and matured later than the improved varieties. Cowpea grain yield per hectare was highly positively correlated with harvest index, shell weight, soil moisture suction measurements, shelling percentage, and grain yield per plant and also significant negative correlation between cowpea grain yield per hectare and number of days to first and 50% flowering, 100-grain weight, number of days to physiological maturity, and pod development period. The results also indicated that fodder yield per hectare was highly positively correlated with photosynthetically active radiation thereby indicating that higher photosynthetically active radiation produced higher yield of fodder.

1. Introduction

Cowpea, *Vigna unguiculata* (L.) Walp, also popularly called "beans" in Nigeria is a legume of vital importance to the livelihood of millions of people in West and Central Africa. It provides nutritious grain and less expensive source of protein for both rural and urban consumers [1]. It is estimated that cowpea supplies 40% of the daily protein requirements to most people in Nigeria [2]. The use of cowpea haulms as fodder is attractive in mixed crop/livestock systems where both grain and fodder can be obtained from the same crop [3]. Some 8 million ha of cowpea are in West and Central Africa, especially in Nigeria, Burkina Faso, Mali, and Senegal. Nigeria is the largest cowpea producer in the world and also has the highest level of consumption [4, 5]. In Nigeria cowpea is largely grown in the northern part of the country which has

savanna type of vegetation and light rainfall [6] and the production trend of cowpea shows a significant improvement with an increase of some 440% in planted area and increase of some 410% in yield over the period of 1961–1995 [7]. Despite the potential for further yield increase, cowpea production faces numerous problems including pest attack, diseases, and drought. As the population continues to increase, new productive cowpea varieties are needed that will overcome these problems [8]. Due to several constraints the average cowpea grain production in West Africa was reported to be as low as 358 kg/ha [4] whereas Singh et al. [9] estimated 240 kg/ha cowpea grain yield as an average for northern Nigeria. In most parts of Borno State, rainfall is unreliable, frequently less and poorly distributed for a good cowpea crop. In semiarid zone, early season and terminal drought conditions are almost an annual event. Improving the yield of cowpea in

Varietal Trials and Physiological Components Determining Yield Differences among Cowpea Varieties in
Semiarid Zone of Nigeria

35

the state requires the use of drought-tolerant and drought-avoidance varieties [10].

The objectives of the study were to evaluate the agronomic performances and to identify physiological traits associated with high grain yieldof some improved cowpea varieties in semiarid zone of Nigeria. The local cowpea varieties are late maturing, low yielding and photosensitive, and very susceptible to drought and heat. Even in the average year, the cowpea cultivars have to rely on moisture stored in the soil after the rains have stopped for grain filling. The crop performs poorly if the rain ends early [11]. The improved varieties have acceptable seed quality for various regions and are resistant to major diseases and parasitic weeds. They also have synchronous flowering and maturity [12]. The improved varieties are therefore early maturing, photoinsensitive and have high yield potential even with less rainfall. In the same vein the improved cowpea varieties have varying degree of yield potentials which could be due to differences in their physiological traits in the dry ecologies of Borno State. Therefore, the need to try these promising cowpea varieties for their adaptability in the semiarid zone of Nigeria is obvious as one of the strategies for improving the productivity of the crop in this region since scanty information is available on the performance of these varieties in this zone. Information on physiological differences of the different cowpea varieties will be valuable for future strategies in the development of high yielding cowpeas for the semiarid zone of Nigeria.

2. Materials and Methods

The study was conducted at the Teaching and Research Farm, Faculty of Agriculture, University of Maiduguri, ($11°47.840'$N; $13°12.021'$E; elevation 319 m asl) in Borno State in the semiarid zone, Nigeria, during the 2010 and 2011 rainy seasons, August to November each year. The gross plot size was 5.0 m × 4.0 m (20 m^2) and the net plot size was 3.6 m × 3.0 m (10.8 m^2). Each plot contained 8 rows of 4.0 m long with spacing of 0.75 m between rows and 0.2 m between plants. The trial consisted of 8 treatments (varieties of cowpea). The treatments included two local varieties, namely, *Kannanado White* and *Borno Brown*, and 6 improved varieties, namely, IT90K-277-2, IT97K-568-18, IT89KD-288, IT97K-499-35, IT98K-131-2, and IT89KD-391. The treatments/varieties were laid out in a randomized complete block design (RCBD) replicated 3 times. Physiological parameters measured are seedling establishment (%) at two weeks after sowing (2 WAS), number of days to first flowering, number of days to 50% flowering, soil moisture suction measurement (centibars), transmitted photosynthetically active radiation, pod development period (days), number of days to physiological maturity, 100-grain weight (g), shelling percentage, harvest index (HI), grain yield per plant (g), grain yield (kg ha^{-1}), shell weight (kg ha^{-1}), and fodder yields. All data were subjected to analysis of variance (ANOVA) using Statistix 8.0 version. Treatment means were compared where *F*-values were significant using Duncan's multiple range test (DMRT) at 5% level of probability [13]. Linear correlation coefficient (*r*) among combined means of two years of cowpea variety and physiological parameters was calculated at 5%.

TABLE 1: Physicochemical characteristics of the soil at the experimental site.

S/No.*	Soil characteristics	Physicochemical properties
	Chemical analysis	
1	pH in H_2O	6.71
2	Organic carbon (g/kg)	4.40
3	Organic matter (g/kg)	7.59
4	Total N (g/kg)	0.05
5	Available potassium (me/100 g)	0.29
6	Available phosphorus (g/kg)	5.30
	Mechanical analysis (0–15 cm depth)	
1	Clay (%)	15.0
2	Sand (%)	70.0
3	Silt (%)	15.0
4	Field texture	Sandy loam

* S/No.: serial number.

3. Results and Discussion

The soil was sandy loam having organic matter content of 7.59 g/kg. The pH of the soil was almost neutral while available phosphorus was 5.30 g/kg (Table 1). Based on the soil properties of the site it was ideal for cowpea growth. Cowpea variety had no significant effect on seedling establishment (%) at 2 WAS (Table 2). The nonsignificance in stand count is a clear indication that there was a good germination of all the varieties; thus seed quality and viability among the varieties were very good. Data on number of days to first and 50% flowering as influenced by cowpea variety as presented in Table 2 showed that *Kannanado White* and *Borno Brown* varieties had significantly the longest number of days to first and 50% flowering compared with the rest of the varieties. The study also revealed that the highest mean soil moisture suction measurements were significantly obtained by IT97K-131-2 compared with the other varieties but only comparable with IT89K-391 and IT97K-499-35 (Table 2). The study shows that the local varieties and IT89KD-288 significantly conserve more moisture. The soil suction reading is a direct measure of the availability of moisture for plant growth. As the soil becomes drier, these films become thinner and the attraction or suction increases. The plant root has to overcome this soil suction, or attraction force, in order to withdraw moisture from the soil [14]. It is noteworthy that these varieties had the highest grain yields compared to the other varieties. This is so because according to Campos et al. [15] and Ogbonnaya et al. [16] cowpea is known to have high stomatal control leading to a rapid closure of stomata under water stress conditions. Table 2 shows the significant effect of cowpea variety on the amount of solar radiation (SR), especially photosynthetically active radiation (PAR), intercepted by the crop where *Kannanado White*, *Borno Brown*, IT97K-499-35, and IT89KD-288 intercepted significantly the highest PAR compared with the rest of the varieties. Also *Borno Brown*, *Kannanado White*, and IT89K-391 varieties significantly took the longest number of days for the individual pods to mature from authesis

TABLE 2: Effect of cowpea variety on physiological parameters at Maiduguri in 2010 and 2011 combined analysis.

Treatment/cowpea variety	Seedling establishment at 2 weeks after sowing	Number of days to first flowering	Number of days to 50% flowering	Soil moisture suction measurements at (centibars)	Transmitted photosynthetic active radiation	Pod development period (days)	Number of days to physiological maturity	100-grain weight (g)	Shelling percentage (%)	Harvest index	Grain yield per plant (g)	Grain yield (kg ha^{-1})	Shell weight (kg ha^{-1})	Fodder yield (kg ha^{-1})
IT90K-277-2	88.50	43.00[d]	52.67[d]	12.83[b]	76.23[bc]	13.64[c]	82.50[cd]	15.50[b]	78.04	34.35[a]	15.92[a]	998.2[a]	291.68[b]	3387.6[cd]
Kannanado White	90.17	64.33[a]	80.00[a]	8.83[c]	92.08[a]	18.08[a]	97.17[a]	19.80[a]	63.13	15.18[e]	7.37[c]	378.3[c]	140.87[d]	4016.1[b]
IT97K-499-35	93.50	39.17[e]	52.17[d]	13.00[ab]	88.53[ab]	17.03[ab]	80.67[cd]	15.17[b]	69.14	34.27[a]	12.27[ab]	990.8[a]	451.37[a]	2720.7[c]
Borno Brown	89.83	64.17[a]	78.33[a]	6.50[cd]	90.75[a]	18.38[a]	98.50[a]	20.42[a]	63.81	14.37[e]	5.59[c]	373.9[c]	167.10[cd]	4902.8[a]
IT89KD-391	96.17	45.17[cd]	55.83[cd]	13.67[ab]	68.68[c]	17.67[a]	78.83[d]	16.07[b]	72.92	27.48[bc]	12.72[ab]	745.6[b]	296.85[b]	3373.0[bc]
IT97K-568-18	87.17	47.67[c]	60.67[bc]	12.50[b]	64.32[c]	16.53[ab]	82.17[c]	14.86[b]	71.15	25.87[c]	9.46[b]	750.4[b]	290.03[b]	3699.1[b]
IT98K-131-2	89.83	44.83[cd]	54.00[cd]	16.00[a]	71.27[c]	15.14[bc]	76.33[e]	15.15[b]	69.63	31.00[ab]	12.46[ab]	938.0[ab]	419.17[a]	3425.6[bc]
IT89KD-288	91.17	56.00[b]	65.33[b]	5.17[d]	94.87[a]	15.13[bc]	91.67[b]	15.97[b]	77.21	20.73[d]	9.47[bc]	784.8[ab]	235.13[bc]	5365.1[a]
SE (±)	2.887	1.201	1.025	1.089	4.891	0.685	0.675	0.419	2.571	1.614	1.586	75.18	32.232	261.46

1: means within a column and treatment followed by similar letter(s) are not significantly different ($P \leq 0.05$) according to Duncan's multiple range test (DMRT).

compared with the other varieties, except varieties IT89KD-391 and IT97K-568-18 (Table 2). The significantly longest number of days to physiological maturity was recorded by the local varieties *Kannanado White* and *Borno Brown* in the combined data (Table 2). Similar results were reported by Elemo [17]. This is probably because they produced most of their flowers and pods at the end of the rain unlike the elite varieties. The results indicated that *Kannanado White* and *Borno Brown* local varieties had significantly the heaviest grains but significantly the lowest grain yields compared with the rest of the varieties (Table 2). Ellis-Jones and Amaza [18] reported lower adoption of IT97K-499-35 (*Striga*-resistant and higher grain yielding) in a study area in North East Nigeria because farmers preferred local varieties that are large-seeded. Efforts should therefore be made to develop cowpea varieties that meet end-user preferences [19].

The data in Table 2 show the differences among the cowpea varieties on shelling percentage and the results indicated that cowpea variety had no significant effect on this parameter. Table 2 also presents that the varieties IT90K-277-2, IT97K-499-35, and IT98K-131-2 had significantly higher harvest index compared with the rest of the varieties. This means that these varieties with higher harvest index have higher ability to partition current assimilates to the grain and the reallocation of stored structural assimilates to the seed [20]. Table 2 further shows the variety IT90K-277-2 had significantly the highest mean yield per plant among the 8 cowpea varieties tested, but only comparable with IT97K-499-35, IT89KD-391, and IT98K-131-2. The highest mean grain yields (kg ha^{-1}) were produced by the varieties IT90K-277-2 and IT97K-499-35 compared with the other varieties, except for IT98K-131-2 and IT89-288. Despite the high yield potential of these varieties, their adoption by farmers may be of some concern. Kamara et al. [8] reported that despite the yield benefits of new varieties, farmers have shown preference for local ones, even when introduced varieties give higher grain yields. The reasons, among others, are ability for relay planting with creeping habit and ability to smother weeds. Also earlier reports showed that seed size is a primary determinant of yield in cowpea [21, 22]; this was not the case in the present study and that of Nakawuka and Adipala [23]. This discrepancy may have been due to the different genotypes used.

The cowpea varieties IT97K-499-35 and IT98K-131-2 had significantly the highest mean shell weight (pod wall) per hectare compared to the rest of the varieties. Data in Table 2 presents the differences among the cowpea varieties on the mean fodder yield (kg ha^{-1}) where the highest fodder yield (5361.1 kg ha^{-1}) was obtained by the semideterminate improved variety (IT89KD-288) but only at par with that of the local variety *Borno Brown* (4902.8 kg ha^{-1}) (Table 2). This observation did not agree with the findings of Kamara et al. [19] who reported that the variety IT97K-499-35 (semierect, determinate) produced 42% more biomass than *Borno Brown* since this variety was more heavily infested with *Striga*. A similar observation was reported by Muli and Saha [24] who found that local cultivars were more productive in terms of leaf yields. This calls for screening efforts to be geared towards high grain yield from indeterminate varieties, while still maintaining a high yield of fodder. The role played by fodder

provision from cowpea to animals during the dry season in the drier northern parts of West Africa is very important [25]. In this study, it is shown that the early maturing cowpea varieties (IT90K-277-2, IT97K-499-35, IT97K-568-18, and IT97K-131-2) produced significantly lower fodder (kg ha^{-1}) and 100-grain (g) weight compared to the other varieties. This is in agreement with the findings of Ntare and Williams [26] who reported that the early maturing cultivars (TVX 3236 and B111-2) produced the smallest grains and fodder yield.

4. Interrelationships among Physiomorphological Parameters

The summary of the correlation coefficients between grain yield per hectare and other physiological parameters average across the two years is presented in Table 3. The correlation among the variables showed many significant values. Cowpea grain yield per hectare was highly positively correlated with harvest index, shell weight, soil moisture suction measurements, shelling percentage, and grain yield per plant. However, the results show significant negative correlation between cowpea grain yield per hectare and number of days to first and 50% flowering, 100-grain weight, number of days to physiological maturity, and pod development period. The significant negative correlation observed between seed yield and duration of reproductive phase in this study implies that an attempt to breed for long duration phase could repress yield, especially in the intermediate cowpea varieties. A similar observation was made by Turk et al. [27] and Ombakho and Tyagi [28] in cowpea. The results also indicated that number of days to 50% flowering, fodder yield per hectare, 100-grain weight, number of days to physiological maturity, and pod development period were significantly positively correlated with number of days to first flowering, while plant stand at harvest (%), harvest index, shell weight, soil moisture suction measurements, and grain yield per plant were significantly negatively correlated with it (Table 3).

Number of days to 50% flowering had a significant positive correlation with fodder yield per hectare, 100-grain weight, number of days to physiological maturity, PAR, and pod development period and a significant negative correlation with harvest index, shell weight, soil moisture suction measurements, shelling percentage, and grain yield per plant (Table 3). Fodder yield per hectare had a significant positive correlation with 100-grain weight, number of days to physiological maturity, and PAR and a significant negative correlation with harvest index, shell weight, and soil moisture suction measurements. The positive correlation of fodder yield per hectare with the percentage transmitted photosynthetically active radiation is in tune with the findings of Gallagher and Biscoe [29] who reported that under nonstressed environmental conditions, the amount of dry matter produced by a crop is linearly related to the amount of solar radiation (SR), especially photosynthetically active radiation (PAR), intercepted by the crop. Also fodder yield per hectare and PAR are highly negatively correlated with soil moisture suction measurements. Therefore, species that intercept a large fraction of PAR are important in dry environments like the semiarid

TABLE 3: Correlation coefficients between cowpea varieties, grain yield, and other parameters tested in 2010 and 2011 at Maiduguri.

	GY	DFF	SE2W	DFPF	FY	HGW	HI	DPM	PAR	PDP	SW	MSHG	SMSM	SP	YPP
GY	1.00														
DFF	-0.66**	1.00													
SE2W	-0.10	-0.24	1.00												
DFPP	-0.78**	0.83**	-0.01	1.00											
FY	-0.25	0.50**	0.04	0.49**	1.00										
HGW	-0.55**	0.76**	-0.08	0.74**	0.32*	1.00									
HI	0.74**	-0.66**	-0.21	-0.79**	-0.61**	0.50**	1.00								
DPM	-0.49**	0.84**	-0.25	0.76**	0.46**	0.81**	-0.46**	1.00							
PAR	-0.10	0.25	0.22	0.30*	0.41**	0.28	-0.40**	0.29*	1.00						
PDP	-0.38**	0.32*	0.18	0.38**	0.05	0.48**	-0.21	0.33*	0.07	1.00					
SW	0.64**	-0.63**	0.25	-0.64**	-0.29	-0.54**	0.34*	-0.69**	0.14	-0.29*	1.00				
MSHG	0.12	0.09	-0.20	-0.05	-0.17	0.11	0.18	0.30*	0.09	0.07	-0.12	1.00			
SMSM	0.40**	-0.45**	-0.26*	-0.53**	-0.74**	-0.34*	0.68**	-0.43**	-0.58**	-0.12	0.23	0.02	1.00		
SP	0.51**	-0.09	-0.42**	-0.40**	-0.06	-0.15	0.62**	0.08	-0.36*	-0.08	-0.16	0.20	0.33*	1.00	
YPP	0.63**	-0.44**	-0.01	-0.61**	-0.24	-0.38**	0.70**	-0.38**	-0.16	-0.19	0.28	-0.01	0.33*	0.52**	1.00

GY: grain yield, SMSM: soil moisture suction measurement, DFF: days to 50% flowering, PAR: photosynthetically active radiation, PDP: pod development period, DPM: days to physiological maturity, HGW: 100-grain weight, SP: shelling percentage, HI: harvest index, SE2W: seedling establishment 2 WAS, YPP: yield per plant, SW: shell weight, FY: fodder yield, DFPF: days to first flowering, **: highly significant at 1% probability level, and *: significant at 5% probability level.

Varietal Trials and Physiological Components Determining Yield Differences among Cowpea Varieties in
Semiarid Zone of Nigeria

39

zone of Nigeria, where sunshine is abundant. 100-grain weights had a significant positive correlation with harvest index, number of days to physiological maturity, and pod development period and a significant negative correlation with shell weight, soil moisture suction measurement, and grain yield per plant (Table 3). Harvest index had a significant positive correlation with shell weight, soil moisture suction measurements, shelling percentage, and grain yield per plant, and a significant negative correlation with number of days to physiological maturity and PAR. Number of days to physiological maturity had a significant positive correlation with PAR and pod development period and a significant negative correlation with shell weight, soil moisture suction measurements, and grain yield per plant. The pod development period had a significant negative correlation with shell weight (Table 3).

5. Conclusions

In the semiarid zone of Nigeria the highest mean grain yields $(kg\,ha^{-1})$ were produced by the varieties IT90K-277-2, IT97K-499-35 compared with the other varieties, except for IT98K-131-2 and IT89-288. Despite the high yield potential of these varieties, their adoption by farmers may be of some concern. Kamara et al. [8] reported that despite the yield benefits of new varieties, farmers have shown preference for local ones, even when introduced varieties give higher grain yields. The reasons, among others, are ability for relay planting with creeping habit and ability to smother weeds. Correlation studies indicate that cowpea grain yield per hectare was highly positively correlated with harvest index, shell weight, soil moisture suction measurements, shelling percentage, and grain yield per plant and negatively correlated with cowpea grain yield per hectare, number of days to first and 50% flowering, 100-grain weight, number of days to physiological maturity, and pod development period. The significant negative correlation observed between seed yield and duration of reproductive phase in this study implies that an attempt to breed for long duration phase could repress yield, especially in the intermediate cowpea varieties.

Conflict of Interests

The authors declare that there is no conflict of interests regarding the publication of this paper.

References

[1] I. Inaizumi, B. B. Singh, P. C. Sanginga, V. M. Manyong, A. A. Adesina, and S. Tarawali, *Adoption and Impact of Dry-Season Dual-Purpose Cowpea in the Semiarid Zone of Nigeria*, Impact (International Institute of Tropical Agriculture), IITA, Ibadan, Nigeria, 1999.

[2] N. C. Muleba, J. B. S. Dabire, I. Drabo, and J. T. Ouedraogo, "Technologies for cowpea production based on genetic and environmental manipulations in the semi-arid tropics," in *Technologies Options for Sustainable Agriculture in Sub-Saharan Africa*, T. Bezuneh, A. M. Emechebe, J. Sedgo, and M. Oeudrago,

Eds., pp. 192–206, Semi-Arid Food Grain Research and Development Agency (SAFGRAD) of the Scientific, Technical and Research Commission of OAU, Ouagadougou, Burkina Faso, 1997.

[3] S. A. Tarawali, B. B. Singh, M. Peters, and S. F. Blade, "Cowpea haulms as fodder," in *Advances in Cowpea Research*, B. B. Singh, D. R. M. Raj, K. E. Dashiel, and L. E. N. Jackai, Eds., pp. 313–325, Copublication of International Institute of Tropical Agriculture (IITA) and Japan International Research Centre for Agricultural Sciences (JIRCAS), Ibadan, Nigeria, 1997.

[4] FAO, 2000, http://www.fao.org/statistics/en/.

[5] B. B. Singh, "Potential and constraints of improved cowpea varieties in increasing the productivity of cowpea-cereal systems in the dry Savannas of West Africa," in *A Plan to Apply Technology in the Improvement of Cowpea Productivity and Utilisation for the Benefit of Farmers and Consumers in Africa: Proceedings of Cowpea Stakeholders Workshop*, P. Majiwa, M. Odera, N. Muchiri, G. Omanya, and P. Werehire, Eds., pp. 14–26, African Agricultural Technology Foundation, Nairobi, Kenya, 2007.

[6] Anonymous, "Cowpea: Abuja Securities and Commodity Exchange PLC," 2008, http://www.abujacomex.com/.

[7] R. Ortiz, "Cowpeas from Nigeria: a silent food revolution," *Outlook on Agriculture*, vol. 27, no. 2, pp. 125–128, 1998.

[8] A. Y. Kamara, J. Ellis-Jones, F. Ekeleme et al., "A participatory evaluation of improved cowpea cultivars in the Guinea and sudan savanna zones of north east Nigeria," *Archives of Agronomy and Soil Science*, vol. 56, no. 3, pp. 355–370, 2010.

[9] B. B. Singh, O. L. Chamblis, and B. Sharma, "Recent advances in cowpea breeding," in *Advances in Cowpea Research*, B. B. Singh, D. R. M. Raj, K. E. Dashiel, and L. E. N. Jackai, Eds., pp. 30–49, Copublication of International Institute of Tropical Agriculture (IITA) and Japan International Research Centre for Agricultural Sciences (JIRCAS), Ibadan, Nigeria, 1997.

[10] J. E. Onyibe, A. Y. Kamara, and L. O. Omoigui, *Guide to Cowpea Production in Borno State, Nigeria*, Promoting Sustainable Agriculture in Borno State (PROSAB), Ibadan, Nigeria, 2006.

[11] A. K. Raheja, "Problems and prospects of cowpea production in the Nigerian Savannas," in *Proceedings of the 1st World Cowpea Research Conference*, Tropical Grain Legume Bulletin, no. 32, pp. 78–87, IITA, November 1986.

[12] B. B. Singh, "Breeding suitable cowpea varieties for West and Central African Savanna," in *Progress in Food Grains Research and Production in Semi-Africa*, J. M. Menyonga, J. B. Bezuneh, J. Y. Yayock, and I. Soumana, Eds., pp. 77–85, OAU/STRC-SAFGRAD, Ouagadougou, Burkina Faso, 1994.

[13] D. B. Duncan, "Multiple range and multiple F tests," *Biometrics*, vol. 11, no. 1, pp. 1–42, 1955.

[14] Soil Moisture Equipment Corp., "Quick draw soil moisture probe," Operating Instructions Model 2900F1, Soil Moisture Equipment Corp., Santa Barbara, Calif, USA, 1989.

[15] P. S. Campos, J. C. Ramalho, J. A. Lauriano, M. J. Silva, and M. D. C. Matos, "Effects of drought on photosynthetic performance and water relations of four *Vigna* genotypes," *Photosynthetica*, vol. 36, no. 1-2, pp. 79–87, 1999.

[16] C. I. Ogbonnaya, B. Sarr, C. Brou, O. Diouf, N. N. Diop, and H. Roy-Macauley, "Selection of cowpea genotypes in hydroponics, pots, and field for drought tolerance," *Crop Science*, vol. 43, no. 3, pp. 1114–1120, 2003.

[17] K. A. Elemo, "Farmer participating in technology testing: a case of agronomic evaluation of cowpea genotypes in the Nigerian Northern Guinea Savanna," *Agricultural Systems in Africa*, vol. 3, no. 1, pp. 39–49, 1993.

[18] J. Ellis-Jones and P. S. Amaza, "PROSAB: promoting sustainable agriculture in Borno State: an adoption and impact assessment of PROSAB's activities over three cropping seasons (2004–2006)," PROSAB Internal Report, 2007.

[19] A. Y. Kamara, D. Chikoye, F. Ekeleme, L. O. Omoigui, and I. Y. Dugje, "Field performance of improved cowpea varieties under conditions of natural infestation by the parasitic weed *Striga gesnerioides*," *International Journal of Pest Management*, vol. 54, no. 3, pp. 189–195, 2008.

[20] N. C. Turner, G. C. Wright, and K. H. M. Siddique, "Adaptation of grain legumes (pulses) to water-limited environments," *Advances in Agronomy*, vol. 71, pp. 193–231, 2001.

[21] B. C. Imrie and R. A. Bray, "Estimates of combining ability and variance components of grain yield and associated characters of cowpea," in *Proceeding of the Australian Plant Breeding Conference*, pp. 202–204, February 1983.

[22] I. O. Obesesan, "Association among grain yield components in cowpea (*Vigna unguiculata* L. Walp.)," *Genetical Agriculture*, vol. 39, no. 4, pp. 377–386, 1985.

[23] C. K. Nakawuka and E. Adipala, "A path coefficient analysis of some yield components interactions in cowpea," *African Crop Science Journal*, vol. 7, no. 4, pp. 327–331, 1999.

[24] M. B. Muli and H. M. Saha, *Participatory Evaluation of Cowpea Cultivars for Adaptation and Yield Performance in Coastal Kenya*, Kenya Agricultural Research Institute, Regional Research Centre, Mtwapa, Kenya, 2008.

[25] N. A. Gworgwor and H. C. Weber, "Studies on biology and control of *Striga*: II. Varietal response of cowpea (*Vigna unguiculata* (L.) Walp.) to *Striga gesnerioides*," *Journal of Agronomy and Crop Science*, vol. 166, no. 2, pp. 136–140, 1991.

[26] B. R. Ntare and J. H. Williams, "Response of cowpea cultivars to planting pattern and date of sowing in intercrops with pearl millet in Niger," *Experimental Agriculture*, vol. 28, no. 1, pp. 41–48, 1992.

[27] K. J. Turk, A. E. Hall, and C. W. Asbell, "Drought adaptation of cowpea. I. Influence of drought on seed yield," *Agronomy Journal*, vol. 72, no. 3, pp. 413–420, 1980.

[28] G. A. Ombakho and A. P. Tyagi, "Correlation and path coefficient analysis for yield and its components in cowpea (*Vigna unguiculata* (L.) Walp.)," *East African Agriculture and Forestry Journal*, vol. 53, no. 1, pp. 23–27, 1987.

[29] J. L. Gallagher and P. V. Biscoe, "Radiation absorption, growth and yield of cereals," *The Journal of Agricultural Science*, vol. 91, no. 1, pp. 47–60, 1978.

5

Response of Maize Seedlings to Cadmium Application after Different Time Intervals

Iqbal Hussain,[1] **Shamim Akhtar,**[1] **Muhammad Arslan Ashraf,**[1] **Rizwan Rasheed,**[1] **Ejaz Hussain Siddiqi,**[2] **and Muhammad Ibrahim**[3]

[1] *Department of Botany, Government College University, Faisalabad 38000, Pakistan*
[2] *Department of Botany, University of Gujrat, Gujarat, Pakistan*
[3] *Department of Applied Chemistry, Government College University, Faisalabad 38000, Pakistan*

Correspondence should be addressed to Iqbal Hussain; iqbalbotanist1@yahoo.com

Academic Editors: M. Arias-Estévez, L. F. Goulao, K. Okuno, and M.-J. Simard

Present study was conducted to appraise the inhibitory effects of cadmium applied at different time intervals on various growth and biochemical parameters in two maize lines, Maize-TargetedMutagenesis 1 and 2 (MTM-1 and MTM-2). Twenty-day-old seedlings were exposed to 0, 3, 6, 9, and 12 mg $CdCl_2$ kg^{-1} sand. Both maize lines exhibited significant perturbations in important biochemical attributes being employed for screening the crops for cadmium tolerance. The results showed that a higher concentration of cadmium (12 mg $CdCl_2$ kg^{-1}) considerably reduced the plant growth in line MTM-1 on the 5th, 10th, and 15th day after the treatment. In contrast, irrespective of exposure time, the plant biomass and leaf area did not show inhibitory effects of cadmium, specifically at 3 mg $CdCl_2$ kg^{-1} in line MTM-2. In addition, MTM-2 was found to be more tolerant than line MTM-1 in terms of lower levels of hydrogen peroxide (H2O2), malondialdehyde (MDA) contents, and relative membrane permeability (RMP). Moreover, H2O2, MDA, RMP, and anthocyanin increased at all levels of cadmium in both lines, but a significant decline was observed in photosynthetic pigments, total free amino acids, and proline contents in all treatments particularly on the 10th and 15th day after treatment.

1. Introduction

Heavy metal stress is one of the major abiotic stresses that cause environmental pollution in recent decades. Cadmium is a toxic pollutant that negatively affects the plant growth. Cadmium is added to the environment by different sources and is persistent for a long time in the environment; it comes into the food chain through plants and threatened the ecosystems [1]. Cadmium is taken up by the plant roots and loaded into the leaves through the phloem and can be accumulated in all parts of the plants [2]. Thus, instead of just reducing the crop productivity and quality [3], it causes a severe health risk to mammals and humans [4].

The cadmium affects the whole life cycle of plants. It inhibits the seed germination [5], disturbs the photosynthetic metabolism and transpiration rate, reduces enzymatic and non enzymatic activities [6], disturbs water homeostasis and ionic relations [7], mineral nutrition [8], induces synthesis of reactive oxygen species, and strongly reduced the biomass production [8]. The cadmium stress causes chlorosis and leaf and root necrosis resulting in stunted growth in the majority of the plants [8]. However, the amount of cadmium deposited into the root, shoot, and interveins of leaves considerably differed among different species [9].

Cadmium induced oxidative stress at the cellular level in different plants [10]. Moreover, cadmium triggers accumulation and/or synthesis of reactive oxygen species like superoxide radical, singlet oxygen, hydrogen peroxides, and hydroxyl radicals [11] that may cause cell death by lipid peroxidation, oxidation of proteins [12], damaging DNA [13], and affect the activities of antioxidant enzymes (superoxide dismutase, guaiacol peroxidase, ascorbate peroxidase, catalase, and glutathione reductase) involved in the oxidative defense system [14]. Thus, a balance between the production of activated oxygen species and quenching activity of antioxidant was disturbed [15]. To repair the inhibitory effects of reactive oxygen species, plants make use of antioxidant defense machinery including enzymatic and nonenzymatic defense system [16].

Maize (*Zea mays* L.) is a suitable crop for tropics and subtropical regions of the world. Maize can be an excellent model plant to study the physiological changes responsible for reduced productivity under stressful conditions. Being a rich source of nutrition (72% starch, 10% protein, 8.5% fiber, and 4.8% edible oil), maize is a major source of food, sugar, cooking oil, and animal feed all over the globe [17].

Keeping the value of maize crop in mind, cadmium stress being the harmful menace to its crop growing, the experiment was performed to assess the effect of various cadmium concentrations on different agronomical, physiochemical attributes of maize seedlings after different time intervals.

2. Materials and Methods

2.1. Plant Material, Treatment, and Plant Growth Conditions. Seeds of two maize lines, Maize-Targeted Mutagenesis 1 and 2 (MTM-1 and MTM-2), were obtained from the Chinese Academy of Agriculture Sciences China and Plant Genetics Resources Institute (PGRI), NARC, Islamabad, Pakistan (Collection Center). An experiment was conducted at the Department of Botany, GC University, Faisalabad, Pakistan. The seeds were disinfected with 0.1% $HgCl_2$ solution for 15 min and washed with distilled water before sowing in the plastic pots containing 10 kg sand. Ten seeds were sown in each pot. The plants were given half strength Hoagland's nutrient solution on a five day basis. After germination, five uniform healthy seedlings were retained in each pot. Twenty-day-old seedlings were exposed to five different levels of cadmium (0.0, 3, 6, 9, and 12 mg $CdCl_2$ kg^{-1} sand) in half strength Hoagland's nutrient solution. The experimental design was completely randomized with three replications per treatment. The sampling was done on the 5th, 10th, and 15th day after cadmium treatment.

2.2. Growth Determinations. Leaf surface area was measured using a leaf area meter. After drying in an oven at 70°C for about 72 h, shoot and root dry weights were recorded. The other plant samples were preserved in cooling chamber at −20°C.

2.3. Chlorophyll a, b and Total Carotenoids Contents. Chlorophyll (Chl) a and b and total carotenoids contents were determined after homogenizing fresh leaves (0.1 g) in 80% acetone (10 mL) and centrifuging at 3000 ×g for 15 min. The absorbance from supernatant was determined at 480 nm, 645 nm, and 663 nm using a spectrophotometer (Hitachi U-2001, Tokyo, Japan). The amounts of chlorophyll contents and total carotenoids were calculated as described by Yoshida et al. [18] and Davies [19].

2.4. Relative Membrane Permeability (RMP). Leaf tissues were collected in the test tubes containing distilled water (10 mL), vortexed for 5 s, and the value of EC_0 was measured. Then EC_1 of the filtrate was measured after 24 h by keeping

them at 4°C. The filtrate was autoclaved for 15 min for measuring EC_2. The percentage of ions leakage was calculated from the equation of Yang et al. [20].

2.5. Hydrogen Peroxide (H2O2) Contents. These were estimated using the method of Velikova et al. [21]. Leaf tissues (0.1 g) were homogenized with 1 mL of 0.1% (w/v) trichloroacetic acid (TCA) on an ice bath. Then homogenate was centrifuged at 12,000 ×g for 15 min. The reaction mixture consisted of supernatant (0.5 mL), 0.5 mL potassium phosphate buffer (10 mM; pH 7.0), and 1 mL KI (1 M), then vortexed, and the absorbance was measured at 390 nm, while 0.1% TCA used as blank. The H2O2 content was determined from a standard curve, and the values are expressed as μmol g^{-1} fresh weight.

2.6. Malondialdehyde (MDA) Contents. The MDA was assayed according to the method of Heath and Packer [22]. Fresh leaf tissues (0.1 g) were homogenized in 5% (w/v) TCA (1 mL). The homogenate was centrifuged at 12,000 ×g for 15 min. After centrifugation, 1 mL of an aliquot of the supernatant was mixed with 20% TCA (1 mL) containing 0.5% (w/v) thiobarbituric acid. The mixture was warmed at 95°C for 30 min, cooled on ice for a while, and then centrifuged at 7500 ×g for 5 min. The absorbance was recorded at 532 nm and 600 nm, whilst 5% TCA used as blank. MDA contents were calculated using an extinction coefficient of 155,000 $nmol\ mol^{-1}$:

$$MDA\left(nmol\ mL^{-1}\right) = \left[\frac{\left(A_{532} - A_{600}\right)}{155000}\right]10^6. \quad (1)$$

2.7. Free Proline Contents. Free proline was determined using the method of Bates et al. [23] Leaf tissue (0.1 g) was homogenized in 3% of aqueous sulphosalicylic acid (5 mL). Filtrate (1 mL) was mixed with acid ninhydrin (1 mL) and glacial acetic acid (1 mL) in a test tube. The mixture was cooled after heating for 10 min at 100°C in an ice bath. The mixture was extracted with toluene (4 mL) and vortexed for 20 s and cooled. The absorbance was measured at 520 nm. The amount of free proline was calculated from the standard curve at 520 nm and expressed as μmol g^{-1} fresh weight.

2.8. Total Free Amino Acid Contents. Total free amino acids were determined by using the ninhydrin method of Hamilton and Slyke [24]. 1.0 g of fresh plant material was extracted by using phosphate buffer (pH 7.0). In 25 mL test tube, 1 mL extract was taken, then 1 mL of 10% pyridine and 1 mL of 2% ninhydrin solution was added in each test tube. And test tubes were heated in a water bath for 30 minutes. The volume was maintained up to 50 mL by using distilled water in each tube. Optical density was measured at 570 nm by using UV-VIS spectrophotometer (Hitachi U-2910, Tokyo, Japan). The amount of total free amino acids was calculated from the standard curve of Lucine at 570 nm and expressed as mg g^{-1} fresh weight.

2.9. Total Anthocyanin Contents. Total anthocyanin contents were determined by the method of Hodges and Nozzolillo [25]. Fresh leaves (0.1 g) of sample were crushed in acidified methanol (2 mL) with the help of pestle and mortar. Then materials transferred to the centrifuge tubes, and heated them in water bath at 50°C for one hour. Centrifuge the materials at 12000 ×g in the centrifuge machine for 15 minutes. The absorbance was measured at 540 nm and 600 nm by using UV-VIS spectrophotometer (Hitachi U-2910, Tokyo, Japan). The amount of total anthocyanin contents was calculated in the original sample at 520 nm and 600 nm, expressed as mgL^{-1} fresh leaves.

2.10. Statistical Analysis. The data collected was subjected to analysis of variance technique (ANOVA) by using a computer software CoStat version 6.2, CoHort Software, 2003 (Monterey, CA, USA). The least significant difference among means was computed. The data in the tables is represented as means ± SE ($n = 3$) for each parameter.

3. Results and Discussion

Cadmium stress significantly ($P \leq 0.001$) decreased different growth attributes of both maize lines (Tables 1 and 2). The data showed that MTM-1 line exhibited markedly lowest growth rate in terms of root and shoot dry weights than MTM-2 at all cadmium levels on the 5th, 10th, and 15th day of sampling. Increasing cadmium levels up to 3 mg $CdCl_2$ kg^{-1} increased shoot and root dry weights and leaf area in MTM-2 than MTM-1 line, while a decrease was noted afterwards on the 5th, 10th, and 15th day of sampling.

Although cadmium toxicity depends upon the type of species and the plant growth stage, it has been shown to reduce crop productivity and quality severely [14]. In the present study, results indicated the differential responses of maize lines to cadmium stress at different intervals of time. Earlier studies have shown that higher level of cadmium changed the pattern of growth and inhibited the plant growth significantly in cucumber at seedling stage [26]. Even at low levels, cadmium is reported to modify the plant metabolism [27]. Changes in plant metabolism could affect the growth patterns as have been reported in some other crops like mung bean [28]. Upon exposure to cadmium, it appeared to inhibit the cell division and thus disrupted root cell expansion and enlargement. In addition, a decreased carbohydrate synthesis due to the inhibitory effect of cadmium on carbohydrate metabolism has also been shown to inhibit shoot and root growth [29]. Uptake of cadmium by living cells causes many drastic changes, leading to cell death depending on the cadmium quantity and time duration of exposure [30]. However, in the present study, a similar decreasing trend in growth was observed in both lines at the 5th, 10th, and 15th day of the cadmium treatment.

A significant ($P \leq 0.001$) variation in leaf chlorophyll a and b, total chlorophyll, and carotenoid contents were recorded in both lines under cadmium stress both on the 5th, 10th, and 15th day of sampling (Tables 1 and 2). Overall, MTM-2 had more photosynthetic pigments than MTM-1 on the 5th, 10th, and 15th day of sampling under cadmium stress. However, the maximum reduction in chlorophyll a and b, total chlorophyll, and total carotenoids were observed at 9 and 12 mg $CdCl_2$ kg^{-1} in both lines, particularly on the 10th and 15th day of sampling in MTM-1.

The cadmium stress has been shown to enhance the stomatal closure and inhibit the photosynthesis via chlorophyll degradation in plants [31]. The cadmium damages the photosynthetic machinery of plants, particularly light harvesting complex-II and photosystem-I (PS-I) and PS-II [32]. Similarly, higher level of cadmium has been shown to reduce the synthesis of chlorophyll a, b and total chlorophyll contents in gram and sorghum [33, 34]. In line with previous studies, we found the maximum reduction in chlorophyll a, b contents at higher cadmium level (9 and 12 mg $CdCl_2$ kg^{-1}) in both lines, particularly on the 15th day of the treatment of cadmium. This implied that the light harvesting system was damaged based on the cadmium exposure time in maize. Plants possessing higher concentrations of chlorophyll and other accessory pigments tend to have higher shoot and root dry weights as these pigments are directly involved in the process of photosynthesis which is closely linked to plant growth and dry matter production [35]. In the present investigation, we observed a significant positive correlation of chlorophyll a ($r = 0.450^{***}$; 0.410^{***}), chlorophyll b ($r = 0.037$ ns; -0.03 ns), and carotenoids ($r = 0.259^*$; 0.265^*) with the shoot and root dry weights that was recorded and presented in Table 3.

Cadmium stress significantly altered leaf RMP, H2O2, MDA, and the total anthocyanin contents in both lines on 5th, 10th, and 15th day of sampling. Although leaf RMP, H2O2, MDA, and total anthocyanin contents increased significantly ($P \leq 0.001$) in both lines at all levels of cadmium stress, MTM-1 line had higher leaf RMP, H2O2, MDA, and total anthocyanin contents on the 5th, 10th, and 15th day of sampling, particularly at 12 mg $CdCl_2$ kg^{-1} on the 15th day of the sampling (Tables 1 and 2).

Cadmium has been shown to damage cell membranes, induce oxidative stress, and thus it enhances the ionic leakage in plants [36]. Cell membrane injury is associated with the production of reactive oxygen species (ROS), which indicated the production of MDA [3]. However, plants have both enzymatic and nonenzymatic antioxidant systems to hinder the production of ROS. Plants having the ability to synthesize these compounds are considered as tolerant under different abiotic stresses. Irrespective of the exposure time, in the present study, cadmium enhanced RMP, H2O2, and MDA contents in both lines. This indicated that the production of ROS is the major factor related to cadmium injury in maize plants. Cadmium-induced oxidative stress-mediated enhanced production of MDA contents has already been reported in maize and *Solanum* [36, 37]. When plants are exposed to abiotic stress, membrane integrity is greatly hampered [38]. Loss of membrane integrity is usually measured as a rise in cellular levels of MDA, a by-product of lipid peroxidation. Researchers have shown that MDA levels are negatively linked with plant growth [39]. Similarly in our study, MDA contents exhibited a negative correlation

T̲ABLE̲ 1: Mean sum of square from ANOVA of data for growth and biochemical attributes of two maize (*Zea mays* L.) lines under different treatments of cadmium after different time intervals.

S.O.V.	df	Shoot dry weight	Root dry weight	Leaf area	Chlorophyll a
Lines (*L*)	1	15.56***	3.97***	2074.52***	4.64***
Cadmium (Cd)	4	5.88***	2.97***	227.06***	5.68***
Time (*T*)	2	38.97***	13.64***	2109.14***	54.73***
$L \times$ Cd	4	1.33***	0.20***	53.27***	0.12***
$L \times T$	2	0.61***	0.31***	2.85*	1.71***
Cd $\times T$	8	0.21***	0.16***	1.69*	0.90***
$L \times$ Cd $\times T$	8	0.44***	0.16***	18.06***	0.31***
Error	60	0.021	0.024	0.713	0.009
LSD 0.05	—	0.237	0.256	1.379	0.153

S.O.V.	df	Chlorophyll b	Total Chlorophyll	Carotenoids	RMP
Lines (*L*)	1	2.19***	13.26***	0.57***	366.91***
Cadmium (Cd)	4	10.97***	32.45***	0.77***	3665.94***
Time (*T*)	2	14.40***	112.56***	4.94***	14370.68***
$L \times$ Cd	4	0.13***	0.25***	0.01***	22.34***
$L \times T$	2	0.12***	2.40***	0.16***	181.05***
Cd $\times T$	8	0.77***	2.78***	0.12***	239.16***
$L \times$ Cd $\times T$	8	0.10***	0.57***	0.04***	55.36***
Error	60	0.011	0.020	0.00026	2.578
LSD 0.05	—	0.171	0.233	0.026	2.622

S.O.V.	df	MDA	H2O2	TAA	Free proline
Lines (*L*)	1	554.23***	117.31***	17.90***	102.86***
Cadmium (Cd)	4	270.68***	310.40***	3.93***	86.99***
Time (*T*)	2	651.96***	834.04***	6.84***	89.23***
$L \times$ Cd	4	42.74***	3.53***	0.77***	1.46***
$L \times T$	2	79.58***	105.45***	4.81***	6.63***
Cd $\times T$	8	3.69**	6.84***	0.07***	2.95***
$L \times$ Cd $\times T$	8	6.78***	7.03***	0.07***	0.31***
Error	60	1.030	0.143	0.017	0.061
LSD 0.05	—	1.657	0.618	0.211	0.403

S.O.V.	df	Anthocyanin		
Lines (*L*)	1	101.98***		
Cadmium (Cd)	4	351.97***		
Time (*T*)	2	193.48***		
$L \times$ Cd	4	8.04***		
$L \times T$	2	9.26***		
Cd $\times T$	8	3.16***		
$L \times$ Cd $\times T$	8	8.35***		
Error	60	0.254		
LSD 0.05	—	0.824		

*,**,*** : significant at 0.05, 0.01, and 0.001 levels, respectively; ns: nonsignificant.
RMP: relative membrane permeability; MDA: malendialdehyde; H2O2: hydrogen peroxide;
TAA: total free amino acids.

($r = -300$**; -0.306**) with shoot and root dry weights, respectively, presented in Table 3. In addition, the production of reactive oxygen species is also a great threat to plant growth. ROS are known for inducing substantial damage to pigments as is evident from the present investigation, where H2O2 exhibited a negative correlation ($r = -0.744$***; -0.817***, -0.846***) with chlorophyll a, b and carotenoids, respectively, presented in Table 3.

H2O2 is also known to inhibit the Calvin cycle that ultimately results in reduced photosynthetic rates. This could have been the major factor for negative correlation of H2O2 with the shoot ($r = -0.502$***) and root ($r = -0.236$*) dry weights presented in Table 3. Measures of relative membrane permeability (RMP) in plants exposed to various environmental stresses including cadmium stress are taken as indicators for stress-induced damage in plants. Plants under

TABLE 2: Means comparison data for growth and biochemical attributes of two maize (*Zea mays* L.) lines under different treatments of cadmium after different time intervals.

Lines	Time (d)	Cd (mg kg^{-1})	SDW	RDW	Leaf area	Chl a	Chl b	Tot-Chl	Car.	RMP	MDA	H2O2	TAA	Proline	Anth.
	5	0	2.23 ± 0.01^{no}	2.23 ± 0.04^{kl}	20.47 ± 0.06^{op}	3.61 ± 0.22^{c}	2.57 ± 0.26^{c}	6.18 ± 0.35^{b}	1.36 ± 0.00^{a}	32.72 ± 0.33^{s}	8.04 ± 0.13^{tu}	6.14 ± 0.01^{s}	0.97 ± 0.02^{lmno}	5.99 ± 0.06^{j}	2.44 ± 0.03^{o}
		3	2.08 ± 0.06^{o}	1.81 ± 0.06^{no}	19.15 ± 1.47^{p}	3.20 ± 0.08^{d}	2.38 ± 0.05^{de}	5.58 ± 0.10^{c}	1.20 ± 0.02^{b}	35.39 ± 0.67^{n}	9.46 ± 0.06st	8.20 ± 0.06^{q}	1.12 ± 0.01^{jklm}	6.20 ± 0.04^{j}	6.62 ± 0.28^{k}
		6	1.73 ± 0.04^{pq}	1.41 ± 0.05^{pq}	17.68 ± 0.09^{q}	2.71 ± 0.04^{e}	2.09 ± 0.04^{f}	4.79 ± 0.07^{de}	0.97 ± 0.00^{d}	39.40 ± 0.33^{n}	12.66 ± 0.32^{qr}	11.45 ± 0.09^{no}	1.17 ± 0.00^{jkl}	7.77 ± 0.04^{g}	9.13 ± 0.28^{i}
		9	1.56 ± 0.02^{q}	1.33 ± 0.04^{qr}	16.67 ± 0.81^{qr}	2.10 ± 0.06^{f}	1.19 ± 0.03^{i}	3.29 ± 0.07^{g}	0.78 ± 0.01^{f}	45.06 ± 1.00^{l}	15.93 ± 0.59^{klmn}	13.46 ± 0.26^{lm}	1.42 ± 0.16^{hi}	9.60 ± 0.19^{d}	10.70 ± 0.15^{gh}
		12	1.30 ± 0.03^{r}	1.13 ± 0.06^{r}	15.70 ± 0.32^{r}	0.55 ± 0.04^{klm}	0.41 ± 0.17^{op}	0.96 ± 0.13^{mn}	0.56 ± 0.01^{h}	56.39 ± 0.33^{j}	18.90 ± 0.57^{gh}	16.14 ± 0.16^{gh}	1.61 ± 0.09^{gh}	12.57 ± 0.22^{b}	13.30 ± 0.28^{e}
	10	0	4.76 ± 0.02^{cd}	2.82 ± 0.08^{fg}	37.27 ± 0.43^{e}	1.63 ± 0.08^{g}	3.03 ± 0.01^{b}	4.66 ± 0.09^{e}	0.85 ± 0.02^{e}	32.55 ± 0.64^{n}	8.21 ± 0.39^{tu}	6.87 ± 0.00^{r}	0.87 ± 0.05^{no}	4.05 ± 0.16^{n}	4.12 ± 0.28^{lm}
		3	3.95 ± 0.10^{f}	2.68 ± 0.05^{ghi}	34.74 ± 0.11^{hi}	1.02 ± 0.00^{i}	2.36 ± 0.03^{de}	3.38 ± 0.03^{fg}	0.65 ± 0.01^{g}	46.93 ± 1.20^{l}	17.94 ± 0.52^{ghij}	11.85 ± 0.06^{n}	1.00 ± 0.04^{lmno}	4.55 ± 0.01^{m}	9.13 ± 0.28^{i}
MTM-1		6	3.42 ± 0.02^{g}	2.61 ± 0.05^{ghi}	33.46 ± 0.33^{ij}	0.65 ± 0.02^{k}	1.08 ± 0.02^{jj}	1.73 ± 0.04^{j}	0.38 ± 0.01^{k}	67.24 ± 0.56^{i}	19.40 ± 1.20^{fg}	12.87 ± 0.34^{m}	1.31 ± 0.11^{ijk}	5.51 ± 0.48^{k}	10.24 ± 0.09^{gh}
		9	3.05 ± 0.08^{hi}	2.50 ± 0.23^{hijk}	32.53 ± 0.50^{jk}	0.47 ± 0.02^{lm}	0.91 ± 0.04^{jk}	1.38 ± 0.02^{k}	0.31 ± 0.01^{l}	83.60 ± 1.19^{de}	22.13 ± 1.37^{e}	15.04 ± 0.13^{j}	1.36 ± 0.05^{ij}	6.91 ± 0.02^{hi}	15.25 ± 0.02^{d}
		12	2.88 ± 0.13^{ijk}	2.47 ± 0.06^{ijk}	31.96 ± 0.03^{k}	0.27 ± 0.01^{o}	0.21 ± 0.00^{qr}	0.48 ± 0.02^{pq}	0.12 ± 0.00^{o}	91.58 ± 1.35^{b}	26.03 ± 0.80^{cd}	17.63 ± 0.15^{f}	1.51 ± 0.09^{hi}	8.93 ± 0.03^{ef}	17.76 ± 0.28^{b}
	15	0	4.78 ± 0.09^{cd}	3.58 ± 0.08^{bc}	36.96 ± 0.35^{ef}	0.81 ± 0.03^{j}	1.25 ± 0.03^{i}	2.06 ± 0.02^{hi}	0.30 ± 0.01^{l}	66.13 ± 1.73^{i}	16.81 ± 0.16^{ijkl}	15.86 ± 0.10^{hi}	0.80 ± 0.08^{o}	3.70 ± 0.17^{n}	11.07 ± 0.74^{fg}
		3	4.28 ± 0.07^{e}	3.40 ± 0.16^{cd}	35.47 ± 0.38^{fgh}	0.25 ± 0.01^{o}	0.69 ± 0.03^{lm}	0.95 ± 0.04^{mn}	0.14 ± 0.01^{o}	80.85 ± 0.36^{fg}	24.58 ± 0.71^{d}	21.51 ± 0.37^{d}	0.90 ± 0.06^{mno}	4.09 ± 0.01^{n}	11.91 ± 0.28^{f}
		6	3.10 ± 0.10^{hi}	2.88 ± 0.05^{fg}	30.04 ± 0.16^{l}	0.18 ± 0.02^{o}	0.36 ± 0.01^{opq}	0.54 ± 0.02^{op}	0.08 ± 0.00^{p}	84.88 ± 1.32^{de}	27.39 ± 0.25^{bc}	26.19 ± 0.55^{b}	0.97 ± 0.03^{lmno}	4.96 ± 0.02^{lm}	13.86 ± 0.48^{e}
		9	3.01 ± 0.07^{hij}	2.38 ± 0.09^{jk}	29.28 ± 0.33^{l}	0.15 ± 0.01^{o}	0.26 ± 0.01^{pqr}	0.41 ± 0.01^{pq}	0.07 ± 0.00^{pqr}	93.24 ± 1.54^{b}	28.34 ± 0.56^{b}	28.05 ± 0.25^{a}	1.29 ± 0.07^{jk}	5.18 ± 0.09^{kl}	16.42 ± 0.36^{c}
		12	2.49 ± 0.11^{lm}	1.95 ± 0.09^{mn}	25.23 ± 0.16^{n}	0.12 ± 0.01^{o}	0.16 ± 0.01^{r}	0.28 ± 0.01^{q}	0.06 ± 0.00^{pqr}	99.72 ± 1.54^{a}	30.47 ± 1.20^{a}	28.33 ± 0.04^{a}	1.48 ± 0.06^{hi}	6.67 ± 0.15^{i}	19.43 ± 0.28^{a}

TABLE 2: Continued.

Lines	Time (d)	Cd (mg kg⁻¹)	SDW	RDW	Leaf area	Chl a	Chl b	Tot-Chl	Car.	RMP	MDA	H2O2	TAA	Proline	Anth.
	5	0	2.70 ± 0.11^{kl}	2.22 ± 0.06^{klm}	30.43 ± 0.35^{l}	4.42 ± 0.10^{a}	3.20 ± 0.01^{ab}	7.63 ± 0.11^{a}	1.34 ± 0.01^{a}	23.74 ± 0.33^{p}	7.59 ± 0.30^{u}	3.44 ± 0.05^{t}	2.05 ± 0.13^{f}	6.86 ± 0.03^{i}	2.73 ± 0.28^{o}
		3	2.78 ± 0.01^{jk}	2.48 ± 0.07^{ijk}	35.41 ± 0.34^{gh}	3.92 ± 0.04^{b}	2.49 ± 0.01^{cd}	6.41 ± 0.06^{b}	1.13 ± 0.02^{c}	29.06 ± 0.58^{o}	9.29 ± 0.30^{stu}	7.45 ± 0.03^{r}	2.86 ± 0.04^{d}	7.91 ± 0.02^{g}	4.80 ± 0.13^{l}
		6	2.38 ± 0.09^{mn}	2.03 ± 0.01^{lmn}	27.41 ± 0.57^{m}	3.13 ± 0.02^{d}	2.27 ± 0.03^{e}	5.40 ± 0.04^{c}	0.75 ± 0.01^{f}	33.40 ± 0.88^{m}	10.45 ± 0.13^{s}	10.46 ± 0.30^{p}	3.13 ± 0.12^{c}	9.17 ± 0.06^{e}	7.84 ± 0.54^{j}
		9	2.16 ± 0.09^{no}	1.90 ± 0.03^{n}	25.41 ± 0.37^{n}	3.10 ± 0.00^{d}	1.76 ± 0.03^{g}	4.86 ± 0.03^{de}	0.49 ± 0.01^{i}	40.07 ± 0.58^{m}	12.47 ± 0.54^{r}	13.68 ± 0.25^{kl}	3.48 ± 0.06^{b}	11.21 ± 0.11^{c}	10.68 ± 0.16^{gh}
		12	1.83 ± 0.08^{p}	1.61 ± 0.06^{op}	21.37 ± 0.28^{o}	2.61 ± 0.03^{e}	0.86 ± 0.00^{kl}	3.47 ± 0.04^{fg}	0.43 ± 0.01^{j}	47.44 ± 1.84^{l}	14.58 ± 0.35^{mnop}	15.44 ± 0.08^{ij}	3.79 ± 0.09^{a}	14.67 ± 0.13^{a}	11.78 ± 0.13^{f}
MTM-2	10	0	4.42 ± 0.12^{e}	3.00 ± 0.08^{ef}	42.94 ± 0.55^{c}	1.76 ± 0.06^{g}	3.23 ± 0.02^{a}	4.99 ± 0.08^{d}	0.26 ± 0.01^{m}	47.39 ± 1.18^{l}	12.20 ± 0.36^{r}	10.91 ± 0.27^{op}	0.87 ± 0.03^{no}	6.85 ± 0.03^{i}	3.02 ± 0.01^{no}
		3	5.56 ± 0.09^{a}	3.31 ± 0.09^{cd}	51.12 ± 0.98^{a}	1.33 ± 0.01^{h}	2.27 ± 0.00^{e}	3.60 ± 0.02^{f}	0.19 ± 0.00^{n}	53.07 ± 1.41^{k}	13.20 ± 0.31^{pqr}	12.02 ± 0.02^{n}	1.08 ± 0.05^{klmn}	7.30 ± 0.08^{h}	6.07 ± 0.56^{k}
		6	4.75 ± 0.06^{cd}	3.20 ± 0.06^{de}	45.04 ± 0.58^{b}	1.29 ± 0.03^{h}	2.08 ± 0.02^{f}	3.37 ± 0.02^{fg}	0.13 ± 0.01^{o}	65.12 ± 0.57^{i}	14.33 ± 0.17^{nopq}	13.49 ± 0.07^{lm}	1.77 ± 0.04^{g}	8.65 ± 0.13^{f}	9.96 ± 0.04^{hi}
		9	3.98 ± 0.08^{f}	2.50 ± 0.03^{hijk}	39.43 ± 0.32^{d}	0.62 ± 0.02^{kl}	1.25 ± 0.02^{i}	1.87 ± 0.03^{ij}	0.12 ± 0.01^{o}	78.65 ± 0.80^{g}	15.25 ± 0.54^{lmno}	14.96 ± 0.03^{j}	2.47 ± 0.10^{e}	11.03 ± 0.02^{c}	11.74 ± 0.08^{f}
		12	3.16 ± 0.09^{h}	2.06 ± 0.01^{lmn}	35.95 ± 0.57^{efgh}	0.19 ± 0.00^{o}	0.67 ± 0.03^{m}	0.86 ± 0.03^{mn}	0.05 ± 0.00^{qr}	85.83 ± 0.29^{cd}	17.47 ± 0.76^{hijk}	16.10 ± 0.56^{gh}	2.77 ± 0.13^{d}	12.23 ± 0.34^{b}	13.02 ± 0.05^{e}
	15	0	4.38 ± 0.09^{e}	3.75 ± 0.10^{ab}	39.65 ± 0.34^{d}	0.64 ± 0.01^{k}	1.47 ± 0.03^{h}	2.12 ± 0.03^{h}	0.25 ± 0.01^{m}	56.16 ± 0.36^{j}	13.99 ± 0.23^{opqr}	11.71 ± 0.24^{n}	0.76 ± 0.04^{o}	4.72 ± 0.11^{m}	3.76 ± 0.05^{mn}
		3	5.32 ± 0.04^{b}	3.98 ± 0.24^{a}	45.70 ± 0.08^{b}	0.50 ± 0.01^{klm}	0.77 ± 0.01^{klm}	1.27 ± 0.00^{kl}	0.14 ± 0.00^{o}	72.17 ± 0.61^{h}	15.37 ± 0.27^{lmno}	14.15 ± 0.08^{k}	0.97 ± 0.03^{lmno}	4.88 ± 0.06^{lm}	6.07 ± 0.28^{k}
		6	4.98 ± 0.14^{c}	3.36 ± 0.05^{cd}	44.67 ± 0.33^{b}	0.44 ± 0.01^{mn}	0.61 ± 0.01^{mn}	1.06 ± 0.01^{lm}	0.07 ± 0.00^{pq}	82.48 ± 0.88^{ef}	16.32 ± 0.27^{jklm}	16.74 ± 0.19^{g}	1.28 ± 0.05^{jk}	6.07 ± 0.03^{j}	14.97 ± 0.28^{d}
		9	4.66 ± 0.14^{d}	3.30 ± 0.08^{cd}	39.33 ± 0.33^{d}	0.29 ± 0.01^{no}	0.48 ± 0.00^{no}	0.78 ± 0.01^{no}	0.07 ± 0.00^{pqr}	87.71 ± 0.33^{c}	18.37 ± 0.33^{ghi}	20.08 ± 0.06^{e}	1.83 ± 0.01^{g}	7.91 ± 0.04^{g}	15.53 ± 0.10^{d}
		12	4.03 ± 0.09^{f}	2.78 ± 0.07^{fgh}	36.77 ± 0.24^{efg}	0.27 ± 0.01^{o}	0.22 ± 0.01^{qr}	0.49 ± 0.00^{pq}	0.04 ± 0.00^{r}	92.82 ± 0.66^{b}	20.97 ± 0.02^{ef}	24.72 ± 0.03^{c}	2.05 ± 0.04^{f}	9.29 ± 0.09^{de}	17.47 ± 0.28^{b}

All the values represent mean + S.E. and values that show the same superscript letter in the same column are not significantly different at $P \leq 0.05$.
SFW: shoot fresh weight; RDW: root dry weight; Chl. a: chlorophyll a; Chl. b: chlorophyll b; Car.: carotenoids; Anth.: anthocyanin.
MDA: malendialdehyde; RMP: relative membrane permeability; H2O2: hydrogen peroxide; TAA: total free amino acids.

Table 3: Correlation among growth and biochemical attributes of two maize (*Zea mays* L.) lines under different treatments of cadmium after different time intervals.

Attributes	SDW	RDW	LA	Chl. a	Chl. b	Car	RMP	MDA	H2O2	TAA	Proline	Anth.
SDW	1											
RDW	0.908***	1										
LA	0.938***	0.852***	1									
Chl a	−0.450***	−0.410***	−0.390***	1								
Chl b	0.037 ns	−0.031 ns	0.057 ns	0.748***	1							
Car.	−0.259*	−0.265*	−0.213 ns	0.953***	0.913***	1						
RMP	−0.506***	−0.465***	−0.531***	−0.873***	−0.702***	−0.854***	1					
MDA	−0.300**	−0.306**	−0.298***	−0.886***	−0.853***	−0.931***	−0.862***	1				
H2O2	−0.502***	−0.236**	−0.012 ns	−0.774***	−0.817***	−0.846***	−0.690***	0.855***	1			
TAA	0.062 ns	0.094 ns	0.023 ns	−0.742***	−0.845***	−0.839***	−0.758***	0.863***	0.902***	1		
Proline	−0.397***	−0.432***	−0.203 ns	−0.361***	−0.041 ns	0.203 ns	0.058 ns	−0.151 ns	−0.194 ns	−0.056 ns	1	
Anth.	−0.490***	−0.602***	−0.329**	0.162 ns	−0.184 ns	0.016 ns	−0.044 ns	−0.042 ns	−0.100 ns	0.008 ns	0.786***	1

*,**,***: significant at 0.05, 0.01, and 0.001 levels, respectively; ns: nonsignificant.
SFW: shoot fresh weight; RDW: root dry weight; Chl. a: chlorophyll a; Chl. b: chlorophyll b; Car.: carotenoids; Anth.: anthocyanin.
MDA: malendialdehyde; RMP: relative membrane permeability; H2O2: hydrogen peroxide; TAA: total free amino acids.

cadmium stress exhibit increase in the permeability of membranes that ultimately leads to loss of membrane integrity. Researchers take the ability of plasma membranes to control the movement of ions across the cell as a potential selection criterion to measure the extent of damage to a great variety of tissues [40]. Likewise in the present investigation, we have observed a negative association of RMP ($r = -0.506^{***}$, -0.465^{***}, -0.531^{**}, -0.873^{***}, -0.702^{***}, -0.854^{***}) with shoot and root dry weights, leaf area, chlorophyll a, b, and carotenoids, respectively, presented in Table 3. This could have been due to the inability of plasma membrane to control the movement of ions across the cell.

Leaf free proline, total free amino acid, and anthocyanin contents significantly ($P \leq 0.001$) differed in both lines under cadmium stress. The plants of MTM-2 line accumulated more free proline, total free amino acid, and anthocyanin contents as compared with the MTM-1 on the 5th, 10th, and 15th day of sampling under cadmium stress (Tables 1 and 2). Moreover higher free proline and total free amino acid contents were produced on the 5th sampling in both lines, while production of free proline and total free amino acids was decreasing on the 10th and 15th day of sampling at all levels of cadmium stress. Furthermore, total anthocyanin contents increased gradually on the 5th, 10th, and 15th day of sampling at all levels of cadmium stress in both lines.

Free proline accumulation occurred in cadmium-treated plants of MTM-2 line. Thus, the accumulation of proline in response to cadmium stress was due to the production of ROS. Previous studies have shown those plants which produced more proline and free amino acids due to more production of hydroxyl radicals. Proline is the only molecule that protects plants against singlet oxygen and free radical since proline acts as a singlet oxygen quencher and as a scavenger of OH radicals [41, 42]. Thus, proline is not only an important molecule in redox signalling but also an effective quencher of reactive oxygen species formed in all plants against abiotic stress while in others proline was produced as an indication of stress [43]. Similarly, in our investigation, we have recorded

that proline did not contribute to a great extent in conferring tolerance to plants against cadmium stress. This could have been due to the fact that endogenous levels of proline were negatively correlated ($r = -0.397^{***}$, -0.432^{***}, -0.361^{***}, -0.194 ns) to shoot and root dry weights, chlorophyll a, and H2O2 (Table 3). In some reports, it has been shown that proline stabilizes the subcellular compartments of cells [44] which were not the case in the present investigation.

There are many reports which show that anthocyanins contents are able to quench the oxygen radicals [45]. In the present investigation, a significant increase in anthocyanin contents was observed in both maize lines, being the highest in MTM-1 line at the level of cadmium stress. Our results of anthocyanins contents are similar to some earlier reports of increases in endogenous levels of flavonoids and total anthocyanins under abiotic stress [46]. Furthermore, flavonoids act as nonenzymatic antioxidants and protect the plants against ROS-induced oxidative stress. Moreover, loss of membrane integrity in terms of lipid peroxidation has been reported in anthocyanin deficient mutant of *Arabidopsis* [47]. Therefore, anthocyanins are considered as a defensive yardstick against various abiotic stresses in plants [48]. Thus, anthocyanin contents may affect the stress defense mechanism.

4. Conclusions

Irrespective of exposure time, cadmium stress affected morphophysiological attributes in both lines. However, on the 15th day of treatment, the cadmium effects and the response of lines to cadmium stress became more evident. The cadmium stress produced oxidative stress in the plants of both lines which was evident from the increased synthesis of H2O2 and MDA contents and increased RMP. Taken together, the results suggested that cadmium-induced oxidative stress was the main cause of reduced plant growth in both lines. Based on having more accumulation of free amino acids, free proline and total anthocyanin contents at all cadmium levels, and the stability or enhancement in plant biomass and leaf area particularly occurred at 3 mg $CdCl_2$ kg^{-1} concentration, the line MTM-2 was considered tolerant as compared with MTM-1 line.

Acknowledgment

This work was partially supported by the Higher Education Commission (HEC), Islamabad, Pakistan, through Project no. PM-IPFP/HRD/HEC/2011/0579.

References

[1] G. R. MacFarlane and M. D. Burchett, "Photosynthetic pigments and peroxidase activity as indicators of heavy metal stress in the grey mangrove, *Avicennia marina* (Forsk.) Vierh," *Marine Pollution Bulletin*, vol. 42, no. 3, pp. 233–240, 2001.

[2] F. Chen, J. Dong, F. Wang et al., "Identification of barley genotypes with low grain Cd accumulation and its interaction with four microelements," *Chemosphere*, vol. 67, no. 10, pp. 2082–2088, 2007.

[3] M. A. Hossain, P. Piyatida, A. Jaime, T. d. Silva, and M. Fujita, "Molecular mechanism of heavy metal toxicity and tolerance in plants: central role of glutathione in detoxification of reactive oxygen species and methylglyoxal and in heavy metal chelation," *Journal of Botany*, vol. 2012, Article ID 872875, 37 pages, 2012.

[4] S. J. Stohs, D. Bagachi, E. Hassoun, and M. Bagachi, "Oxidative mechanism in the toxicity of chromium and cadmium ions," *Journal of Environmental Pathology. Toxicology Oncology*, vol. 19, no. 3, pp. 201–213, 2000.

[5] A. Larbi, F. Morales, A. Abadia, Y. Gogorcena, J. J. Lucena, and J. Abadia, "Effects of Cd and Pb in sugar beet plants grown in nutrient solution: induced Fe deficiency and growth inhibition," *Functional Plant Biology*, vol. 29, no. 12, pp. 1453–1464, 2002.

[6] Y. X. Chen, Y. F. He, Y. M. Luo, Y. L. Yu, Q. Lin, and M. H. Wong, "Physiological mechanism of plant roots exposed to cadmium," *Chemosphere*, vol. 50, no. 6, pp. 789–793, 2003.

[7] C. Poschenrieder, G. Gunse, and J. Barcelo, "Influence of cadmium on water relations, stomatal resistance and abscisic acid content in expanding bean leaves," *Plant Physiology*, vol. 90, no. 4, pp. 1365–1371, 1989.

[8] M. Prasad, "Cadmium toxicity and tolerance in vascular plants," *Environmental and Experimental Botany*, vol. 35, no. 4, pp. 525–545, 1995.

[9] A. Wahid, A. Ghani, and F. Javed, "Effect of cadmium on photosynthesis, nutrition and growth of mungbean," *Agronomy for Sustainable Development*, vol. 28, no. 2, pp. 273–280, 2008.

[10] I. Hussain, M. Iqbal, S. Qurat-UL-Ain et al., "Cadmium dose and exposure-time dependent alterations in growth and physiology of maize (*Zea mays*)," *International Journal of Agriculture and Biology*, vol. 14, no. 6, pp. 959–964, 2012.

[11] Y. Y. Kim, D. Y. Kim, D. Shim et al., "Expression of the novel wheat gene TM20 confers enhanced cadmium tolerance to bakers' yeast," *Journal of Biological Chemistry*, vol. 283, no. 23, pp. 15893–15902, 2008.

[12] A. Paradiso, R. Berardino, M. C. de Pinto et al., "Increase in ascorbate-glutathione metabolism as local and precocious systemic responses induced by cadmium in durum wheat plants," *Plant and Cell Physiology*, vol. 49, no. 3, pp. 362–374, 2008.

[13] T. Gichner, Z. Patková, J. Száková, and K. Demnerová, "Cadmium induces DNA damage in tobacco roots, but no DNA damage, somatic mutations or homologous recombination in tobacco leaves," *Mutation Research*, vol. 559, no. 1-2, pp. 49–57, 2004.

[14] S. S. Gill, N. A. Khan, and N. Tuteja, "Differential cadmium stress tolerance in five indian mustard (*Brassica juncea* L) cultivars: an evaluation of the role of antioxidant machinery," *Plant Signaling and Behavior*, vol. 6, no. 2, pp. 293–300, 2011.

[15] W. A. Kasim, "Changes induced by copper and cadmium stress in the anatomy and grain yield of *Sorghum bicolor* L. Moench," *Nature of Science*, vol. 8, no. 1, pp. 1–8, 2008.

[16] S. S. Gill and N. Tuteja, "Reactive oxygen species and antioxidant machinery in abiotic stress tolerance in crop plants," *Plant Physiology and Biochemistry*, vol. 48, no. 12, pp. 909–930, 2010.

[17] C. R. Dowswell, R. L. Y. Paliwal, and R. P. Cantrell, *Maize is the Third World*, Westview Press, Boulder, Colo, USA, 1996.

[18] S. Yoshida, D. A. Forno, J. H. Cock, and K. A. Gomez, *Laboratory Manual for Physiological Studies of Rice*, IRRI, Los Banos, Calif, USA, 1976.

[19] B. H. Davies, "Carotenoids," in *Chemistry and Biochemistry of Plant Pigments*, T. W. Goodwin, Ed., pp. 199–217, Academic Press, London, UK, 1976.

[20] G. Yang, D. Rhodes, and R. J. Joly, "Effect of high temperature on membrane stability and chlorophyll fluorescence in glycine betaine-deficient and glycine betaine containing maize lines," *Australian Journal of Plant Physiology*, vol. 23, no. 4, pp. 437–443, 1996.

[21] V. Velikova, I. Yordanov, and A. Edreva, "Oxidative stress and some antioxidant systems in acid rain-treated bean plants protective role of exogenous polyamines," *Plant Science*, vol. 151, no. 1, pp. 59–66, 2000.

[22] R. L. Heath and L. Packer, "*Photoperoxidation in Isolated Chloroplasts Archives of Biochemistry and Biophysics*, vol. 125, no. 1, pp. 189–198, 1968.

[23] L. S. Bates, R. P. Waldren, and I. D. Teare, "Rapid determination of free proline for water-stress studies," *Plant and Soil*, vol. 39, no. 1, pp. 205–207, 1973.

[24] P. B. Hamilton and D. D. Van Slyke, "Amino acid determination and metal accumulation by *Brassica juncea* L.," *International Journal of Plant Production*, vol. 3, no. 1, pp. 1735–8043, 1943.

[25] D. M. Hodges and C. Nozzolillo, "Anthocyanin and anthocyanoplast content of cruciferous seedlings subjected to mineral nutrient deficiencies," *Journal of Plant Physiology*, vol. 147, no. 6, pp. 749–754, 1996.

[26] S. S. Abu-Muriefah, "Growth parameters and elemental status of cucumber (*Cucumus sativus*) seedlings in response to cadmium accumulation," *International Journal of Agriculture and Biology*, vol. 10, no. 3, pp. 261–266, 2008.

[27] S. Kevrešan, S. Kiršek, J. Kandrač, N. Petrović, and D. J. Kelemen, "Dynamics of cadmium distribution in the intercellular space and inside cells in soybean roots, stems and leaves," *Biologial Plantarum*, vol. 46, no. 1, pp. 85–88, 2003.

[28] A. Ghani, "Effect of cadmium toxicity on the growth and yield components of mungbean [*Vigna radiata* (L.) Wilczek]," *World Applied Sciences Journal*, vol. 8, no. 1, pp. 26–29, 2008.

[29] M. M. Al-rumaih, S. S. Rushdy, and A. S. Warsy, "Effect of cadmium chloride on seed germination and growth characteristics of cowpea (*Vigna unguiculata* L.) plants in the presence and absence of gibberellic acid," *Saudi Journal of Biological Sciences*, vol. 8, no. 1, pp. 41–50, 2001.

[30] A. P. Vitória, P. J. Lea, and R. A. Azevedo, "Antioxidant enzyme responses to cadmium in radish tissues," *Phytochemistry*, vol. 57, no. 5, pp. 701–710, 2001.

[31] S. A. Hasan, Q. Fariduddin, B. Ali, S. Hayat, and A. Ahmad, "Cadmium: toxicity and tolerance in plants," *Journal of Environmental Biology*, vol. 30, no. 2, pp. 165–174, 2009.

[32] M. Mobin and N. A. Khan, "Photosynthetic activity, pigment composition and antioxidative response of two mustard (*Brassica juncea*) cultivars differing in photosynthetic capacity subjected to cadmium stress," *Journal of Plant Physiology*, vol. 164, no. 5, pp. 601–610, 2007.

[33] M. S. Tantrey and R. K. Agnihotri, "Chlorophyll and proline content of gram (*Cicer arietinum* L.) under cadmium and mercury treatments," *Research Journal of Agriculture and Biological Sciences*, vol. 1, no. 2, pp. 119–122, 2010.

[34] L. Da-lin, H. Kai-qi, M. Jing-jing, Q. Wei-wei, W. Xiu-ping, and Z. Shu-pan, "Effects of cadmium on the growth and physiological characteristics of sorghum plants," *African Journal of Biotechnology*, vol. 10, no. 70, pp. 15770–15776, 2011.

[35] M. H. Siddiqui, M. H. Al-Whaibi, A. M. Sakran, M. O. Basalah, and H. M. Ali, "Effect of calcium and potassium on antioxidant system of *Vicia faba* L. under cadmium stress," *International Journal of Molecular Science*, vol. 13, no. 6, pp. 6604–6619, 2012.

[36] Y. Cui and Q. Wang, "Physiological responses of maize to elemental sulphur and cadmium stress," *Plant, Soil and Environment*, vol. 52, no. 11, pp. 523–529, 2006.

[37] A. Nasraoui-Hajaji, H. Gouial, E. Carrayol, and C. Haouari-Chaffei, "Ammonium alleviates redox state in solanum seedlings under cadmium stress conditions," *Journal of Environmental and Analytical Toxicology*, vol. 2, article 141, 2012.

[38] M. Bajji, J. M. Kinet, and S. Lutts, "The use of the electrolyte leakage method for assessing cell membrane stability as a water stress tolerance test in durum wheat," *Plant Growth Regulation*, vol. 36, no. 1, pp. 61–70, 2002.

[39] M. A. Ashraf, M. Ashraf, and Q. Ali, "Response of two genetically diverse wheat cultivars to salt stress at different growth stages: leaf lipid peroxidation and phenolic contents," *Pakistan Journal of Botany*, vol. 42, no. 1, pp. 559–565, 2010.

[40] O. Blokhina, E. Virolainen, and K. V. Fagerstedt, "Antioxidants, oxidative damage and oxygen deprivation stress: a review," *Annals of Botany*, vol. 91, pp. 179–194, 2003.

[41] P. Pardha Saradhi, Alia, and P. Mohanty, "Involvement of proline in protecting thylakoid membranes against free radical-induced photodamage," *Journal of Photochemistry and Photobiology B*, vol. 38, no. 2-3, pp. 253–257, 1997.

[42] Alia and P. P. Saradhi, "Suppression in mitochondrial electron transport is the prime cause behind stress induced proline accumulation," *Biochemical and Biophysical Research Communications*, vol. 193, no. 1, pp. 54–58, 1993.

[43] M. Ashraf and M. R. Foolad, "Roles of glycine betaine and proline in improving plant abiotic stress resistance," *Environmental and Experimental Botany*, vol. 59, no. 2, pp. 206–216, 2007.

[44] S. Hayat, Q. Hayat, M. N. Alyemeni, A. S. Wani, J. Pichtel, and A. Ahmad, "Role of proline under changing environments: a review," *Plant Signaling and Behavior*, vol. 7, no. 1, pp. 1456–1466, 2012.

[45] R. Farkhondeh, E. Nabizadeh, and N. Jalilnezhad, "Effect of salinity stress on proline content, membrane stability and water relations in two sugar beet cultivars," *International Journal of AgriScience*, vol. 2, no. 5, pp. 385–392, 2012.

[46] K. L. Hale, H. A. Tufan, I. J. Pickering et al., "Anthocyanins facilitate tungsten accumulation in *Brassica*," *Physiologia Plantarum*, vol. 116, no. 3, pp. 351–358, 2002.

[47] G. Agati, P. Matteini, A. Goti, and M. Tattini, "Chloroplast-located flavonoids can scavenge singlet oxygen," *New Phytologist*, vol. 174, no. 1, pp. 77–89, 2007.

[48] S. Chutipaijit, S. Cha-um, and K. Sompornpailin, "High contents of proline and anthocyanin increase protective response to salinity in *Oryza sativa* L. spp. Indica," *Australian Journal of Crop Sciences*, vol. 5, no. 10, pp. 1191–1198, 2011.

6

Growth Responses and Leaf Antioxidant Metabolism of Grass Pea (*Lathyrus sativus* L.) Genotypes under Salinity Stress

Dibyendu Talukdar

Department of Botany, R.P.M. College, University of Calcutta, Uttarpara, Hooghly, West Bengal 712 258, India

Correspondence should be addressed to Dibyendu Talukdar; dibyendutalukdar9@gmail.com

Academic Editors: A. Berville and F. Volaire

Response of six improved grass pea genotypes to prolonged salinity stress was investigated on seedlings grown in pot experiment using 150 mM NaCl up to 60 days of growth after commencement of treatment (DAC). NaCl exposure significantly reduced growth potential of varieties PUSA-90-2 and WBK-CB-14, but no such effect was observed in varieties B1, BioL-212 and in two mutant lines LR3 and LR4. A time-bound measurement at 15, 30 and 60 DAC revealed significant reduction in plant dry matter production, orchestrated through abnormally low capacity of leaf photosynthesis accompanied by low K^+/Na^+ ratio and onset of oxidative stress in all six genotypes at 15 DAC and the extension of the phenomena in PUSA-90-2 and WBK-CB-14 to 60 DAC. High superoxide dismutase (SOD) activity coupled with low ascorbate redox and declining ascorbate peroxidase (APX) and catalases (CAT) levels led to abnormal rise in H_2O_2 content at reproductive stage (30 DAC) in the latter two genotypes, consequently, resulting in NaCl-induced oxidative damage. H_2O_2 level in the rest of the four genotypes was modulated in a controlled way by balanced action of SOD, APX and CAT, preventing oxidative damage even under prolonged NaCl-exposure. Enzyme isoforms were involved in regulation of foliar H_2O_2-metabolism, which was critical in determining As tolerance of grass pea genotypes.

1. Introduction

Soil salinity is one of the most severe abiotic stresses affecting production of the crops worldwide [1, 2]. This problem is more severe in arid and semiarid regions, and legume plants already face a notable impact of salt stress in these regions [3, 4]. The legume family is the second only to the cereals in their importance to mankind [3], but unfortunately, improvements of this group of plants for their tolerance against soil salinity stress have not kept pace with those of cereals and oil seeds.

Salinity induces oxidative stress through the generation of reactive oxygen species (ROS) within the plant cells [5]. The resultant damage is generally manifested by different alterations at cellular level including membrane lipid peroxidation, electrolyte leakage, and sometimes over accumulation of hydrogen peroxide (H_2O_2). H_2O_2 is a highly diffusible ROS within plant cell and its dual roles as a stress-inducer and at the same time as a signaling molecule to upregulate primary antioxidant defense during oxidative stress have been increasingly recognized in different crops including legumes [6–8]. Among the prominent enzymatic

system involved in ROS scavenging, SOD constitutes the first line of defense, but it steadily generates H_2O_2 during dismutation of superoxide radicals mainly by the action of its membrane bound Cu/Zn isoforms [9, 10]. This H_2O_2 is readily scavenged by ascorbate peroxidase (APX) using AsA as its exclusive cofactor within the AsA-GSH cycle and by catalases (CAT) outside this cycle [11]. These three prominent H_2O_2 metabolizing enzymes hold the key in controlling H_2O_2 level during the onset of salinity-induced oxidative stress in plants [4, 11]. In addition, plants accumulate different osmo-regulatory substances under stress and, among them, the role of proline is being debated most. Besides cytosolic osmotic adjustment, compatible solutes possibly play vital roles in stabilizing the structure and activities of enzymes and protein complexes, scavenging ROS and maintaining the integrity of membranes under dehydration stress conditions [12, 13]. Another function of compatible solutes may be in maintaining cytosolic K^+ homeostasis by preventing NaCl-induced K^+ leakage from the cells [14].

Grass pea (*Lathyrus sativus* L.) is a hardy cool-season legume crop, cultivated for both forage and grain in Indian

Subcontinent, Australia, the Mediterranean regions, North Africa, parts of North Europe, and in South America [14, 15]. This crop is highly valued for its remarkable capacity to grow in almost every agroclimatic condition with marginal or often no input and is a promising source of seed protein, minerals, and antioxidant compounds like flavonoids and other polyphenolic compounds [16–19]. The property of type II diabetes-related enzyme inhibition capacity in this crop has recently been revealed in raw and different processed forms of seeds [19, 20]. In recent times, genetic improvement programs for desirable agronomic features particularly high yield and low antinutritional factors including neurotoxin (β-ODAP) have gained momentum in grass pea with development of robust mutation genetic and cytogenetic stocks [21–29]. However, like many other pulses, grass pea faces diverse types of abiotic stresses such as drought, salinity, metal, and weed-induced toxicity [30–33]. Exposure to NaCl-significantly modulated early seedling growth and leaf biochemical parameters in grass pea genotypes under salinity stress [34], and quite alarmingly, enhanced the seed neurotoxin content in grass pea genotypes [34]. Vaz Patto et al. [30] reported good adaptability to salinity stress in the Mediterranean germplasm, while reduction of growth was known in Iranian germplasms of grass pea [35]. Although vast areas under grass pea cultivation are now salinity-affected and increasing salinity is posing great danger for broader introduction of promising genotypes (high yield with low seed neurotoxin content), virtually nothing is known about growth responses and primary antioxidant defense mechanism of this crop. As salt tolerance mechanisms may vary from species to species and at different developmental stages, understanding of specific physiological and intrinsic biochemical mechanism in relation to plant growth is extremely necessary to develop screening markers for genetic improvement of salinity tolerance in crop plants [36]. Along with growth parameters, the antioxidant defense responses are often regarded as one of the important criteria for determining tolerance level of plants to salinity stress [37]. Hence, the object of the present study was set to evaluate the effects of salt stress on growth responses of different grass pea genotypes and to analyze the primary antioxidant defense response in leaves of control and treated plants for better understanding the mechanisms of salt tolerance in grass pea.

2. Materials and Method

2.1. Plant Materials and Field Location.

The experimental materials comprised six elite grass pea (*Lathyrus sativus* L.) genotypes [34], namely, B1, BioL-212, PUSA-90-2, WBK-CB-14, LR3, and LR4. Cultivars (var.) B1, BioL-212 and PUSA-90-2 were selected in the present study for their high yield and low seed toxin levels (<0.2%) and introduction of cultivation in different parts of India as improved pure line varieties [38]. Cultivar WBK-CB-14 (Coochbehar Local) is a dwarf genotype and has been cultivated as a locally-adapted genotype in sub-Himalayan foothills of Dooars region. The LR3 and LR4 are two induced mutant lines, developed through 300 and 350 Gy gamma radiation of seeds of var. PUSA-90-2 and var. WBK-CB-14, respectively, and have been isolated as NaCl-tolerant mutant lines in grass pea [39]. Fresh and healthy seeds with uniform size and from last season harvest (winter of 2011-2012) were collected from Pulses and Oilseeds Research Station, Berhampur (24.1°N, 88.25°E), West Bengal, India. Fresh seeds from M_2 generation harvest of LR3 and LR4 mutant lines were collected separately and used in the present study.

2.2. Treatment Protocol and Plant Growth.

20 dry, healthy, and uniform-sized seeds genotype^{-1} treatment^{-1} were surface sterilized in 70% ethanol for 2 min, rinsed twice in deionized water, and then placed on water-moistened filter papers in Petri dishes in an incubator at 25°C with 12 h light following the guidelines of ISTA [40]. Germinated seeds were immediately transferred to twelve inches earthen pots containing a mixture (total amount of 6.5 Kg) of this soil, vermiculite, and farmyard manure (1 : 1 : 1). The experimental soil was clay loam in texture (clay 32.67%, silt 49.22%, and sand 18.11%) and neutral (pH 7.0) in reaction and contained 40.7 mg kg^{-1} exchangeable Na and 7.39 mg kg^{-1} water exchangeable Cl. Seedlings were thinned to two per pot after emergence and watered evenly for their uniform growth until 7 days after first emergence. The pots were kept under control condition (temperature 20°C–27°C, humidity of 70–77%, neutral light intensity 330–400 μmol m^{-2} s^{-1}) during October–December. Salt treatment commenced on 20-day-old seedlings. The control plants from each of the six genotypes were irrigated with distilled water, while others were subjected to salinity stress by watering them with 150 mM NaCl-supplemented distilled water (300 mL water in each pot), respectively, thrice a week. Five pots (two plants pot^{-1}) genotype^{-1} were arranged in a randomized complete block design with five replications of each treatment. Salt concentration of 150 mM was found critical in an earlier study for determining tolerance level of grass pea genotypes to salt stress [41] and, thus, was selected for the present study. Salt concentration in pot soil was regularly checked by measuring electrical conductivity with a conductivitimeter (Systronics M-308, Kolkata, India), and evapotranspirational losses were compensated daily with deionized water. Each of the six control plants exhibited nonsignificant ($P > 0.05$) variations for the traits studied, and therefore mean of all controls was presented for comparison with treated genotypes.

2.3. Growth Measurements.

Plants were harvested at 15, 30, and 60 days after the commencement of salt treatments (DAC) and separated into roots and shoots. Plant height (cm) and number of primary branches were recorded at harvest. To determine dry weights, plants were separated into roots and shoots. Roots were washed in tap water to remove soil and rinsed in de-ionised water. Plant materials were oven-dried at 65°C for 48 h and weighed. Fully expanded leaf samples from primary branches of plants were used for analysis of leaf biochemical parameters and antioxidant defense response.

2.4. Measurement of Chlorophyll and Carotenoids Contents and Rates of Photosynthesis. Leaf chlorophyll and carotenoid contents were determined by the method of Lichtenthaler [42]. Leaf tissue (50 mg) was homogenized in 10 mL chilled acetone (80%). The homogenate was centrifuged at 4,000 ×g for 12 min. Absorbance of the supernatant was recorded at 663, 647 and 470 nm for chlorophyll *a*, chlorophyll *b* and carotenoids, respectively. The contents were expressed as mg chlorophyll or carotenoids g^{-1} FW. The chlorophyll stability indices (CSI%) were measured using the following formula: (Total chlorophyll content in stressed leaves/total chlorophyll content in control leaves) × 100. Leaf photosynthetic rate was assayed following the methods of Coombs et al. [43] using a portable photosynthesis system (LI-6400XT, LI-COR, USA).

2.5. Estimation of Na^+ and K^+ Contents. Fully expanded leaves of control and salt-treated plants were analysed for total Na^+ and K^+ contents following the method of Kumar and Sharma [44]. The oven-dried leaf (0.2 g) was ground to fine powder and transferred to a digestion flask (50 mL) containing acid mixture (3 mL) of concentrated H_2SO_4 and $HClO_4$ in the ratio of 9 : 1 (v/v). The flask was heated gently over a hot plate for 10 to 12 min until the solution became colorless. The cooled digest was then diluted by adding double distilled water and volume was made up as required. The estimation of Na^+ and K^+ contents in acid extracts was carried out using an atomic absorption spectrophotometer (Perkin Elmer, AA-100).

2.6. Determination of Leaf Proline Level. Leaf proline content was estimated according to the method of Bates et al. [45] from fully expanded leaf samples collected from first formed primary branches on respective harvest dates (20, 40, and 60 DAS).

2.7. Analysis of H_2O_2 Content. Fresh tissue of 0.1 g was powdered and blended with 3 mL acetone for 30 min at 4°C. Then the sample was filtered through eight layers of gauze cloth. After addition of 0.15 g active carbon, the sample was centrifuged twice at 3,000 ×g for 20 min at 4°C and then 0.2 mL 20% $TiCl_4$ in HCl and 0.2 mL ammonia was added to 1 mL of the supernatant. After reaction, the compound was centrifuged at 3,000 ×g for 10 min, the supernatant was discarded, and the pellet was dissolved in 3 mL of 1 M H_2SO_4 and absorbance was measured at 410 nm. H_2O_2 content was measured from the absorbance at 410 nm using a standard curve, following the methods of Wang et al. [46].

2.8. Estimation of Lipid Peroxidation. Lipid peroxidation rates were determined by measuring the malondialdehyde (MDA) equivalents following the method of Hodges et al. [47]. About 0.5 g of fresh tissue was homogenized in a mortar with 80% ethanol. The homogenate was centrifuged at 3,000 ×g for 12 min at 4°C. The pellet was extracted twice with the same solvent. The supernatants were pooled and 1 mL of this sample was added to a test tube with an equal volume of either the solution comprised of 20% TCA and 0.01% butylated hydroxy toluene (BHT) or solution of 20% TCA,

0.01% BHT, and 0.65% TBA. Samples were heated at 95°C for 25 min and cooled to room temperature. Absorbance was measured at 450, 532, and 600 nm. Level of lipid peroxides was calculated following Hodges et al. [47] and expressed as nmol MDA g^{-1} FW.

2.9. Assay of Electrolyte Leakage. Electrolyte leakage (EL) was assayed by measuring the ions leaching from tissue into deionised water [39]. Fresh samples (100 mg) were cut into small pieces (about 5 mm segments) and placed in test tubes containing 10 mL deionised water. Tubes were kept in a water bath at 32°C for 2 h. After incubation, electrical conductivity (EC_1) of the bathing solution was recorded with an electrical conductivity meter (Systronics M-308, Kolkata, India). The samples were then autoclaved at 121°C for 20 min to completely kill the tissues and release all electrolytes. Samples were then cooled to 25°C and final electrical conductivity (EC_2) was determined. The EL was expressed as a percentage by the formula (EL%) = (EC_1)/(EC_2) × 100.

2.10. Estimation of Foliar Ascorbic Acid. Reduced AsA and oxidized ascorbate (DHA) contents were determined by the method of Law et al. [48]. AsA redox was calculated as AsA/(AsA + DHA).

2.11. Antioxidant Enzyme Assays. Fresh leaf tissue of 250 mg was homogenized in 1 mL of 50 mM potassium phosphate buffer (pH 7.8) containing 1 mM EDTA, 1 mM dithiotreitol, and 2% (w/v) polyvinylpyrrolidone (PVP) using a chilled mortar and pestle kept in an ice bath. The homogenate was centrifuged at 15,000 ×g at 4°C for 20 min. Clear supernatant was used for enzyme assays. For measuring APX (EC 1.11.1.11) activity, the tissue was separately ground in homogenizing medium containing 2.0 mM AsA in addition to the other ingredients. All assays were done at 25°C. Soluble protein content was determined according to Bradford [49] using BSA as a standard.

SOD (EC 1.15.1.1) activity was determined by nitro blue tetrazolium (NBT) photochemical assay following Beyer and Fridovich [50]. In this method, 1 mL of solution containing 50 mM potassium phosphate buffer (pH 7.8), 9.9 mM L-methionine, 57 μM NBT, and 0.025% triton-X-100 were added into small glass tubes followed by 20 μL of enzyme extract. Reaction was started by adding 10 μL of riboflavin solution (0.044 mg mL^{-1}) and placing the tubes in an aluminium foil-lined box having two 20 W fluorescent lamps for 7 min. A parallel control was run where buffer was used instead of sample. After illumination, the absorbance of solution was measured at 560 nm. A nonirradiated complete reaction mixture served as a blank. SOD activity was expressed as U (unit) min^{-1} mg^{-1} protein. One unit of SOD was equal to that amount which causes a 50% decrease of SOD-inhibited NBT reduction. SOD isozymes were individualized by native PAGE on 10% acrylamide gel and were localized by a photochemical method [50]. Activity staining gels were incubated for 30 min in 50 mM K-phosphate buffer, pH 7.5, containing 2 mM KCN or 5 mM H_2O_2. Cu/Zn-SOD is inhibited by both KCN and H_2O_2; Fe SOD is inactivated by

H_2O_2 but resistant to KCN, and Mn SOD is resistant to both inhibitors.

APX activity was assayed following methods adopted by Nakano and Asada [51]. Three milliliters of the reaction mixture contained 50 mM K-phosphate buffer (pH 7.0), 0.5 mM AsA, 0.1 mM EDTA, 0.1 mM H_2O_2, and 0.1 mL enzyme extract. The H_2O_2-dependent oxidation of AsA was followed by a decrease in the absorbance at 290 nm (ε = 2.8 mM^{-1} cm^{-1}). APX activity was expressed as nmol AsA oxidized min^{-1} mg^{-1} protein. Native PAGE of APX isozymes was performed in 4% gel and stained following Mittler and Zilinskas [52] based on the inhibition of NBT reduction by AsA.

CAT (EC 1.11.1.6) activity was measured according to Chance and Maehly [53] with slight modifications. Enzymatic activity was initiated by adding 50 μL of enzyme extract into the reaction mixture containing 500 μL of K-1 (0.1 M, pH 6.5), 250 μL of distilled water, and 200 μL of 75 mM H_2O_2. CAT activity was monitored at 240 nm for 2 min at 25°C after initiation of the reaction and was measured against a blank reaction mixture containing no enzyme extract. CAT specific activity (nmol H_2O_2 degraded min^{-1} mg^{-1} protein) was calculated using the molar absorptivity of 43.6 M^{-1} cm^{-1} for H_2O_2 at 240 nm. CAT isozyme profiling was done on 6% acrylamide gel, following Woodbury et al. [54].

2.12. Statistical Analyses. The results presented here are the mean values ± standard error (SE) of at least five replicates. Means were compared by ANOVA using the SPSS v. 10 (SPS Inc, USA) and evaluated using Duncan's Multiple Range Test at $P \leq 0.05$.

3. Results

3.1. Effect of Salt Stress on Growth. Six grass pea genotypes subjected to NaCl-treatment (150 mM) exhibited significant ($P < 0.05$) variations among themselves for growth parameters. While 15 DAC represented early vegetative growth, 30 DAC denoted flowering stage and 60 DAC reflected pod-bearing stage. Plant height, length of internodes, primary branches/plant, and shoot and root dry weight were decreased significantly compared to those of control in all six genotypes with significantly different magnitudes at 15 DAC (Figures 1(a)–1(e)) and were further reduced in var. PUSA-90-2 and WBK-CB-14 at 30 DAC. Shoot dry weight in saline versus control plants was reduced by about 2-3-fold at 15 DAC with higher magnitude in PUSA-90-2 and WBK-CB-14 than B1 (2.7-fold), BioL-212 (2.5-fold), LR3 (2.2-fold) and LR4 (2-fold). Similar trend was noticed in case of root dry weight. Nearly 4-fold reduction in biomass production (total dry weight) was measured in PUSA-90-2 and WBK-CB-14 at 30 DAC. Growth traits plummeted to lowest level in PUSA-90-2 and WBK-CB-14 at 60 DAC with 5–5.5 reduction in relation to control. By contrast, nearly normal plant height, primary branches and biomass production were noticed in the rest four genotypes at 60 DAC (Figures 1(a)–1(e)).

3.2. Photosynthetic Apparatus, CSI%, and Photosynthesis. Significant variation was observed among six genotypes regarding pigment composition during progression of NaCl-treatment. Within photosynthetic apparatus, chl *a* content was changed significantly (Figure 1(f)) but carotenoid content varied nonsignificantly between 1.49 mg g^{-1} FW and 1.61 mg g^{-1} FW among the six genotypes under NaCl-treatment (data not shown). Compared to control plants, chl *a* content was measured markedly lower at 15 DAC across the genotypes but was further reduced only in PUSA-90-2 and WBK-CB-14 by 2-fold at 30 DAC and by another 2-fold at 60 DAC (Figure 1(f)). Chl *a* content was increased substantially and became quite normal in B1, BioL-212, LR3, and LR4 at 60 DAC. Chl *a/b* ratio was also changed, accordingly. This value was significantly lower in all genotypes at 15 DAC but was dropped to <1 in PUSA-90-2 and WBK-CB-14 at 30 DAC and was further reduced in these two varieties at 60 DAC (Figure 2(a)). CSI% was 100% in control plants but was reduced substantially across the six genotypes at 15 DAC with severest effect on PUSA-90-2 (24.78%) and WBK-CB-14 (25.03%). It became normal in the rest four genotypes, varying 98–100% at 60 DAC. Compared to control, leaf photosynthetic rate was decreased by about 2-fold in LR3 and LR4, 3-fold in B1, 3.3-fold in BioL-212, and nearly 4-fold in PUSA-90-2 and WBK-CB-14 till 30 DAC. Photosynthetic rate was significantly low in PUSA-90-2 and WBK-CB-14 at 60 DAC also, but it was quite normal in B1, BioL-212, LR3, and LR4 at 60 DAC (Figure 2(b)).

3.3. Salt Stress Effect on Leaf Proline Content. Compared to control (3.05–3.17 μg g^{-1} fresh weight) no significant change in leaf proline content was observed in six genotypes at 15 DAC. The content, however, was increased by about 2-fold in B1 and BioL-212 and about 3-fold in the two mutant lines at 30 DAC. Proline level remained normal in PUSA-90-2 and WBK-CB-14 and it continued till 60 DAC (data not shown).

3.4. Effect on Leaf Na$^+$ and K$^+$ Ions. Changes in leaf Na$^+$ and K$^+$ content were significant ($P < 0.05$) across six genotypes (Figures 3(a) and 3(b)). This was due to significant accumulation of Na$^+$ and reduction of K$^+$ concentration in all six genotypes at 15 DAC and in PUSA-90-2 and WBK-CB-14 at 30 DAC and 60 DAC. Compared to control, Na$^+$ concentration was markedly decreased while K$^+$ was considerably enhanced in B1, BioL-212, LR3, and LR4 genotypes at 30 DAC, pushing K$^+$/Na$^+$ ratio significantly higher than control in B1 and BioL-212 and close to normal level in both mutant lines at 30 DAC. It was nonsignificantly changed at 60 DAC (Figure 3(c)).

3.5. As-Induced H_2O_2 Accumulation, Lipid Peroxidation, and Electrolyte Leakage (EL%). Compared to control, accumulation of H_2O_2 and MDA was significantly higher in B1, BioL-212, LR 3, and LR4 at 15 DAC but was significantly different in different genotypes (Figures 4(a)–4(c)). The increase of H_2O_2 was 3.5-4.5-fold while MDA level was increased by about 3-fold across the four genotypes at 15 DAC. Statistically significant rise was also observed for EL% (Figure 4(c)). By

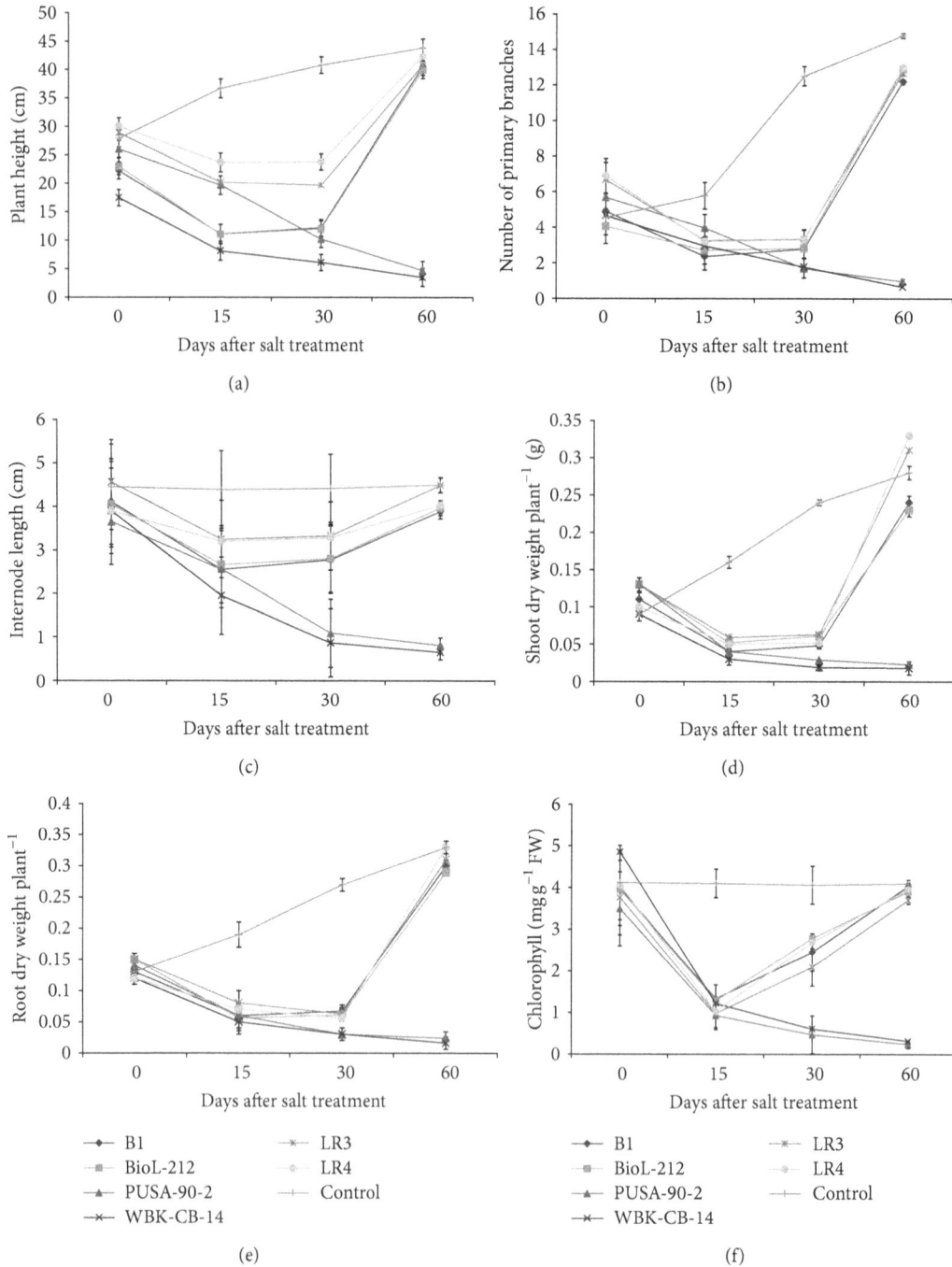

FIGURE 1: Changes in (a) plant height, (b) primary branches/plant, (c) internode length, (d) shoot dry weight (g), (e) root dry weight (g), and (f) chlorophyll *a* content in B1, BioL-212, PUSA-90-2, WBK-CB-14, LR3, and LR4 along with mean control value of grass pea (*Lathyrus sativus* L.) at 15, 30, and 60 days after salt (150 mM NaCl) treatment. Data are means ± SE of four replicates. Means of respective genotypes with different small letters are significantly different at $P \leq 0.05$ by ANOVA followed by Duncan's Multiple Range Test. The control of each of the six genotypes being without any significant differences ($P > 0.05$) among themselves; mean value of all controls was presented.

contrast, H_2O_2 content was markedly low but MDA content and EL% was quite high in PUSA-90-2 and WBK-CB-14 at 15 DAC. At 30 DAC, H_2O_2-level was decreased by about 2–2.2-fold in B1, BioL-212, LR3, and LR4 from the level recorded at 15 DAC, but the content was still significantly higher than that in control (Figure 4(a)). MDA content and EL%, however,

were normal in these four genotypes at 30 DAC and also, at 60 DAC. Both H_2O_2 and MDA content were considerably enhanced over control in PUSA-90-2 and WBK-CB-14 at 30 DAC, and remained unchanged at 60 DAC. EL% followed the same trend of change in MDA level in all genotypes (Figures 4(b) and 4(c)).

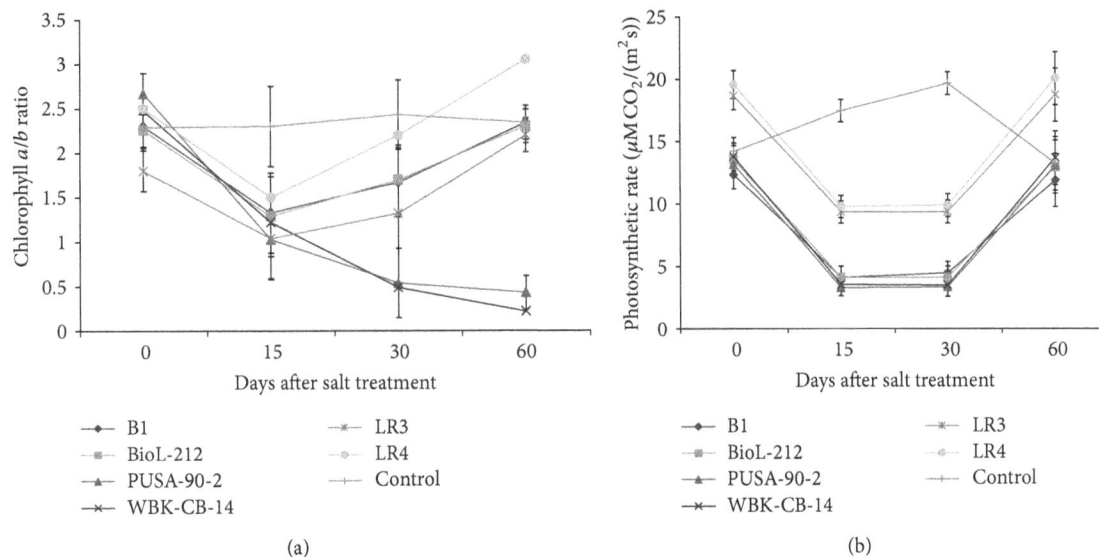

FIGURE 2: Effects of salt treatment (150 mM NaCl) on leaf (a) chlorophyll (chl) a/b ratio and (b) photosynthesis in six grass pea genotypes along with mean control value at 15, 30, and 60 days after treatment. Data are means ± SE of four replicates. Means of respective genotypes with different small letters are significantly different at $P \leq 0.05$ by ANOVA followed by Duncan's Multiple Range Test. The control of each of the six genotypes being without any significant differences ($P > 0.05$) among themselves; mean value of all controls was presented.

3.6. Antioxidant Defense Responses to NaCL-Treatment. Reduced ascorbate (AsA) content as well as AsA-redox was significantly lower in all six genotypes at 15 DAC (Table 1). At 30 DAC, AsA content was increased considerably from its 15 DAC level in B1, BioL-212, LR3, and LR 4 genotypes and became normal (Table 1). Further increase in AsA level was measured in these four genotypes at 60 DAC. AsA level and its redox state were the highest in LR4, followed by BioL-212, LR3 and then by B1 at 60 DAC (Table 1). AsA content and AsA-redox was significantly lower in PUSA-90-2 and WBK-CB-14 cultivars during 30 and 60 DAC (Table 1).

Among the H_2O_2-metabolizing enzymes, statistically significant higher SOD activity over the control was observed in B1, BioL-212, LR3, and LR4 seedlings since 15 DAC and the trend was continued till 60 DAC (Table 1). Compared to control, SOD activity was quite low in PUSA-90-2 (3-fold) and WBK-CB-14 (3.5-fold) at 15 DAC, but bounced back at 30 DAC registering an increase of 4-fold in PUSA-90-2 and of 6-fold in WBK-CB-14 and the level was maintained at 60 DAC (Table 1). APX activity was increased in LR3 and LR4 by about 2–2.5-fold but was quite low in B1 and BioL-212 at 15 DAC. At the same period, APX activity was normal in PUSA-90-2 and WBK-CB-14. Completely reverse situation was encountered at 30 DAC and 60 DAC. APX activity was significantly declined in LR3, LR4, PUSA-90-2 and WBK-CB-14 but was enhanced by about 2-fold in B1 and BioL-212 at 30 DAC, and nonsignificant ($P > 0.05$) changes of respective values occurred at 60 DAC (Table 1). CAT activity was markedly higher in B1 and BioL-212 and normal in PUSA-90-2 and WBK-CB-14 but was reduced substantially in LR3 and LR4 mutant lines at 15 DAC (Table 1). At 30 DAC, CAT activity was decreased significantly in B1, BioL-212, LR3,

and LR4 but was increased in two mutant lines by about 2-3-fold (Table 1). Similar trend was noticed at 60 DAC, also (Table 1).

3.7. In-Gel Activity of Antioxidant Enzymes Under As-Exposure. Altogether five activity bands were resolved for SOD on the basis of their increasing mobility and sensitivity to H_2O_2 and KCN and zymograms obtained at 60 DAC were presented (Figures 5(a) and 5(b)). Two Cu/Zn isoforms (I and II) were consistently resolved by native PAGE in the leaf extract of control plants (Figure 5(a)). At 15 DAC, Cu/Zn-SOD I and II were visualized in leaves of all the six genotypes but with much higher intensity in B1, BioL-212, LR3, and LR4 (figure not shown). At 60 DAC, in addition to Cu/Zn I and II, two Mn -SOD isoforms (Mn SOD I and II) was distinctly visualized in PUSA-90-2 and WBK-CB-14. At the same period, one Mn SOD (Mn SOD I) in LR3 and one Fe SOD in LR4 mutant line appeared in addition to Cu/Zn SOD I and II (Figures 5(a) and 5(b)). Similar pattern was maintained at 30 DAC in all the genotypes (Figures not presented).

For APX, a total of three isoforms (APX 1, APX 2, and APX 3) were clearly resolved in the leaf extract at 60 DAC in control and treated genotypes but at different intensity (Figure 6(a)). At 15 DAC, APX 1 was resolved as faint band in B1 and BioL-212 but along with APX 2 and APX 3 it was quite normal in the rest of the genotypes (Figure not shown). All the three isoforms were visualized as strong bands in B1 and BioL-212 at 30 DAC and 60 DAC but were diminished as faint bands in the rest of the genotypes (Figure 6(a)).

CAT activity was uniformly resolved as a single zone across the genotypes (Figure 6(b)). Band intensity was much stronger in LR3 and LR4 but was considerably lower in the

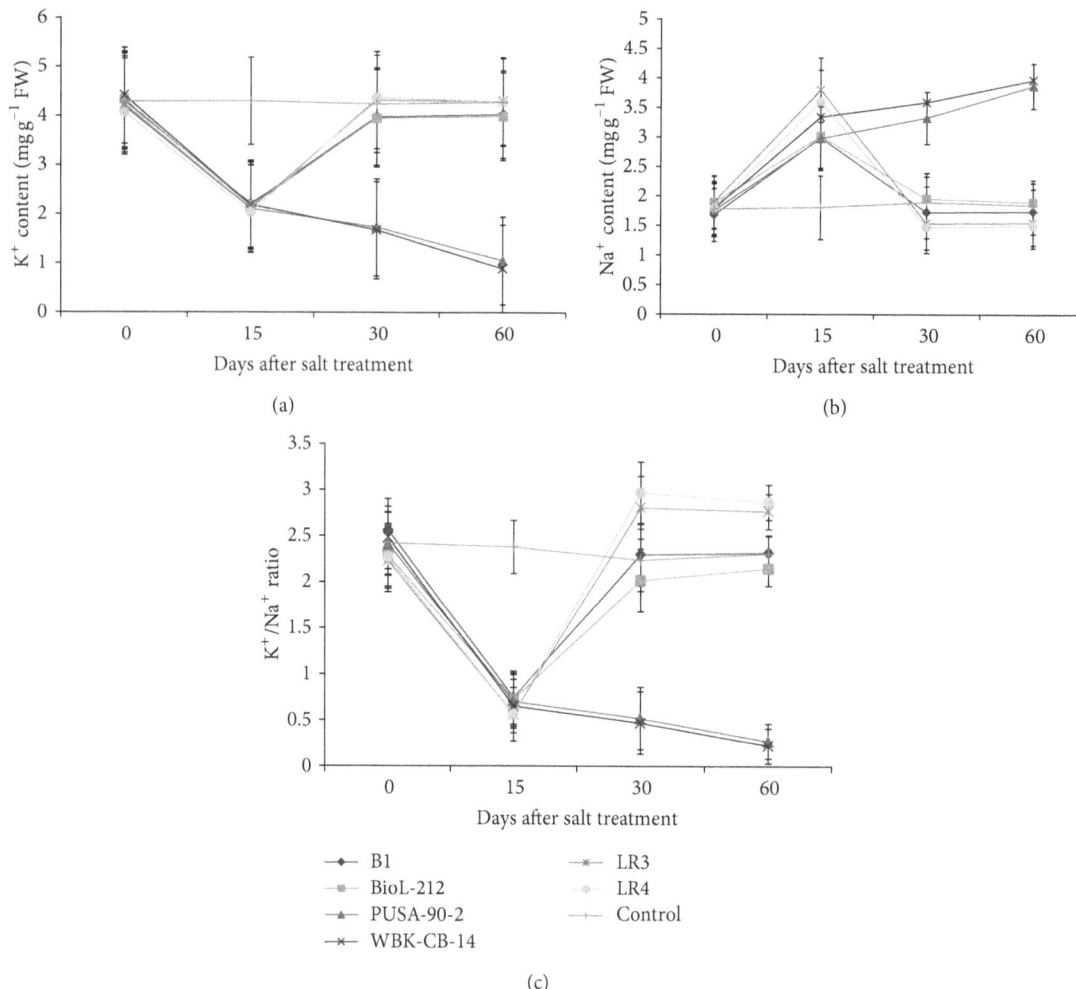

FIGURE 3: Changes in leaf (a) K^+, (b) Na^+ content, and (c) K^+/Na^+ ratio in six grass pea genotypes along with mean control value at 15, 30, and 60 days after 150 mM NaCl-treatment. Data are means ± SE of four replicates. Means of respective genotypes with different small letters are significantly different at $P \leq 0.05$ by ANOVA followed by Duncan's multiple range test. The control of each of the six genotypes being without any significant differences ($P > 0.05$) among themselves mean value of all controls was presented.

rest of the genotypes with faintest appearance was resolved in case of PUSA-90-2 and WBK-CB-14 genotypes at 30 and 60 DAC (Figure 6(b)). Low to moderate in-gel activity was noticed at 15 DAC (Figure not shown).

4. Discussion

Dry weight of plants has been considered as one of the integrative criteria in determining salt responses in plants [36]. In the present study, distinctly different response was obtained in six genotypes during progression of salt treatment. Significant reduction in plant height and primary branches might be due to marked reduction in internodes length and attributed to lower shoot dry weight in all six genotypes at 15 DAC, indicating effect of salt-treatment at early vegetative growth stages. This situation got significant twist at later stages of growth. Biomass production as measured by dry weight of shoots and roots was further reduced in var. PUSA-90-2 and WBK-CB-14 due to significantly low plant

height and branches/plant at 30 DAC and 60 DAC. However, despite increase in treatment duration, growth retardation was checked in the rest four genotypes at 30 DAC, and remarkably enough, growth traits bounced back to normal in these four genotypes at 60 DAC. Results illustrated that the greatest reduction in plant growth occurred during the first period of salt treatment (vegetative growth stage) in cases of B1, BioL-212, LR3, and LR4 but it was extended to its severest inhibitory effect on growth of PUSA-90-2 and WBK-CB-14 at later stages (reproductive) of growth. The normal (control like) dry matter production in four genotypes at flowering stage (30 DAC) and its maintenance up to pod-bearing stage (60 DAC) despite initial blow at 15 DAC was unique in grass pea and strongly suggested effective prevention of early NaCl-stress at reproductive period to maintain salinity-tolerance in the present experimental conditions. A good positive correlation between plant height ($r = 0.95$, $n = 10$) as well as primary branches/plant ($r = 0.78$, $n = 10$) and shoot dry weight was indicative that salt tolerance was

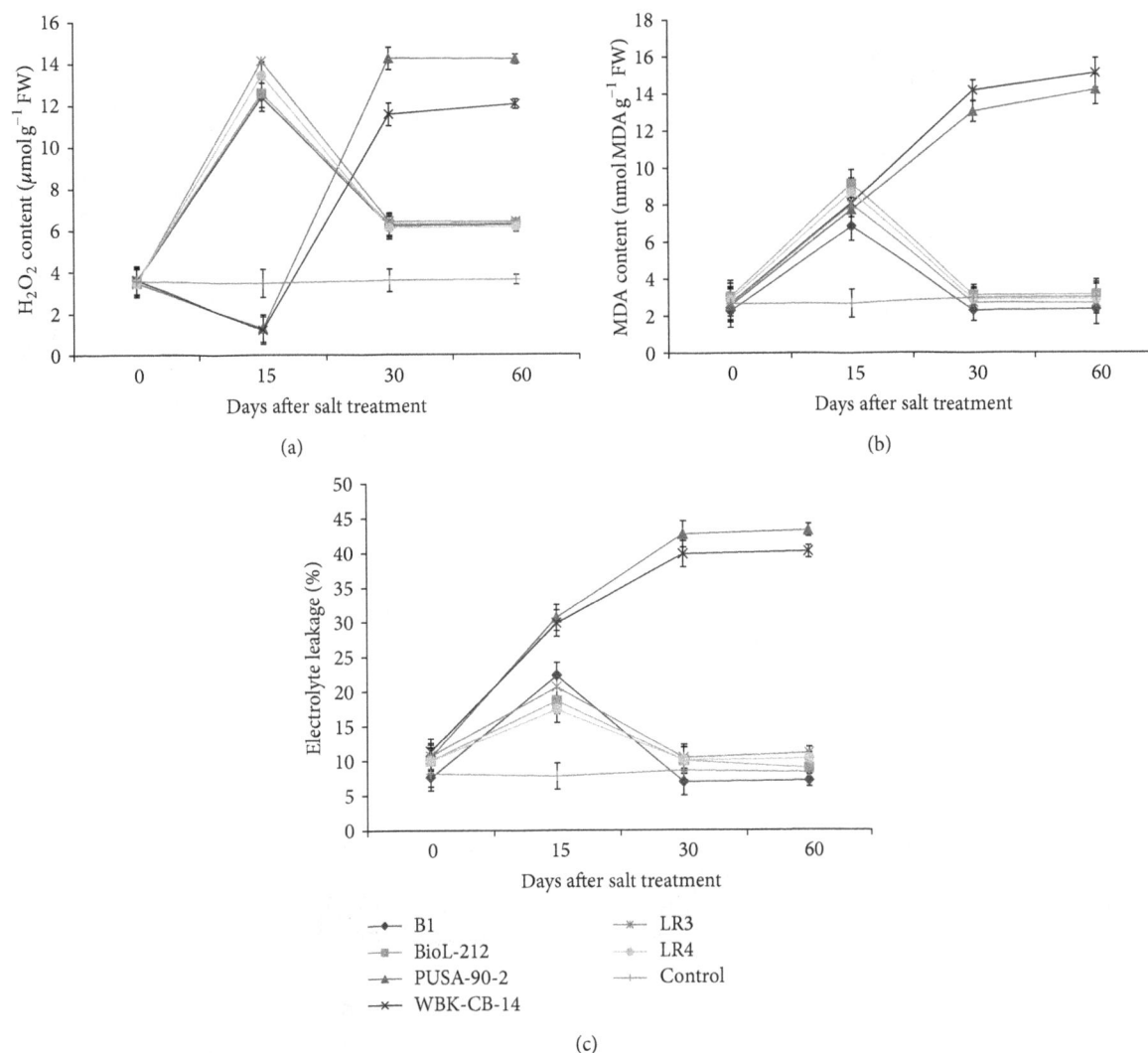

(a)

(b)

(c)

FIGURE 4: Changes in foliar (a) accumulation of H_2O_2, (b) malondialdehyde (MDA) content, and (c) electrolyte leakage % in six grass pea genotypes along with mean control value at 15, 30, and 60 days after 150 mM NaCl-treatment. Data are means ± SE of four replicates. Means of respective genotypes with different small letters are significantly different at $P \leq 0.05$ by ANOVA followed by Duncan's multiple range test. The control of each of the six genotypes being without any significant differences ($P > 0.05$) among themselves, mean value of all controls was presented.

positively associated with normal plant dry weight, which was increased by positive contribution from other components, least affected by salt induced injury in tolerant lines. In grass pea, high tolerance of dwf1 and dwf2 dwarf mutant lines to 170 mM NaCl-treatment was indicated by their normal (control like) plant dry weight, while low plant biomass accumulation in the third dwarf line, dwf3, was associated with symptoms of increased salt sensitivity [55]. Reduction in plant dry weight under high salt treatment was reported in pea, grass pea, *Phaseolus* and lentil [1, 35, 41, 56, 57] while increasing dry weight over control was reported in grass pea mutant lines showing high tolerance to salinity stress [39].

The toxic effect of salt inside the plant triggers the "phase II" response of growth, and the most affected organ is leaf [36]. Changes in plant biomass production under salinity may be due to many reasons such as lack of maintenance of turgor,

sodium/chloride ion toxicity and disturbances in metabolic pathways. Since these factors disturb the functioning of gas exchange attributes they ultimately lead to a decline in activity of photosynthetic apparatus [5, 57]. In the present study, reduction in chl *a* content was mainly responsible for decrease in chl *a/b* ratio, suggesting NaCl-induced disruption in photosynthetic apparatus of all six genotypes at early stages of growth. The magnitude of this disruption was also manifested by markedly low chlorophyll stability index, consequently resulting in low rate of photosynthesis in salt-affected genotypes. This situation were significantly improved at later stages of growth in B1, BioL-212, LR3, and LR4 with normal chl *a/b* ratio, chlorophyll stability and normal (control like) rate of photosynthesis, as the marks of salt-tolerance. Contrastingly, significant reduction in all these parameters in PUSA-90-2 and WBK-CB-14 at 30

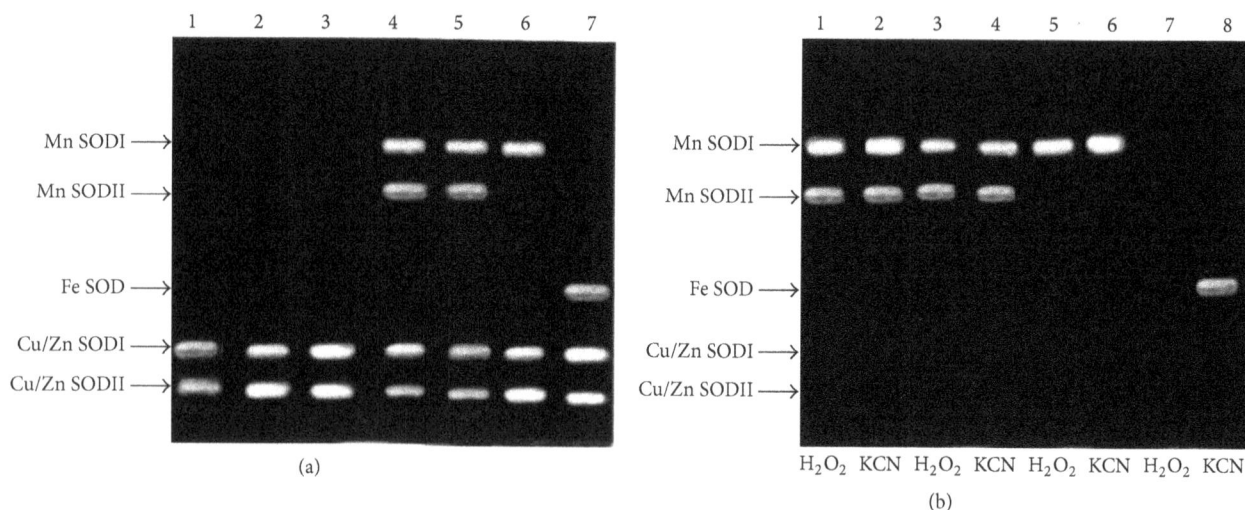

FIGURE 5: (a) Activity gel of superoxide dismutase (SOD) in native PAGE on 10% acrylamide gels of leaf extracts of six grass pea (*Lathyrus sativus* L.) genotypes at 60 days after commencement of 150 mM NaCl-treatment; Lane 1—control plants (Cu/Zn SOD I and II), lane 2—variety B1 (Cu/Zn SOD I and II), lane 3—variety BioL-212 (Cu/Zn SOD I and II), lane 4—PUSA-90-2 (Mn SOD I and II, Cu/Zn SOD I and II), lane 5—WBK-CB-14 (Mn SOD I and II, Cu/Zn SOD I and II), lane 6—LR3 mutant (Mn SOD I, Cu/Zn SOD I and II), and lane 7—LR4 mutant (Fe SOD, Cu/Zn SOD I and II), and (b) inhibitor studies with H_2O_2 and KCN and visualization of SOD isoforms in native PAGE of leaf extracts of grass pea (*Lathyrus sativus* L.) variety PUSA-90-2 (lanes 1 and 2), WBK-CB-14 (lanes 3 and 4), LR3 mutant (lanes 5 and 6), and LR 4 mutant (lanes 7 and 8) at 60 days after commencement of salt treatment. The control plants, variety B1 and BioL-212 containing only Cu/Zn SOD I and II showed no bands in inhibitor study and thus not presented.

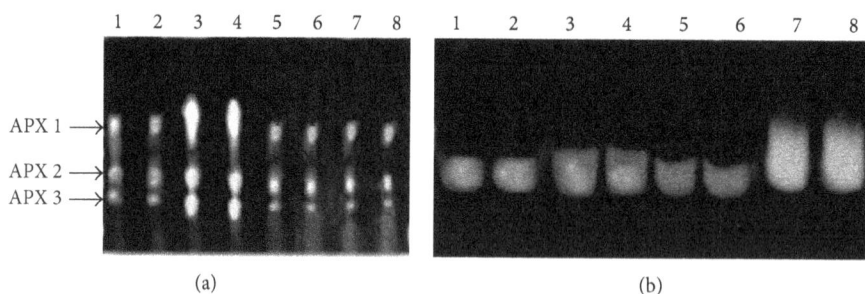

FIGURE 6: Effects of 150 mM NaCl-treatments on isozymes banding of (a) APX activity in native PAGE on 4% acrylamide gels of leaf extracts of grass pea (*Lathyrus sativus* L.) genotypes at 60 days after commencement of treatment: lanes 1 and 2—control plant, lane 3—variety B1, lane 4—BioL-212, lane 5—PUSA-90-2, lane 6—WBK-CB-14, lane 7—LR3 mutant and lane 8—LR4 mutant: (b) CAT activity in native PAGE on 6% acrylamide gels of leaf extracts of grass pea (*Lathyrus sativus* L.) genotypes at 60 days after commencement of treatment; lanes 1 and 2—control plants, lane 3—variety B1, lane 4—BioL-212, lane 5—PUSA-90-2, lane 6—WBK-CB-14, lane 7—LR3, and lane 8—LR4 mutant.

DAC and at 60 DAC severely impeded plant photosynthesis at reproductive stage which might be instrumental in low biomass production in these two genotypes. This is despite the fact that chl *b* and carotenoids were quite normal (control like) throughout the treatment period. Carotenoid has been regarded as a primary nonenzymatic antioxidant defense components protecting plants from adverse effect of ROS generated under NaCl-induced oxidative stress in plants including legumes [5]. Increase in chl *a* and *b* and carotenoid content, however, was observed in grass pea genotypes and mutant lines exhibiting tolerance to salt treatment [39, 41]. Presumably, chl *a* content played more vital role than chl *b* and carotenoids in maintaining normal photosynthesis in the present genotypes under prolonged salt treatment. The reduction in chlorophyll contents could have been due to

the displacement of Mg^{2+} by toxic Na^+ ions, which caused the degradation of green pigments through the disruption of ultrastructure of pigment-protein-lipid complex by ion toxicity, as explained in salt-stressed cotton plants [58]. Disruption of chlorophyll and carotenoid contents under salinity stress was also reported in *Lens culinaris* [56], *Pisum sativum* [57] and *Vigna radiata* [59].

Conflicting reports are available regarding specific role of proline in salt-tolerant genotypes. A positive correlation between proline over accumulation and increasing salinity/drought tolerance has been found in different crop plants including transgenics that were engineered for overproduction of proline [12]. Increase in proline content under NaCl-stress was manifested as one of the mechanisms of salt tolerance in *Lathyrus sativus* [39, 41], *Pisum sativum* [37,

TABLE 1: Effects of NaCl-treatments (150 mM) on reduced ascorbate (AsA, μmol g^{-1} FW), redox state of ascorbate (AsA/AsA + DHA), activities of superoxide dismutase (SOD, U min^{-1} mg^{-1} protein), ascorbate peroxidase (APX, nmol AsA oxidized min^{-1} mg^{-1} protein), and catalases (CAT, nmol H$_2$O$_2$ degraded min^{-1} mg^{-1} protein) in leaves of control and six treated genotypes [varieies B1, BioL-212, PUSA-90-2, WBK-CB-14, and two mutant lines, namely, Lathyrus resistant mutant 3 (LR3) and Lathyrus resistant mutant 4 (LR4)] at control (0) and 15, 30, and 60 days after commencement of treatment (DAC) of Lathyrus sativus L.

Traits	DAC	B1	BioL-212	PUSA-90-2	WBK-CB-14	LR3	LR4	Mean control
AsA	0	1.81 ± 1.0$^{ab'}$	1.75 ± 1.0$^{ab'}$	1.89 ± 0.97$^{aa'}$	1.79 ± 0.89$^{aa'}$	1.87 ± 0.87$^{aa'}$	1.82 ± 0.88$^{ab'}$	1.83 ± 0.9$^{aa'}$
	15	1.09 ± 0.9$^{cc'}$	1.08 ± 0.8$^{cc'}$	0.98 ± 0.3$^{cb'}$	1.01 ± 0.8$^{cb'}$	1.12 ± 0.3$^{bb'}$	1.03 ± 0.8$^{cc'}$	1.89 ± 0.9$^{aa'}$
	30	1.80 ± 0.3$^{ab'}$	1.78 ± 0.8$^{ab'}$	0.57 ± 0.4$^{bc'}$	0.73 ± 0.2$^{bc'}$	1.82 ± 0.8$^{aa'}$	1.79 ± 0.8$^{ab'}$	1.75 ± 0.9$^{aa'}$
	60	1.91 ± 7.9$^{ba'}$	2.28 ± 10.0$^{aa'}$	0.43 ± 7.9$^{dd'}$	0.61 ± 8.1$^{cd'}$	2.14 ± 10.8$^{aa'}$	2.42 ± 9.9$^{aa'}$	1.81 ± 0.9$^{ba'}$
AsA redox	0	0.887 ± 0.09$^{aa'}$	0.893 ± 0.09$^{aa'}$	0.911 ± 0.08$^{aa'}$	0.915 ± 0.10$^{aa'}$	0.905 ± 0.1$^{aa'}$	0.90 ± 0.1$^{aa'}$	0.891 ± 0.09$^{aa'}$
	15	0.632 ± 0.11$^{ab'}$	0.643 ± 0.10$^{ab'}$	0.557 ± 0.12$^{bb'}$	0.619 ± 0.09$^{ab'}$	0.645 ± 0.1$^{ab'}$	0.647 ± 0.1$^{ac'}$	0.860 ± 0.09$^{ba'}$
	30	0.842 ± 0.08$^{ba'}$	0.901 ± 0.09$^{aa'}$	0.468 ± 0.09$^{bc'}$	0.497 ± 0.11$^{bc'}$	0.970 ± 0.11$^{aa'}$	0.891 ± 0.1$^{ab'}$	0.855 ± 0.09$^{ba'}$
	60	0.861 ± 0.09$^{ca'}$	0.958 ± 0.11$^{aa'}$	0.305 ± 0.09$^{dd'}$	0.315 ± 0.12$^{dd'}$	0.930 ± 0.11$^{ba'}$	0.977 ± 0.1$^{aa'}$	0.859 ± 0.09$^{ca'}$
SOD	0	111.5 ± 5.1$^{ac'}$	118.9 ± 4.9$^{ac'}$	113.3 ± 4.7$^{ab'}$	111.5 ± 5.5$^{ab'}$	123.9 ± 4.9$^{ac'}$	118.7 ± 4.9$^{ac'}$	116.5 ± 4.9$^{ad'}$
	15	130.9 ± 5.8$^{bb'}$	147.8 ± 4.8$^{ab'}$	46.7 ± 5.1$^{cc'}$	37.6 ± 7.7$^{dc'}$	143.9 ± 4.8$^{ab'}$	147.2 ± 4.8$^{ab'}$	136.5 ± 4.9$^{bc'}$
	30	158.8 ± 4.8$^{ca'}$	191.8 ± 10.5$^{aa'}$	192.6 ± 9.8$^{aa'}$	179.8 ± 8.5$^{ba'}$	189.8 ± 10.5$^{aa'}$	194.8 ± 10.5$^{aa'}$	176.5 ± 4.9$^{bb'}$
	60	161.6 ± 11.6$^{ab'}$	190.6 ± 9.8$^{aa'}$	195.1 ± 6.9$^{aa'}$	187.5 ± 10.2$^{aa'}$	191.89 ± 10.5$^{aa'}$	199.8 ± 10.5$^{aa'}$	200.5 ± 4.9$^{aa'}$
APX	0	175.8 ± 7.8$^{ab'}$	179.9 ± 8.1$^{ab'}$	179.5 ± 8.9$^{aa'}$	181.7 ± 10.0$^{aa'}$	175.9 ± 8.1$^{ab'}$	182.9 ± 8.1$^{ab'}$	181.3 ± 8.1$^{aa'}$
	15	94.7 ± 8.6$^{cc'}$	87.7 ± 12.6$^{cc'}$	178.3 ± 4.3$^{ba'}$	183.73 ± 4.5$^{ba'}$	338.8 ± 10$^{aa'}$	375.1 ± 10.0$^{aa'}$	181.0 ± 9.2$^{ba'}$
	30	350.9 ± 4.9$^{aa'}$	329.6 ± 9.3$^{aa'}$	100.4 ± 18.5$^{cb'}$	98.8 ± 13.2$^{cb'}$	97.9 ± 11.2$^{cd'}$	108.8 ± 10.8$^{cc'}$	176.3 ± 6.6$^{ba'}$
	60	358.7 ± 6.1$^{aa'}$	311.3 ± 5.5$^{aa'}$	111.5 ± 11.3$^{bb'}$	107.1 ± 10.9$^{bb'}$	101.9 ± 11.6$^{bc'}$	107.9 ± 11.8$^{bc'}$	161.6 ± 7.1$^{ab'}$
CAT	0	39.1 ± 4.7$^{ab'}$	40.1 ± 6.7$^{ab'}$	37.7 ± 5.1$^{aa'}$	39.8 ± 5.6$^{aa'}$	44.8 ± 5.3$^{ab'}$	49.8 ± 5.9$^{ab'}$	42.1 ± 4.9$^{aa'}$
	15	60.5 ± 3.8$^{aa'}$	68.8 ± 6.4$^{aa'}$	38.6 ± 4.9$^{ba'}$	40.9 ± 5.7$^{ba'}$	20.8 ± 5.9$^{cc'}$	24.9 ± 4.8$^{cc'}$	42.8 ± 5.2$^{ba'}$
	30	18.5 ± 4.1$^{cc'}$	20.3 ± 4.7$^{cc'}$	11.7 ± 5.3$^{cb'}$	13.3 ± 6.0$^{cb'}$	133.3 ± 7.9$^{aa'}$	151.3 ± 8.8$^{aa'}$	49.1 ± 4.9$^{ba'}$
	60	16.7 ± 5.0$^{cc'}$	18.9 ± 5.1$^{cc'}$	8.6 ± 6.1$^{db'}$	10.7 ± 5.9$^{db'}$	137.5 ± 8.9$^{aa'}$	163.5 ± 10.9$^{aa'}$	50.1 ± 4.9$^{ba'}$

Data were presented as means ± SE of four replicates. Different small letters in each row indicate significant differences among genotypes and letters with prime in each column indicate significant differences among treatments (for a particular trait) at $P \leq 0.05$ by ANOVA followed by Duncan's multiple range test. Controls of each genotype exhibited nonsignificant ($P > 0.05$) differences and thus mean of all controls is presented here.

57] and Phaseolus aureus [60]. Present result is in partial agreement with these results. Increase in proline level was observed in B1, BioL-212, LR3, and LR4 from 30 DAC and it was maintained till 60 DAC. However, it is noteworthy that seedlings from all genotypes faced severe growth inhibition at 15 DAC even when proline level was close to control value. Significantly enough, proline level was control like in both PUSA-90-2 and WBK-CB-14 throughout the growth period although the genotypes continuously suffered growth inhibition. Proline accumulation in plants was regarded as a symptom of stress in less-salinity-tolerant species, and its contribution to osmotic adjustment was found negligible as compared with K$^+$ [2]. Certainly, this is not the situation in the present case. Higher accumulation of proline in four genotypes showing good biomass accumulation at 30 DAC and at 60 DAC suggested its responsiveness during reproductive development of plant conferring protection against salt-induced osmotic disruption in tolerant genotypes. Its usual level during salt-induced growth inhibition, however, requires further study.

Equilibrium of cellular Na$^+$ and K$^+$ content is absolutely essential in imparting salinity tolerance in plants [36, 41]. Excessive accumulation of Na$^+$ in leaves has been considered highly harmful for normal metabolism of plant, and tolerant genotype has the capacity of successful salt exclusion [13, 36]. Salt stress also impairs K$^+$ uptake of plants, and it has been suggested that K$^+$ deficiency might be a contributing factor to salt-induced growth inhibition through induction of oxidative stress and related cell damage [57, 61, 62]. The K$^+$: Na$^+$ ratio has been used as a discriminating factor between tolerant and sensitive genotypes with greater capacity of former to block or reduce the uptake or exclude the excess amount of Na$^+$ and associated increase in K$^+$ content [36]. In the present material, reduction in K uptake and transport to aerial part of plants under salinity in competition with higher Na absorption resulted in decrease in K$^+$: Na$^+$ ratio and low plant biomass production in all the six genotypes at 15 DAC, and in PUSA-90-2 and WBK-CB-14 at 30 DAC and also, at 60 DAC. Growing evidences indicate that abnormal accumulation of Na$^+$ and/or K$^+$ deficiency severely disrupts photosynthetic capability and impeded translocation of photosynthates from leaves into sink organs [36, 57, 63], and might be one of the prime reasons for reduction in photosynthetic capacity during salt-sensitivity of the present genotypes. In salt-sensitive Phaseolus species, K$^+$ deficiency combined with salt stress induced a reduction in CO$_2$ photo assimilation and stomata closure [62]. Similar situation was encountered in salt-sensitive genotypes and mutant lines of grass pea experiencing different NaCl-treatment regimes [39, 41, 55]. By contrast, higher level of K^+ content than Na^+ pushed $K^+ : Na^+$ ratio to quite normal level and ensured better plant growth in B1, BioL-212, LR3, and LR4 genotypes at later stages of growth despite prolonged NaCl-treatment. The result indicated that early surge in Na^+ accumulation and low $K^+ : Na^+$ ratio may be detrimental to vegetative growth but its normal (control like) level is crucial in reproductive stages to maintain flowering and subsequent grain-filling stage under salt-exposure. Normal (control like) concentration of K^+ and $K^+ : Na^+$ ratio at 30 DAC and also at 60 DAC in these four genotypes might be orchestrated through a compensation over time, presumably through translocation of K^+ from roots and stems to leaves [61], a sustained acquisition despite appreciable overall Na^+ uptake [1, 62], a high K^+ selectivity and/or K^+/Na^+ exchange across the plasmalemma of the root epidermis [34, 39, 57]. The maintenance of higher leaf K^+ concentrations and $K^+ : Na^+$ ratio in the present grass pea genotypes showing good growth under salinity strongly indicated their ability to withdraw Na^+ and to retranslocate K^+ during reproductive stages which seems to be highly crucial for superior salt tolerance [61], as explained in salt-stressed grass pea [39], beans [62] and barley seedlings [61].

Perhaps, one of the most intriguing questions in the present investigation is the remarkable capacity of four grass pea genotypes to restore their normal growth at flowering stage (30 DAC) and its maintenance through early pod-bearing stages (60 DAC) despite them experiencing pro-longed salt exposure. In order to ascertain this apparent conflicting situation, response of four primary defense components involved in H_2O_2-metabolism and concomitant oxidative stress response were studied. These four components play vital roles in primary defense mechanism of plants during ROS-detoxification [11]. H_2O_2 is a highly diffusible ROS across cellular membranes and inflicts oxidative damage to thiol-containing enzymes [6–8, 11, 64]. On the other hand, it can induce antioxidant defense to be upregulated against oxidative stress as a signaling molecule [6, 7, 10]. Therefore, regulation of its concentration at a particular level within cellular environment is the most vital during plant growth and development [8, 10, 64]. In the present study, a time-bound measurements at 15, 30 and 60 DAC revealed huge increase of H_2O_2 level in leaves of B1, BioL-212, LR3, and LR4 but a significant decline in PUSA-90-2 and WBK-CB-14 at 15 DAC, followed by a remarkable rise in PUSA-90-2 and WBK-CB-14 and decrease in the rest of the four genotypes at 30 DAC and 60 DAC. Interestingly enough, the level of H_2O_2 in B1, BioL-212, LR3, and LR4 at 30 DAC was still higher than that in control and this level was maintained at 60 DAC. This situation can be better explained if enzymatic regulation of H_2O_2-metabolism is taken into account. High SOD activity strongly suggested NaCl-induced generation of excess superoxide radicals and thus, formed huge H_2O_2 in B1, BioL-212, LR3, and LR4 genotypes at 15 DAC, and in the rest of the genotypes up to 60 DAC. In many studies, up-regulation of SOD, APX and CAT activities was cited as prime reason for salt-tolerance while their decreased level was associated with onset of salt-induced oxidative stress in sensitive genotypes [39, 57, 65]. However, the

intrinsic relationship among these three prominent H_2O_2-metabolizing enzymes is not straightforward in the present case. Both APX and CAT activities were quite normal in PUSA-90-2 and WBK-CB-14 genotypes at 15 DAC but it reduced significantly at 30 DAC and 60 DAC. Completely reverse response, however, was noticed between activity of APX and CAT in the rest of the four genotypes throughout the treatment period. Declining APX level in B1 and BioL-212 was accompanied by high CAT activity at 15 DAC which went just reverse during later stages of growth. Similarly, high APX activity in the two mutant lines was associated with abrupt fall in CAT level at 15 DAC, followed by decline in APX activity and concomitant increase in CAT level at 30 DAC and 60 DAC. The results clearly indicated that low SOD activity coupled with normal level of both APX and CAT might be responsible for significant fall in H_2O_2 generation in PUSA-90-2 and WBK-CB-14 during early stages of treatment. At the same period, low APX level accompanied with high SOD activity might have attributed to over-accumulation of H_2O_2 in B1 and BioL-212, indicating failure of H_2O_2-scavenging machinery within AsA-GSH cycle, and CAT alone could not cope up with rising level of H_2O_2 at early growth stages. By contrast, rise in H_2O_2 level in LR3 and LR4 mutant lines was presumably due to abnormal fall in CAT activity and despite high APX level, H_2O_2 was generated in excess amount due to high SOD activity. This situation got remarkable twist once the treated plants entered flowering and subsequent pod-bearing stage. High APX activity coupled with low CAT level in B1 and BioL-212 and low APX combined with high CAT activity in LR3 and LR4 mutant lines in the backdrop of enhanced SOD activity led to decrease in H_2O_2 level in a particular concentration. This level was still higher than control but was not toxic to plant growth and development at reproductive period under NaCl-exposure. Presumably, leaf H_2O_2 concentration in these four genotypes was regulated in controlled way by opposite mechanisms of prominent H_2O_2-metabolizing enzymes, balancing it to a particular level during later stages of growth. This critical balance, however, was lost when APX activity coupled with CAT began to decline and SOD activity was enhanced in PUSA-90-2 and WBK-CB-14 during 30 DAC and 60 DAC, resulting in un-regulated generation and accumulation of H_2O_2 in treated genotypes. Obviously, a well-integrated catalase-peroxidase system along with SOD was instrumental in regulation of H_2O_2-metabolism in stressed plants in different manners, and this is distinctly functional at reproductive stages of growth. Declining level of APX might be due to significant reduction in AsA level and its redox state. Besides serving as cofactor of APX, AsA itself can detoxify ROS through nonenzymatic mechanism [64, 66] and its availability and redox state plays key roles in signaling network through controlled metabolism of H_2O_2 during NaCl-stress [66]. Pro-longed NaCl-exposure, presumably, led to crippling of AsA-mediated antioxidant defense in PUSA-90-2 and WBK-CB-14, resulting in significant rise of H_2O_2 content under NaCl-treatment. The decline in CAT activity at 30 and 60 DAC further aggravated the situation. H_2O_2 has a longer half-life than superoxides [7], and therefore its initial surge in leaves of four tolerant genotypes indicated early "oxidative burst" in the photosynthetic organs which might trigger enhanced defense response against excess ROS at later stages of growth. Rise in either APX activity powered by high AsA content or AsA-redox and increasing CAT level at 30 DAC and 60 DAC in B1, BioL-212, LR3, and LR4 was the strong enough to stimulate antioxidant response, and might be instrumental to maintain H_2O_2 at a particular level. Presumably, H_2O_2 played dual roles in stress perception of NaCl-treated grass pea genotypes plants in a concentration-dependent manner; it promoted plant growth in B1, BioL-212, LR3, and LR4 at 30 DAC and 60 DAC in a concentration higher than control but induced toxicity in these four genotypes at the far higher level during 15 DAC. Certainly, the status of cellular H_2O_2 informed the plant cell about the severity of the oxidative stress and hence the appropriate level of antioxidant enzyme activities through induction [6, 8, 10]. Present results indicates that the H_2O_2 level is highly critical in stress perception, which can be regulated/adjusted in favor of plant growth and yield performances during prolonged exposure to high NaCl.

Obviously, extremely low H_2O_2 level in PUSA-90-2 and WBK-CB-14 at 15 DAC was not enough to upregulate antioxidant defense against NaCl-induced oxidative stress. On the other hand, its huge surge was associated with significant accumulation of MDA content in B1, BioL-212, LR3, and LR4 at 15 DAC and in PUSA-90-2 and WBK-CB-14 at 30 DAC and 60 DAC. MDA along with high H_2O_2 level was commonly used as cellular markers of salinity-induced oxidative stress in sensitive genotypes [5, 37, 39, 57, 59]. However, the relationship between H_2O_2 content and oxidative stress is not straightforward in the present case. The absence of oxidative stress symptom in shoots of B1, BioL-212, LR3, and LR4 lines at 60 DAC despite the high H_2O_2 content might be due to the significantly low level of membrane lipid peroxidation and electrolyte leakage. Similarly, low H_2O_2 level in PUSA-90-2 and WBK-CB-14 at 15 DAC did not guarantee low oxidative damage, as MDA as well as EL% was quite high in these two genotypes at this stage. Rising MDA content and EL% in treated genotypes might be instrumental in lowering of photosynthesis by damaging the photosynthetic pigment composition and stability along with low K^+ content is probably responsible for lower pigment levels in NaCl-treated B1, BioL-212, LR3, and LR4 at 15 DAC and in PUSA-90-2 and WBK-CB-14 varieties throughout the treatment period. Similar situation was also observed in salt-affected cereals, legumes and other vegetable crops [34, 39, 57, 61, 67]. Significant increase in MDA content and loss in photosynthetic apparatus have been recognized as the marks of oxidative stress in plants [5, 6, 10, 39, 57, 64, 68, 69] and may be one of the prime reasons for NaCl-induced growth inhibition in the present material. It is also clear that oxidative damage at initial growth stages was effectively prevented in four genotypes by mitigating membrane damage at reproductive stage, but it was quite impossible for the PUSA-90-2 and WBK-CB-14 varieties to recover from NaCl-induced oxidative damage due to complete failure of antioxidant defense throughout the treatment period.

Changes in antioxidant enzyme activities coincided with a variable increase or decrease of their individual isoform expression. Analysis of isoforms and inhibitor studies revealed that the increase in SOD activity in B1, BioL-212, LR3, and LR4 lines over control was purely due to over-activity of both Cu/Zn I and II isoforms, as suggested by high staining intensity. On the other hand, significant enhancement in SOD activity at 30 DAC was mainly due to origin of a new Mn SOD I isoform in LR3, one Fe SOD isoform in LR4 line and two new Mn SOD isoforms (I and II), in addition to existing Cu/Zn isoforms. No new isoforms were observed at 60 DAC. Thus, further increase in activity of SOD might be attributed to enhanced expression of existing isoforms, visualized as stronger intensity. The supply of NaCl reportedly enhanced the activity of Cu/Zn-SOD II in wheat seedlings [70]. The present result revealed origin of three novel isoforms namely two Mn SODs (I and II) and one Fe SOD at 30 DAC and it was observed at 60 DAC, also. It seems likely that increased activity of SOD at 15 DAC was mainly mediated through existing isoforms but as the treatment progressed induction of SOD expression was required in different cellular compartments to combat the elevated generation of superoxide radicals due to prolonged NaCl-exposure. While Cu/Zn isoforms are predominantly present in chloroplast and cytosol, Mn SODs are located in peroxisomes and mitochondria and Fe SODs are mainly chloroplastidic [9], suggesting participation of SOD isoforms in different cellular compartments to combat NaCl-induced generation of free radicals in the present material. For APX, the increased activity was mainly due to enhanced expression of APX 1, APX 2 and APX 3 isozymes, as was evidenced by strongly intense bands in zymograms. On the other hand, the decreased activity of APX was manifested by diminishing intensity of different APX bands. Leaf APX isoforms are rapidly deactivated by over-accumulating H_2O_2 at low AsA pool [71], and this was, perhaps, one of the prime reasons for reduced APX level in treated genotypes at different growth stages. CAT activity was uniformly visualized as a single zone across the treatments. However, the increasing staining intensity confirmed the enhanced level of CAT expression in treated genotypes. Induction of new SOD isoforms at 30 DAC and their retaining till 60 DAC, however, is interesting and has immense significance as the plants were then at flowering and pod-bearing stages and major changes (increase or decrease) in H_2O_2 level occurred in this period. Present results clearly indicated criticality of blooming stage in determining regulation of H_2O_2-metabolism through NaCl-induced enzyme expression. The variations in the isozyme pattern and their correspondence to total assayed activity suggested that the three H_2O_2-metabolizing enzymes responded strongly to NaCl-induced oxidative stress in grass pea genotypes.

5. Conclusions

For the first time, effect of prolonged NaCl-treatment was studied in six promising grass pea genotypes. Results revealed significant genotypic differences in response to

150 mM NaCl-treatment and even differences between growth stages of a particular genotype. NaCl-treatment significantly reduced dry matter production in all six genotypes at early vegetative stage by inhibiting photosynthetic capacity probably orchestrated through diminishing level of K^+, low K^+/Na^+ ratio and onset of severe oxidative stress. The negative impact of NaCl-stress was ameliorated in B1, BioL-212, LR3, and LR4 at later stages of growth through modulation of H_2O_2-metabolism in fine tune by balanced action of SOD, APX and CAT. This along with healthy ionic balance in favor of K^+ helped these four genotypes to maintain normal growth through restoration of normal photosynthetic capacity. This balance, however, was completely lost in PUSA-90-2 and WBK-CB-14 genotypes at the most crucial stages of reproductive growth, leading to un-regulated accumulation of H_2O_2 and high rate of lipid peroxidation as the marks of oxidative stress. Present result also suggested differential response of CAT/APX system during scavenging of H_2O_2 in four tolerant genotypes. Isozyme analysis revealed origin of unique isoforms of SOD in response to As treatment and increase in activity of existing isoforms of SOD, APX, and CAT, conferring enhanced activity of the enzymes. Considering the overall response of growth and leaf antioxidant metabolism of six grass pea genotypes, it can be finally concluded that var. B1 and BioL-212 along with LR3 and LR 4 mutant lines are tolerant to NaCl-stress while var. PUSA-90-2 and WBK-CB-14 are highly salt sensitive. H_2O_2-metabolism holds the key in determining sensitivity of grass pea genotypes to prolonged exposure of NaCl-induced oxidative stress. Understanding the intrinsic modulation of physiological and antioxidant metabolism by H_2O_2 through different growth stages would be an important step in formulating effective breeding strategies for improving NaCl-tolerance in crop plants.

Conflict of Interests

The author declares that there is no conflict of interests regarding the publication of this paper.

Acknowledgments

The author is thankful to University Grants Commission, ERO, Calcutta, India, for granting financial assistance in the form of research project (Grant no. PSW-047/11-12 ERO) to carry out present work. Technical assistance from Dr. Tulika Talukdar and guidance from Dr. Zahed Hossain during the experimentation are gratefully acknowledged. Thanks are also due to Dr. Madan Srivastava, Joint Director Pulses (in Pulses and Oilseeds Research Station, Berhampur, West Bengal) for kindly providing the verified seed samples.

References

[1] J. S. Bayuelo-Jiménez, D. G. Debouck, and J. P. Lynch, "Salinity tolerance in *Phaseolus* species during early vegetative growth," *Crop Science*, vol. 42, no. 6, pp. 2184–2192, 2002.

[2] W. Wang, B. Vinocur, and A. Altman, "Plant responses to drought, salinity and extreme temperatures: towards genetic engineering for stress tolerance," *Planta*, vol. 218, no. 1, pp. 1–14, 2003.

[3] P. H. Graham and C. P. Vance, "Legumes: importance and constraints to greater use," *Plant Physiology*, vol. 131, no. 3, pp. 872–877, 2003.

[4] Z. Hossain, A. K. A. Mandal, R. Shukla, and S. K. Datta, "NaCl stress: its chromotoxic effects and antioxidant behavior in roots of *Chrysanthemum morifolium* Ramat," *Plant Science*, vol. 166, no. 1, pp. 215–220, 2004.

[5] J. A. Hernández, A. Jiménez, P. Mullineaux, and F. Sevilla, "Tolerance of pea (*Pisum sativum* L.) to long-term salt stress is associated with induction of antioxidant defences," *Plant, Cell and Environment*, vol. 23, no. 8, pp. 853–862, 2000.

[6] D. Talukdar, "Exogenous calcium alleviates the impact of cadmium-induced oxidative stress in *Lens culinaris* Medic. seedlings through modulation of antioxidant enzyme activities," *Journal of Crop Science and Biotechnology*, vol. 15, no. 4, pp. 325–334, 2012.

[7] S. Neill, R. Desikan, and J. Hancock, "Hydrogen peroxide signalling," *Current Opinion in Plant Biology*, vol. 5, no. 5, pp. 388–395, 2002.

[8] X. Hu, M. Jiang, J. Zhang, A. Zhang, F. Lin, and M. Tan, "Calcium-calmodulin is required for abscisic acid-induced antioxidant defense and functions both upstream and downstream of H_2O_2 production in leaves of maize (*Zea mays*) plants," *New Phytologist*, vol. 173, no. 1, pp. 27–38, 2007.

[9] R. G. Alscher, N. Erturk, and L. S. Heath, "Role of superoxide dismutases (SODs) in controlling oxidative stress in plants," *Journal of Experimental Botany*, vol. 53, no. 372, pp. 1331–1341, 2002.

[10] D. Talukdar, "Arsenic-induced changes in growth and antioxidant metabolism of fenugreek," *Russian Journal of Plant Physiology*, vol. 60, no. 5, pp. 652–660, 2013.

[11] C. H. Foyer and G. Noctor, "Redox sensing and signalling associated with reactive oxygen in chloroplasts, peroxisomes and mitochondria," *Physiologia Plantarum*, vol. 119, no. 3, pp. 355–364, 2003.

[12] N. Anoop and A. K. Gupta, "Transgenic indica rice cv IR-50 over-expressing *Vigna aconitifolia* Δ1-Pyrroline-5-carboxylate synthetase cDNA shows tolerance to high salt," *Journal of Plant Biochemistry and Biotechnology*, vol. 12, no. 2, pp. 109–116, 2003.

[13] T. J. Flowers, "Improving crop salt tolerance," *Journal of Experimental Botany*, vol. 55, no. 396, pp. 307–319, 2004.

[14] C. Chen, C. Tao, H. Peng, and Y. Ding, "Genetic analysis of salt stress responses in asparagus bean (*Vigna unguiculata* (L.) ssp. sesquipedalis Verdc.)," *Journal of Heredity*, vol. 98, no. 7, pp. 655–665, 2007.

[15] D. Talukdar, "Dwarf mutations in grass pea (*Lathyrus sativus* L.): origin, morphology, inheritance and linkage studies," *Journal of Genetics*, vol. 88, no. 2, pp. 165–175, 2009.

[16] M. T. Jackson and A. G. Yunus, "Variation in the grass pea (*Lathyrus sativus* L.) and wild species," *Euphytica*, vol. 33, no. 2, pp. 549–559, 1984.

[17] D. Talukdar, "Recent progress on genetic analysis of novel mutants and aneuploid research in grass pea (*Lathyrus sativus* L.)," *African Journal of Agricultural Research*, vol. 4, no. 13, pp. 1549–1559, 2009.

[18] E. Pastor-Cavada, R. Juan, J. E. Pastor, M. Alaiz, and J. Vioque, "Antioxidant activity of seed polyphenols in fifteen wild *Lathyrus* species from South Spain," *Food Science and Technology*, vol. 42, no. 3, pp. 705–709, 2009.

[19] D. Talukdar, "Antioxidant potential and type II diabetes related enzyme inhibition properties of raw and processed legumes in Indian Himalayas," *Journal of Applied Pharmaceutical Science*, vol. 3, no. 3, pp. 013–019, 2013.

[20] D. Talukdar, "*In Vitro* antioxidant potential and type II diabetes related enzyme inhibition properties of traditionally processed legume-based food and medicinal recipes in Indian Himalayas," *Journal of Applied Pharmaceutical Science*, vol. 3, no. 1, pp. 026–032, 2013.

[21] D. Talukdar, "Cytogenetic characterization of seven different primary tetrasomics in grass pea (*Lathyrus sativus* L.)," *Caryologia*, vol. 61, no. 4, pp. 402–410, 2008.

[22] S. Kumar, G. Bejiga, S. Ahmed, H. Nakkoul, and A. Sarker, "Genetic improvement of grass pea for low neurotoxin (β-ODAP) content," *Food and Chemical Toxicology*, vol. 49, no. 3, pp. 589–600, 2011.

[23] D. Talukdar and A. K. Biswas, "Induced seed coat colour mutations and their inheritance in grass pea (*Lathyrus sativus* L.)," *Indian Journal of Genetics and Plant Breeding*, vol. 65, no. 2, pp. 135–136, 2005.

[24] D. Talukdar and A. K. Biswas, "Seven different primary trisomics in grass pea (*Lathyrus sativus* L.). I. Cytogenetic characterisation," *Cytologia*, vol. 72, no. 4, pp. 385–396, 2007.

[25] D. Talukdar, "Cytogenetics of a reciprocal translocation integrating distichous pedicel and tendril-less leaf mutations in *Lathyrus sativus* L.," *Caryologia*, vol. 66, no. 1, pp. 21–30, 2013.

[26] D. Talukdar, "Reciprocal translocations in grass pea (*Lathyrus sativus* l.): pattern of transmission, detection of multiple interchanges and their independence," *Journal of Heredity*, vol. 101, no. 2, pp. 169–176, 2010.

[27] D. Talukdar, "Genetics of pod indehiscence in grass pea (*Lathyrus sativus* L.)," *Journal of Crop Improvement*, vol. 25, no. 2, pp. 161–175, 2011.

[28] D. Talukdar, "Cytogenetic analysis of a novel yellow flower mutant carrying a reciprocal translocation in grass pea (*Lathyrus sativus* L.)," *Journal of Biological Research*, vol. 15, pp. 123–134, 2011.

[29] D. Talukdar, "Flavonoid-deficient mutants in grass pea (*Lathyrus sativus* L.): genetic control, linkage relationships, and mapping with aconitase and S nitrosoglutathione reductase isozyme loci," *Scientific World Journal*, Article ID 345983, 11 pages, 2012.

[30] M. C. Vaz Patto, B. Skiba, E. C. K. Pang, S. J. Ochatt, F. Lambein, and D. Rubiales, "Lathyrus improvement for resistance against biotic and abiotic stresses: from classical breeding to marker assisted selection," *Euphytica*, vol. 147, no. 1-2, pp. 133–147, 2006.

[31] D. Talukdar, "Effect of arsenic-induced toxicity on morphological traits of *Trigonella foenum-graecum* L. and *Lathyrus sativus* L. during germination and early seedling growth," *Current Research Journal of Biological Sciences*, vol. 3, no. 2, pp. 116–123, 2011.

[32] D. Talukdar, "A glutathione-overproducing mutant in grass pea (*Lathyrus sativus* L.): alterations in glutathione content, modifications in antioxidant defense response to cadmium stress and genetic analysis using primary trisomics," *International Journal of Recent Scientific Research*, vol. 3, no. 4, pp. 234–243, 2012.

[33] D. Talukdar, "Selenium priming selectively ameliorates weed-induced phytotoxicity by modulating antioxidant defense components in lentil (*Lens culinaris* Medik.) and grass pea (*Lathyrus sativus* L.)," *Annual Review and Research in Biology*, vol. 3, no. 3, pp. 195–212, 2013.

[34] D. Talukdar, "Modulation of plant growth and leaf biochemical parameters in grass pea (*Lathyrus sativus* L) and fenugreek (*Trigonella foenum-graecum* L.) exposed to NaCl treatments," *Indian Journal of Fundamental and Applied Life Sciences*, vol. 2, no. 3, pp. 20–28, 2012.

[35] B. Mahdavi and S. A. M. M. Sanavy, "Germination and seedling growth in grasspea (*Lathyrus sativus*) cultivars under salinity conditions," *Pakistan Journal of Biological Sciences*, vol. 10, no. 2, pp. 273–279, 2007.

[36] R. Munns, "Genes and salt tolerance: bringing them together," *New Phytologist*, vol. 167, no. 3, pp. 645–663, 2005.

[37] P. Ahmad, R. John, M. Sarwat, and S. Umar, "Responses of proline, lipid peroxidation and antioxidative enzymes in two varieties of *Pisum sativum* L. under salt stress," *International Journal of Plant Production*, vol. 2, no. 4, pp. 353–366, 2008.

[38] I. M. Santha and S. L. Mehta, "Development of low ODAP somaclones of *Lathyrus sativus*," *Lathyrus Lathyrism Newsletter*, vol. 2, article 42, 2001.

[39] D. Talukdar, "Isolation and characterization of NaCl-tolerant mutations in two important legumes, *Clitoria ternatea* L. and *Lathyrus sativus* L.: induced mutagenesis and selection by salt stress," *Journal of Medicinal Plant Research*, vol. 5, no. 16, pp. 3619–3628, 2011.

[40] ISTA, *International Rules for Seed Testing*, International Seed Testing Association, 2008.

[41] D. Talukdar, "Morpho-physiological responses of grass pea (*Lathyrus sativus* L.) genotypes to salt stress at germination and seedling stages," *Legume Research*, vol. 34, no. 4, pp. 232–241, 2011.

[42] H. K. Lichtenthaler, "Chlorophylls and carotenoids: pigments of photosynthetic biomembranes," *Methods in Enzymology*, vol. 148, pp. 350–382, 1987.

[43] J. Coombs, D. O. Hall, S. P. Long, and J. M. O. Scurlock, *Techniques in Bioproductivity and Photosynthesis*, Pergamon Press, Oxford, UK, 1985.

[44] V. Kumar and D. R. Sharma, "Isolation and characterization of sodium chloride-resistant callus culture of *Vigna radiata* (L.) wilczek var. *radiata*," *Journal of Experimental Botany*, vol. 40, no. 1, pp. 143–147, 1989.

[45] L. S. Bates, R. P. Waldren, and I. D. Teare, "Rapid determination of free proline for water-stress studies," *Plant and Soil*, vol. 39, no. 1, pp. 205–207, 1973.

[46] C.-Q. Wang, M. Chen, and B.-S. Wang, "Betacyanin accumulation in the leaves of C_3 halophyte *Suaeda salsa* L. is induced by watering roots with H_2O_2," *Plant Science*, vol. 172, no. 1, pp. 1–7, 2007.

[47] D. M. Hodges, J. M. DeLong, C. F. Forney, and R. K. Prange, "Improving the thiobarbituric acid-reactive-substances assay for estimating lipid peroxidation in plant tissues containing anthocyanin and other interfering compounds," *Planta*, vol. 207, no. 4, pp. 604–611, 1999.

[48] M. Y. Law, S. A. Charles, and B. Halliwell, "Glutathione and ascorbic acid in spinach (*Spinacia oleracea*) chloroplasts. The effect of hydrogen peroxide and of Paraquat," *Biochemical Journal*, vol. 210, no. 3, pp. 899–903, 1983.

[49] M. M. Bradford, "A rapid and sensitive method for the quantitation of microgram quantities of protein utilizing the principle of protein dye binding," *Analytical Biochemistry*, vol. 72, no. 1-2, pp. 248–254, 1976.

[50] W. F. Beyer Jr. and I. Fridovich, "Assaying for superoxide dismutase activity: some large consequences of minor changes in conditions," *Analytical Biochemistry*, vol. 161, no. 2, pp. 559–566, 1987.

[51] Y. Nakano and K. Asada, "Hydrogen peroxide is scavenged by ascorbate-specific peroxidase in spinach chloroplasts," *Plant and Cell Physiology*, vol. 22, no. 5, pp. 867–880, 1981.

[52] R. Mittler and B. A. Zilinskas, "Detection of ascorbate peroxidase activity in native gels by inhibition of the ascorbate-dependent reduction of nitroblue tetrazolium," *Analytical Biochemistry*, vol. 212, no. 2, pp. 540–546, 1993.

[53] B. Chance and A. C. Maehly, "Assay of catalases and peroxidases," *Methods in Enzymology*, vol. 2, pp. 764–817, 1955.

[54] W. Woodbury, A. K. Spencer, and M. A. Stahmann, "An improved procedure using ferricyanide for detecting catalase isozymes," *Analytical Biochemistry*, vol. 44, no. 1, pp. 301–305, 1971.

[55] D. Talukdar, "Flower and pod production, abortion, leaf injury, yield and seed neurotoxin levels in stable dwarf mutant lines of grass pea (*Lathyrus sativus* L.) differing in salt stress responses," *International Journal of Current Research*, vol. 2, no. 1, pp. 46–54, 2011.

[56] M. A. Turan, V. Katkat, and S. Taban, "Variations in proline, chlorophyll and mineral elements contents of wheat plants grown under salinity stress," *Journal of Agronomy*, vol. 6, no. 1, pp. 137–141, 2007.

[57] M. A. Shahid, R. M. Balal, M. A. Pervez et al., "Differential response of pea (*Pisum sativum* L.) genotypes to salt stress in relation to the growth, physiological attributes antioxidant activity and organic solutes," *Australian Journal of Crop Science*, vol. 6, no. 5, pp. 828–838, 2012.

[58] D. A. Meloni, M. A. Oliva, C. A. Martinez, and J. Cambraia, "Photosynthesis and activity of superoxide dismutase, peroxidase and glutathione reductase in cotton under salt stress," *Environmental and Experimental Botany*, vol. 49, no. 1, pp. 69–76, 2003.

[59] P. Saha, P. Chatterjee, and A. K. Biswas, "NaCl pretreatment alleviates salt stress by enhancement of antioxidant defense system and osmolyte accumulation in mungbean (*Vigna radiata* l. wilczek)," *Indian Journal of Experimental Biology*, vol. 48, no. 6, pp. 593–600, 2010.

[60] N. Misra and A. K. Gupta, "Effect of salt stress on proline metabolism in two high yielding genotypes of green gram," *Plant Science*, vol. 169, no. 2, pp. 331–339, 2005.

[61] E. Degl'Innocenti, C. Hafsi, L. Guidi, and F. Navari-Izzo, "The effect of salinity on photosynthetic activity in potassium-deficient barley species," *Journal of Plant Physiology*, vol. 166, no. 18, pp. 1968–1981, 2009.

[62] J. S. Bayuelo-Jimenez, N. Jasso-Plata, and I. Ochoa, "Growth and physiological responses of *Phaseolus* species to salinity stress," *International Journal of Agronomy*, vol. 2012, Article ID 527673, 13 pages, 2012.

[63] V. Arbona, A. J. Marco, D. J. Iglesias, M. F. López-Climent, M. Talon, and A. Gómez-Cadenas, "Carbohydrate depletion in roots and leaves of salt-stressed potted *Citrus clementina* L," *Plant Growth Regulation*, vol. 46, no. 2, pp. 153–160, 2005.

[64] D. Talukdar, "Ascorbate deficient semi-dwarf *asfL1* mutant of *Lathyrus sativus* exhibits alterations in antioxidant defense," *Biologia Plantarum*, vol. 54, no. 4, pp. 675–682, 2012.

[65] Z. Noreen and M. Ashraf, "Assessment of variation in antioxidative defense system in salt-treated pea (*Pisum sativum*) cultivars and its putative use as salinity tolerance markers," *Journal of Plant Physiology*, vol. 166, no. 16, pp. 1764–1774, 2009.

[66] N. Smirnoff, "Ascorbic acid: metabolism and functions of a multi-facetted molecule," *Current Opinion in Plant Biology*, vol. 3, no. 3, pp. 229–235, 2000.

[67] C. Pandolfi, S. Mancuso, and S. Shabala, "Physiology of acclimation to salinity stress in pea (*Pisum sativum*)," *Environmental and Experimental Botany*, vol. 84, pp. 44–51, 2012.

[68] D. Talukdar, "Arsenic-induced oxidative stress in the common bean legume, *Phaseolus vulgaris* L. seedlings and its amelioration by exogenous nitric oxide," *Physiology and Molecular Biology of Plants*, vol. 19, no. 1, pp. 69–79, 2013.

[69] D. Talukdar, "Studies on antioxidant enzymes in *Canna indica* plant under copper stress," *Journal of Environmental Biology*, vol. 34, no. 1, pp. 93–98, 2013.

[70] F. (Inci) Eyidoğan, H. A. Öktem, and M. Yücel, "Superoxide dismutase activity in salt stressed wheat seedlings," *Acta Physiologiae Plantarum*, vol. 25, no. 3, pp. 263–269, 2003.

[71] A. N. P. Hiner, J. N. Rodríguez-López, M. B. Arnao, E. L. Raven, F. García-Cánovas, and M. Acosta, "Kinetic study of the inactivation of ascorbate peroxidase by hydrogen peroxide," *Biochemical Journal*, vol. 348, pp. 321–328, 2000.

7

Effect of Management of Sulfonylurea Resistant *Stellaria media* on Barley Yield

Tuomas Uusitalo,[1,2] **Asmo Saarinen,**[3] **and Pirjo S. A. Mäkelä**[1]

[1] *Department of Agricultural Science, P.O. Box 27, University of Helsinki, 00014 Helsinki, Finland*
[2] *Raisioagro Oy, P.O. Box 101, 21201 Raisio, Finland*
[3] *Berner Oy, Eteläranta 4B, 00130 Helsinki, Finland*

Correspondence should be addressed to Pirjo S. A. Mäkelä; pirjo.makela@helsinki.fi

Academic Editors: M. Ruiz and S. S. Xu

Sulfonylureas represent one of the largest herbicide groups that have been widely used since 1980s. Their continuous use has resulted in development of sulfonylurea resistance in weeds. The aim of this research was to investigate options to manage putative sulfonylurea-resistant chickweed in barley stands and to evaluate the effect of chickweed and its management on barley yield. A field experiment was arranged as a randomized complete block design and included 14 herbicide treatments applied at two different times. Tribenuron-methyl (sulfonylurea) affected minimal control of chickweed. A bromoxynil-ioxynil (photosystem II inhibitor) mix did not control chickweed efficiently. However, nearly total control was achieved with fluroxypyr, mecoprop, and their mixtures (synthetic auxins and photosystem II inhibitors). Chickweed had no effect on barley yield whether controlled or uncontrolled. Therefore, further evaluation of the chickweed management threshold would be needed. It seems that even in the boreal region, typified by a cold climate, limited solar radiation, a very short growing season, and relatively low-intensity cropping systems, unilateral use of sulfonylureas might lead to herbicide resistance. Although resistant weed populations can be controlled with herbicides of groups other than the sulfonylureas, this represents an increasing problem when planning weed management, especially when including sulfonylurea-resistant crops.

1. Introduction

Acetolactate synthase (ALS) inhibitors were commercialized in 1982, chlorsulfuron being one of the first active ingredients used. They quickly spread around the world and, due to their selectivity, low application rate, and broad-spectrum effectiveness, in many cases came to represent a key component of weed management [1]. Sulfonylureas belong to the group of ALS inhibitors. Widespread reliance solely on sulfonylureas has led to a situation where by 1998 large numbers of weed species were reported to be resistant to sulfonylureas [2]. Sulfonylurea-resistant weed species currently number over 100 [1]. The first reported sulfonylurea-resistant weed, prickly lettuce (*Lactuca serriola* L.), in 1987 was resistant to chlorsulfuron [3]. This was followed in 1988 by a common chickweed [*Stellaria media* (L.) Vill.] population in Canada [4]. In the northern temperate region (Norway and Sweden)

ALS-resistant chickweed has been reported from Sweden in 1995 and Norway in 2002 (www.weedscience.com) but not in the northern boreal region (e.g., Finland).

Sulfonylurea resistance derives mainly from a single-point mutation in the ALS gene, which leads to substitutions in branched-chain amino acids [1]. In some cases the mutation can lead to cross-resistance against other ALS-inhibitor herbicides [5]. Some weeds, such as rigid ryegrass (*Lolium rigidum* Gaud) and blackgrass (*Alopecurus myosuroides* Huds.), have also developed non-target-site cross-resistance across several herbicide modes of action, including ALS inhibitors and other herbicide groups never used in the areas in which the resistant weeds have been found [1].

Further development and spread of sulfonylurea resistance in weeds could be prevented, or at least reduced, through the use of herbicides belonging to groups with different modes of action, as well as by applying herbicide mixtures [6]. Effective crop management practices, such as attempts to reduce the weed seed bank and diversified weed management systems, are feasible options to reduce the problem of development of herbicide resistance [7]. Increased attention should be paid, however, to the use of clean seed and equipment, and use of crop rotations and cover crops [8].

Chickweed, originally from Europe, is distributed globally. In Northern Europe chickweed is one of the most common weeds in spring and winter cereal stands, with a cool and humid climate favoring its growth [9]. Chickweed is adapted to low light intensities and thus grows well under the shade of a crop canopy [10]. Moreover, root growth of chickweed is faster than, for example, that of barley (*Hordeum vulgare* L.) and chickweed takes up nitrogen more effectively than barley [11] and can out-compete cereal and oilseed crops for resources, resulting in yield and quality losses [12].

The aims of the work were to evaluate the effect of sulfonylurea-resistant chickweed and its management options on barley stands in most northern agricultural areas in the boreal region [13].

2. Materials and Methods

A field experiment was conducted in 2012 in Somero ($60°70, 63'$ N, $23°23, 59'$ E, 80 m asl). The soil type was a rich loamy silt loam, pH 5.9. The field had a long history of cereal and oilseed production, and sulfonylureas were mainly used to control weeds. In 2010 and 2011, substantial chickweed populations developed following herbicide treatment. Seed samples were collected in 2010 and analyzed at DuPont, Germany, for herbicide resistance. Results indicated that the chickweed populations had developed sulfonylurea resistance.

Barley (cv. NCF-Tipple) was sown on May 10, 2012, at 500 seeds/m^2 and was fertilized with 108 kg/ha N (N-P-K : 27-2-3; Pellon Y1, Yara Finland) at sowing. Herbicides were applied either early, 28 days after sowing (DAS), at growth stage 21 of barley [14] and growth stage 10–16 for chickweed, or late, 41 DAS, at growth stage 30 of barley and growth stage 20 for chickweed. There were 14 specific herbicide treatments (Table 1) and water was applied as a control. Herbicides were chosen to represent the ALS-inhibiting group (both sulfonylureas and nonsulfonylureas), synthetic auxins, and the photosystem II-inhibiting group. Application rates ranged from the lowest to the highest recommendations. Diseases were controlled twice during the growing season, first (stage 31) with 0.5 L/ha Prosaro EC 250 (prothioconazole 125 g/L and tebuconazole 125 g/L, Bayer Crop Science) and later (stage 39) with 0.25 L/ha Comet (pyraclostrobin 250 g/L, BASF) and 0.25 L/ha Sportak 45 HF (prochlorazine 450 g/L, BASF). A growth regulator (0.6 L/ha Terpal, mepiquat chloride 305 g/L, and etefon 155 g/L, BASF) was applied after the initial fungicide application (stage 32).

Botanical analysis of weeds was conducted before treatments were applied. Since chickweed was practically the only weed present (Table 2), no further attention was paid to other weeds. The population density of chickweed was assessed from an area of 0.1 m^2 from each plot one day before each herbicide treatment and 14 and 28 days after herbicide treatment. Biomass samples of chickweed were collected 35 days after the late herbicide treatment. An area of 0.25 m^2 was cut above the soil surface and chickweed plants were dried at $65°C$ for two days, weighed, ground, and stored at room temperature for further analysis. The nitrogen (N) content of chickweed was analyzed from ground biomass samples using the Dumas combustion method (Elementar Vario Max C/N, Elementar Analysensysteme GmbH, Hanau, Germany). The height of the chickweed population was measured, and the weed coverage and treatment efficiency were evaluated visually 42 days after the late herbicide treatment using an 11-step scale (e.g., 0, no weeds; 3, 1–5%; 15, 10–22%; 30, 22–40%; 50, 40–60%; 70, 60–78%; 85, 78–90%; 97, 95–99%; 100, full coverage). At maturity, barley was harvested. Grain was sorted and weighed, and the test weight and moisture content were recorded. Protein content was analyzed using a near infrared spectrometer (DA7200 NIR Analyser, Perten Instruments, Sweden). The experiment was arranged in a completely randomized block design with four replications. Plot size was 33 m^2 (3×11 m).

Data for analyzed traits were subjected to ANOVA using the PASW 18.0 program (IBM Chicago, IL, USA) to compare the effects of herbicide treatments. Statistically significant differences among treatment means were established using Tukey's test.

3. Results and Discussion

The population density of chickweed decreased substantially following herbicide treatments, to a greater extent when evaluated 28 days after treatment (Table 3) than 14 days after treatment (data not shown). The only exception among herbicide treatments was tribenuron-methyl, which did not have a markedly different effect from the control, bromoxynil-ioxynil mix, and a low application rate of florasulam, which reduced the population density by only approximately 50% (Table 3). A similar decreasing trend was noted for chickweed biomass, since it was only approximately 25% lower following tribenuron-methyl and bromoxynil-ioxynil mix treatment in comparison with the control (Table 3). However, tribenuron-methyl should be more than 90% efficient against chickweed [5]. It can therefore be suggested that the chickweed population developed resistance against sulfonylureas, especially with the earlier confirmation of existence of sulfonylurea resistance by DuPont seed testing. Since the efficiency of tribenuron-methyl against chickweed was about 50%, it seems that the population contains resistant and nonresistant plants.

Similarly to the reports of Kudsk et al. [5] and Marshall et al. [15], fluroxypyr and mecoprop-P were effective against chickweed (Table 3). Fluroxypyr was effective even when applied at low application rates and at late growth stages on larger plants (Table 3). The best chickweed control was at the highest application rate of fluroxypyr and mecoprop-P, after the application of which practically all chickweed plants were destroyed (Table 3). The lowest application rate (54 g/ha)

TABLE 1: Trade names, active ingredients and their concentrations, and the concentrations of active ingredients applied in the field experiment in 2012. All herbicides were applied as an aqueous solution at 150 L/ha.

Active ingredient	Trade name	Active ingredient applied	
		g/L or g/g	g/ha
Bromoxynil : ioxynil	Oxitril EC	200 : 200	100 : 100
Bromoxynil : ioxynil	Oxitril EC	200 : 200	150 : 150
Bromoxynil : ioxynil	Oxitril EC	200 : 200	200 : 200
Dichlorprop-P : MCPA : mecoprop-P	K-Trio SL	310 : 160 : 130	310 : 160 : 130
Dichlorprop-P : MCPA : mecoprop-P	K-Trio SL	310 : 160 : 130	465 : 240 : 195
Dichlorprop-P : MCPA : mecoprop-P	K-Trio SL	310 : 160 : 130	620 : 320 : 260
Florasulam	Primus SC	50	2.5
Florasulam	Primus SC	50	3.75
Florasulam	Primus SC	50	5
Fluroxypyr	Starane 180 EC	180	54
Fluroxypyr	Starane 180 EC	180	108
Fluroxypyr	Starane 180 EC	180	162
MCPA : clopyralid : fluroxypyr	Ariane S SL	200 : 20 : 40	400 : 20 : 80
Tribenuron-methyl	Express 50 SX	0.5	10

TABLE 2: Botanical analysis and number of weeds in barley plant stands before herbicide treatments in 2012.

Scientific name	Common name	Number of plants/m^2
Stellaria media (L.) Vill.	Common chickweed	237
Lamium purpureum L.	Red deadnettle	2
Fumaria officinalis L.	Common fumitory	1
Chenopodium album L.	Common lamb's quarter	<1
Fallopia convolvulus L.	Wild buckwheat	<1

TABLE 3: Reduction in common chickweed population density, coverage of common chickweed, and treatment efficiency against common chickweed evaluated after herbicide treatments in a field experiment in 2012.

Applied treatment		Density reduction, %		Treatment efficiency, %	Coverage, %	Biomass, g/m^2
Active ingredient	g/ha	28 DAS	41 DAS			
Control	0	4	+22	0	99	61.37
Bromoxynil : ioxynil	100 : 100	42	38	54	67	13.02
Bromoxynil : ioxynil	150 : 150	49	44	76	41	5.34
Bromoxynil : ioxynil	200 : 200	65	64	82	42	3.20
Dichlorprop-P : MCPA : mecoprop-P	310 : 160 : 130	83	94	97	5	0.25
Dichlorprop-P : MCPA : mecoprop-P	465 : 240 : 195	86	98	98	4	0.17
Dichlorprop-P : MCPA : mecoprop-P	620 : 320 : 260	89	97	99	2	1.82
Florasulam	2.5	54	84	97	5	0.46
Florasulam	3.75	62	89	98	4	0.33
Florasulam	5	68	91	98	3	0.03
Fluroxypyr	54	63	75	89	30	6.03
Fluroxypyr	108	82	95	98	4	0.31
Fluroxypyr	162	86	100	99	2	0.02
MCPA : clopyralid : fluroxypyr	400 : 20 : 80	84	97	97	4	0.16
Tribenuron-methyl	10	28	20	47	69	20.01
S.E.M.		5.9***	6.0***	6.9***	8.7***	3.847***

*** Statistically significant at $P = 0.001$.
Data shown are combined means of early and late herbicide treatments ($n = 8$), except for reduction data means ($n = 4$), since there were no significant differences between timings of treatment.

TABLE 4: Barley grain yield and its protein content following herbicide treatments in the field experiment of 2012.

Applied treatment		Yield, kg/ha	Test weight, hL	Protein content, %
Active ingredient	g/ha			
Control	0	5930	62.3	11.3
Bromoxynil : ioxynil	100 : 100	6044	62.0	11.4
Bromoxynil : ioxynil	150 : 150	6089	62.1	11.5
Bromoxynil : ioxynil	200 : 200	6037	62.5	11.4
Dichlorprop-P : MCPA : mecoprop-P	310 : 160 : 130	6059	62.0	11.5
Dichlorprop-P : MCPA : mecoprop-P	465 : 240 : 195	6140	62.5	11.4
Dichlorprop-P : MCPA : mecoprop-P	620 : 320 : 260	6233	62.6	11.4
Florasulam	2.5	6045	62.5	11.5
Florasulam	3.75	5951	61.9	11.6
Florasulam	5	6056	61.1	11.6
Fluroxypyr	54	6192	63.1	11.2
Fluroxypyr	108	6188	62.8	11.4
Fluroxypyr	162	5986	61.8	11.6
MCPA : clopyralid : fluroxypyr	400 : 20 : 80	6050	61.7	11.6
Tribenuron-methyl	10	6133	62.8	11.2
	S.E.M.	152.8[ns]	0.68[ns]	0.15[ns]

[ns]Not statistically significant, $P > 0.05$.
Data shown are combined means of early and late herbicide treatments ($n = 8$), since there were no significant differences between timings of treatment.

of fluroxypyr would not have resulted in satisfactory control in practice. The bromoxynil-ioxynil mix should be efficient against chickweed [16], but only the highest application rate resulted in acceptable control (Table 3). Nonetheless, it would represent an additional option to control sulfonylurea-resistant chickweed.

The N content of the chickweed increased to 3.2% following tribenuron-methyl treatment, whereas it was 2.6% in control plants (data not shown). Thus, the N uptake of untreated chickweed was 16 kg/ha. However, the N uptake of tribenuron-methyl-treated chickweed was only 6 kg/ha, while for other treatments it was generally around 0.1 kg N/ha. Even though the N content in chickweed following tribenuron-methyl treatment probably increased due to the lower degree of intraspecific competition, as suggested by Cahill Jr. [17], lower number of plants accumulated less N due to lower total biomass (Table 3). A low level of intraspecific competition could also result in decreased root growth at low root densities [18], thus limiting the root surface area to a smaller volume of nitrogen-containing soil [19]. Although chickweed accumulated substantial amounts of nitrogen without weed control, this was not reflected in the barley grain yield, which was approximately 6 100 kg/ha, its protein content, which was approximately 11.4%, and its test weight, which was approximately 62.2 (Table 4). The different herbicide treatments did not affect barley yield or quality (Table 4). Hamouz et al. [20] also concluded that various weed control treatments had no effect on winter wheat (*Triticum aestivum* L.) yield. Earlier, Salonen and Erviö [21] reported that chemical weed control in general increased the grain yield of spring cereals. Moreover, a simple increased number of chickweed plants within the crop stand have led to significant decrease in yield of winter wheat [22].

Thus, it seems that the low chickweed canopy, below 100 mm, might have negatively affected competition with barley. Even though the chickweed population did not affect barley yield during the year of the experiment, over the long term problems could arise, especially if less competitive, shorter, lodging sensitive, and slower growing crops are grown in the area without adequate weed management. Moreover, the chickweed seed bank was reported to respond directly to weed control treatments [23] and thus the species could increase so as to cause severe yield reductions if not managed properly.

4. Conclusions

In conclusion, the most effective active herbicide ingredients tested to control putative sulfonylurea-resistant chickweed populations were fluroxypyr, MCPA-clopyralid-fluroxypyr and dichlorprop-PMCPA-mecoprop-P mixtures. Good chickweed control was obtained at application rates of 80 g/ha for fluroxypyr, and 130 g/ha for mecoprop. Although, at least in the case of sulfonylurea resistance, other active ingredients can be used to control the weeds successfully, it is possible to avoid development of resistance through good crop management practices, including crop rotations and the use of herbicides mixtures using components from different groups. Rotations could be limited by sulfonylurea-resistance developing in crops, such as *Brassica* species, and the limited range of herbicides on sale. Since the chickweed did not decrease barley yield or its quality, a more profound investigation should evaluate its management threshold taking also into account the economy. In this experiment, chickweed merely served as a nitrogen trap.

Conflict of Interests

The authors declare that there is no conflict of interests regarding the publication of this paper.

References

[1] S. B. Powles and Q. Yu, "Evolution in action: plants resistant to herbicides," *Annual Review of Plant Biology*, vol. 61, pp. 317–347, 2010.

[2] P. J. Tranel and T. R. Wright, "Resistance of weeds to ALS-inhibiting herbicides: what have we learned?" *Weed Science*, vol. 50, no. 6, pp. 700–712, 2002.

[3] C. A. Mallory-Smith, D. C. Thill, and M. J. Dial, "Identification of sulfonylurea herbicide-resistant prickly lettuce (*Lactuca serriola*)," *Weed Technology*, vol. 4, pp. 163–168, 1990.

[4] J. T. O'Donovan, G. M. Jeffers, D. Maurice, and M. P. Sharma, "Investigation of a chlorsulfuron-resistant chickweed [*Stellaria media* (L.) Vill.] population," *Canadian Journal of Plant Science*, vol. 74, no. 4, pp. 693–697, 1994.

[5] P. Kudsk, S. K. Mathiassen, and J. C. Cotterman, "Sulfonylurea resistance in *Stellaria media* [L.] Vill," *Weed Research*, vol. 35, no. 1, pp. 19–24, 1995.

[6] H. J. Beckie and X. Reboud, "Selecting for weed resistance: herbicide rotation and mixture," *Weed Technology*, vol. 23, no. 3, pp. 363–370, 2009.

[7] J. K. Norsworthy, S. M. Ward, D. R. Shaw et al., "Reducing the risks of herbicide resistance: best management practices and recommendations," *Weed Science*, vol. 60, pp. 31–62, 2012.

[8] W. K. Vencill, R. L. Nichols, T. M. Webster et al., "Herbicide resistance: toward an understanding of resistance development and the impact of herbicide-resistant crops," *Weed Science*, vol. 60, pp. 2–30, 2012.

[9] L. G. Holm, *The Worlds' Worst Weeds: Distribution and Biology*, University Press of Hawaii, Honolulu, Hawaii, USA, 1977.

[10] H. Fogelfors, "The competition between barley and five weed species as influenced by MCPA treatment," *Swedish Journal of Agricultural Research*, vol. 7, no. 3, pp. 147–151, 1977.

[11] H. H. Mann and T. W. Barnes, "The competition between barley and certain weeds under controlled conditions IV. Competition with *Stellaria media*," *Annals of Applied Biology*, vol. 37, pp. 139–148, 1950.

[12] P. J. W. Lutman, P. Bowerman, G. M. Palmer, and G. P. Whytock, "Prediction of competition between oilseed rape and *Stellaria media*," *Weed Research*, vol. 40, no. 3, pp. 255–269, 2000.

[13] L. Hämet-Ahti, "The boreal zone and its biotic subdivision," *Fennia*, vol. 159, pp. 69–75, 1981.

[14] U. Meier, Ed., *Growth Stages of Mono-and Dicotyledonous Plants: BBCH Monograph*, Federal Biological Research Centre for Agriculture and Forestry, 2001, http://www.jki.bund.de/fileadmin/dam_uploads/_veroeff/bbch/BBCH-Skala_englisch.pdf.

[15] R. Marshall, R. Hull, and S. R. Moss, "Target site resistance to ALS inhibiting herbicides in *Papaver rhoeas* and *Stellaria media* biotypes from the UK," *Weed Research*, vol. 50, no. 6, pp. 621–630, 2010.

[16] P. Kudsk and S. K. Mathiassen, "Herbicid resistens—status," *DJF Rapport*, vol. 99, pp. 127–139, 2004.

[17] J. F. Cahill Jr., "Lack of relationship between below-ground competition and allocation to roots in 10 grassland species," *Journal of Ecology*, vol. 91, no. 4, pp. 532–540, 2003.

[18] P. J. Kramer and J. S. Boyer, *Relations of Plants and Soils*, Academic Press, San Diego, Calif, USA, 1995.

[19] A. H. Fitter, "Effects of nutrient supply and competition from other species on root growth of *Lolium perenne* in soil," *Plant and Soil*, vol. 45, no. 1, pp. 177–189, 1976.

[20] P. Hamouz, K. Hamouzová, J. Holec, and L. Tyšer, "Impact of site-specific weed management on herbicide savings and winter wheat yield," *Plant, Soil and Environment*, vol. 59, pp. 101–107, 2013.

[21] J. Salonen and L. R. Erviö, "Efficacy of chemical weed control in spring cereals in Finland," *Weed Research*, vol. 28, no. 4, pp. 231–235, 1988.

[22] E. Stasinskis, "Effect of preceding crop, soil tillage and herbicide application on weed and winter wheat yield," *Agronomy Research*, vol. 7, pp. 103–112, 2009.

[23] A. Légère, F. C. Stevenson, D. L. Benoit, and N. Samson, "Seedbank-plant relationships for 19 weed taxa in spring barley-red clover cropping systems," *Weed Science*, vol. 53, no. 5, pp. 640–650, 2005.

8

Effects of Media Formulation on the Growth and Morphology of Ectomycorrhizae and Their Association with Host Plant

Ferzana Islam and Shoji Ohga

Department of Agro-Environmental Sciences, Faculty of Agriculture, Kyushu University, Fukuoka 811-2415, Japan

Correspondence should be addressed to Shoji Ohga; ohga@forest.kyushu-u.ac.jp

Academic Editors: A. D. Arencibia, J. S. Swanston, and I. Vasilakoglou

Tricholoma matsutake and *Rhizopogon roseolus* form ectomycorrhizal (ECM) association with their host plant on natural habitats. The main objective of this study was to test mycelial growth, morphology, and host plant survival both *in vitro* and *in vivo* when treated with enriched media. Aseptically germinated seedlings of *Pinus densiflora* and *P. thunbergii* were inoculated with the strains of *T. matsutake* and *R. roseolus*, respectively. Under *in vitro* conditions mycelial growth rates performed best on pH 5 and were better on Modified-Melin-Norkrans-(MMN) based medium and Potato Dextrose Agar (PDA); addition of micronutrients and vitamins in MMN mycelial growth rates had 6–27% differences. Without ECM, plant survival rates on standard media were 30% to below 30% and by inclusion of elements they were 50% to 80%. On *in vivo*, soil containing different media with ECM allowed successful mycorrhizal association and increased seedling survival rates approximately 100%. Our findings confirm that MMN and PDA allowed higher mycelial growth but poor plant survival (<30%); however, enriched media supported 100% plant survival with successful ECM associations. The present method is advantageous in terms of giving objectivity for ECM by employing suitable media for strains and host plant, and making it possible for mass production of ECM-infected seedlings.

1. Introduction

The development and survival of many forest trees and the success of a reforestation programme depend on the symbiosis involving host tree and ectomycorrhizae—their growth and establishment. Mycorrhizal symbiosis develops capabilities of the host root system by extending the plant's ability to tolerate biological and environmental stresses such as phytopathogenic attacks, nutritional insufficiencies, pollution of heavy metal, extensive erosion, drought, and different pH [1]. These positive effects of the ectomycorrhizal symbiosis on the establishment and growth of forest plants have made the ectomycorrhizal inoculation a valuable technical tool for plant production in forestry [2]. *Pinus densiflora* and *P. thunbergii* have received extra attention due to their potential usage in pine forest reforestation programs. However, the pine forest in Japan has been under threat over the recent decades and now is facing a serious crisis to survive. The recent decline in pine forest has been aggravated by many interactive disfavourable growing conditions both for the host and the mycorrhizal fungi [2, 3]. This is ultimately reducing the production of edible mushrooms growing in pine forests which have significant economic importance and cultural value in Japan [3].

Mushrooms have become attractive as a functional food and are important as a source for the development of drugs and nutraceuticals [3, 4], especially antioxidants [5, 6] and antimicrobial compounds [7]. Alternative or substitute mushroom products are mycelia which are used as food and food-flavoring material, and also for the formulation of nutraceuticals and functional foods [8]. In the culture of Japan, *Tricholoma matsutake* and *Rhizopogon roseolus* have long been prized for its flavor, distinct taste and holds its exceptional commercial and cultural value as highly sought edible mushrooms [3, 9]. Besides, a number of bioactive compounds, antioxidants, and antifat properties have been identified in *T. matsutake* which gives this mushroom a special importance for containing medicinal properties [10, 11]. Nevertheless, *T. matsutake* has been also studied that, mycelia preparation in bulk quantity was proven to have anti-tumor activity as well as preventive activity against the formation of azoxymethane-induced precancerous lesions in case of the colon organs [12, 13]. Besides *T. matsutake*, *R. roseolus* also holds an important position in the culture of Japan.

Some ECM mushroom fruiting bodies are difficult to grow on a large production scale. Therefore, growing mushroom mycelium on defined nutrient medium could be an alternative method for the production of ECM fungal biomass [14, 15]. For ectomycorrhizal (ECM) fungi previous research studies show that pH level plays a very important role and Modified Melin Norkrans (MMN) usually offers the best results for this group of fungi [16]. Genetic variation within species and within the strains can influence both the degree of root colonization by ECM fungi and the response to the plant to mycorrhizal symbiosis [17]. To improve forest productivity the ECM symbiosis requires fungal inoculants in a large scale level, for these reasons it is necessary to define the optimal composition of the culture medium for each fungus accounting different strains and their host plant establishment on a large variation of soil conditions.

The overall objective of this study was therefore to assess the improvement of the media formulations favorable for ECM mycelial growth, host plant survival rate *in vitro* and suitable for ectomycorrhizal association with *P. densiflora* and *P. thunbergii in vivo* conditions with best plant survival. For this, firstly the experiment had been extrapolated on *in vitro* conditions for validating the growth and development of ECM and host plant survival rates on different media adjoining the nutrients and vitamins, which sharpen the focus on the composition of a medium suitable for both ECM and their host plants without their associations. Secondly, on *in vivo* conditions inoculation of *T. matsutake* and *R. roseolus* strains were established with *P. densiflora* and *P. thunbergii* to evaluate the invariability of the improved media formulations for successful ECM association with host plant and their survival rates.

2. Materials and Methods

2.1. Under In Vitro Conditions

ECM Samples. Three different strains of *T. matsutake* and one strain of *R. roseolus* were used in this experiment. The strains of *T. matsutake* were NBRC 109050, NBRC 109051, and NBRC 109052 {NITE (National Institute of Technology and Evaluation) Biological Research Centre}. For NBRC 109050 and NBRC 109051 strains, the origin of the sources was Kyoto and strain NBRC 109052 was collected from Iwate, Japan. In case of *R. roseolus* strain RR, the origin was Fukuoka, Japan. The collected specimens were first cultured on Potato Dextrose Agar (PDA) (Wako Pure chemical Industries Ltd., Osaka, Japan) in petri dishes and to get the actively growing mycelium, mycelium plugs were cut and transferred to a fresh PDA medium every 4 weeks and pregrown there.

The Effect of Different Culture Medium and pH on Mycelial Growth. The effect of the culture media and pH on mycelial growth of *T. matsutake* and *R. roseolus* was observed on petri dishes containing 10 mL of solid medium. The following different nutritive solid mediums (Table 1) were tested: Modified-Melin-Norkrans-(MMN-) based medium,

L-MMN, G-MMN {MMN, L-MMN, and G-MMN were Adapted modified media composition based on the media formulation used by Langer et al. [18]} and PDA. Each medium was adjusted to three pH levels: 4, 5, and 6 with 1 N KOH solution, and was autoclaved for sterilization. For inoculation, 4-week-old mycelial discs were cut from colonies of different strains and culture in petri dishes on different solid medium having three replicates of each and were incubated at $23 \pm 2°C$ for 90 days in dark. Mycelial growth (colony) was measured weekly at 4 right angles during the experiment. As our research results showed that the applied pH values had no effect on the morphological characteristics on the mycelial growth so we documented these characteristics on pH 5. Meanwhile the culture media had a lot of influences on the morphological characteristic of mycelial structure. Mycelial morphological identifications were guided by the method by Barros et al. and were summarized in Table 2 [19].

Preparation of Aseptic Seedlings of Pine. Seeds of *P. densiflora* and *P. thunbergii* were collected from the University forest at Sasaguri (Kyushu University of Japan) in 2010. Growing of aseptic seedlings was guided according to the methods of Lagutte et al. and [15]. We selected 160 and 80 healthy seedlings of *P. densiflora* and *P. thunbergii*, respectively.

Testing of Improved Media for Host Plant Survival Rate. Among germinated seedlings 40 of each *Pinus* sp. seedlings were transferred to test tubes (13×100 mm), which contained MMN, L-MMN, G-MMN, and PDB (Potato Dextrose Broth, Wako, Japan) semisolid media (50% agar that were used in the original composition used for each media agar) with pH 5, to determine the favorable media composition for plant survival rate. For each medium composition 10 test tubes were used and each of them contained one seedling. The lower portions (root) of the seedlings were inserted into the medium and the upper portions (stem) were kept out of the test tube. The opening portions of the test tubes were sealed with parafilm tape keeping the stem out of the test tube. Seedlings were kept (incubated at 15–25°C 10–30,000 lux fluorescent light, $25 \pm 2°C$, and 16 hrs. photoperiod) for 7 weeks followed by the methods used by Guerin-Lagutte et al. [15] and Park et al. [20]; some of the seedlings started to turn yellowish colour with shedding their needle (pine leaf). Plant survival rates were counted by selecting the green and healthy seedlings.

Preparation of ECM Inoculum for Mixed Soil. Our finding that on *in vitro* condition pH 5 was the best condition for mycelial growth rates, so we preferred pH 5 to continue our following experimental steps. Glass flasks of 200 mL containing 100 mL of different liquid media in each of them were autoclaved. After that twenty to thirty pieces of each strain were cut from PDA medium and were transferred for culturing on different liquid media for three months at $23 \pm 2°C$ in darkness. Before inoculation to the mixer soil containing container, the mycelial suspension from each flask was homogenized with autoclavable blender guided by Guerin-Lagutte et al. [15].

Preparation Mixed Soil with Infection Medium. Autoclave proof 160 culture containers was used. The soil used a mixture

TABLE 1: Composition of culture media used in this study.

Elements	Compounds	MMN[a]	L-MMN[b]	G-MMN[c]	PDA
Macroelements (mg/L)	KH_2PO_4	500.00	500.00	500.00	
	$(NH_4)_2SO_4$		250.00		
	$(NH_4)_2HPO_4$	250.00		250.00	
	$MgSO_4 \cdot 7H_2O$	150.00	150.00	150.00	
	$CaCl_2 \cdot 2H_2O$	50.00	50.00	50.00	
	NaCl	25.00	25.00	25.00	
Microelements (mg/L)	$FeCl_3 \cdot 6H_2O$	12.00	12.00	12.00	
	H_3BO_3		15.458	15.458	
	$MnSO_4 \cdot H_2O$		9.295	9.295	
	$CuSO_4 \cdot 5H_2O$		1.310	1.310	
	$ZnSO_4 \cdot 7H_2O$		5.750	5.750	
	$Na_2MnO_4 \cdot 2H_2O$		0.003	0.003	
Vitamins (mg/L)	Thiamine HCl	1.00	10.00	0.100	
	Myo-Inositol		100.00		
	Nicotinic acid		1.00		
	Pyridoxine HCl		1.00		
Potato (g/L)					4
Carbohydrate source (g/L)	Dextrose				20
	Glucose	2.5	5.0	5.0	
	Malt extract	10.0		3.0	
Solidification agent (g/L)		9.0	9.0	9.0	15

[a,b, and c]Adapted modified media composition based on the media formulation used by Langer et al. [18].

of perlite and *Sphagnum* peatmoss at a ratio of 100 : 7–10, followed by Park et al. [20]. The bed soil was autoclaved in culture containers having 250 g of soil each for 30 min. twice per day for three days. After that, soil was mixed with 100 mL of different liquid media and was autoclaved. Each of the 40 containers contained the same medium composition. Each liquid medium, containing three month's old (cultured) mycelial suspension (of different strains) was poured into the mixed soil, wherein it was carried out on a clean bench. From total 160 container 40 of each contained same formulated medium, again 10 out of 40 contained same strain with same medium composition.

Planting of Aseptic Pine Seedlings into the Infection Soil Medium. Seedlings were aseptically immersed in the suspension of mycelia for 5 hours and their roots were carefully placed into the container with mixed soil prepared in previous step; the remaining suspension of mycelia was applied closely to the root using a sterile syringe as guided by Park et al. [20]. Each container contained 1 seedling. The open surfaces of the container were then sealed with parafilm carefully keeping the stem portion exposed to the outer environment. Following the same day they were kept in the room temperature under dark condition for hardening.

2.2. Under In Vivo Conditions. All containers were placed outside, under *in vivo* conditions, in the natural environment of Kyushu University Sasaguri forest nursery in Fukuoka,

Japan. Each container was supplied with different specific liquid media containing specific strain of mycelial inoculum with the help of sterile syringe at five-day-interval. For specific medium composition, specific strain and specific *Pinus* sp. were used. Each of the containers was irrigated 50 mL of distilled water using sterile syringe every day. During the experimental period some of the seedlings started to turn yellowish with shedding their needle (pine leaf) and plant survival rates were counted by selecting the green and healthy seedlings [18]. Formation of ectomycorrhizal roots was counted in the plant roots with naked eyes which were guided according to the methods of Chung et al. [21]. Experiments were continued for seven weeks.

2.3. Statistical Analysis. Our present experiment was carried out using 3 replicates (4 radius values for each one). Data were statistically analyzed using SPSS for windows version 15 (SPSS Inc., Chicago, IL, USA).

3. Results

3.1. Under In Vitro Conditions

3.1.1. The Effect of Culture Media on the Mycelial Growth. The radial growth rates of mycelia on different nutritive solid culture media (MMN, L-MMN, G-MMN, and PDA) were studied. On MMN medium the mycelial growth rates recorded were near about 3.73, 4.57, 4.48, and 4.76 cm

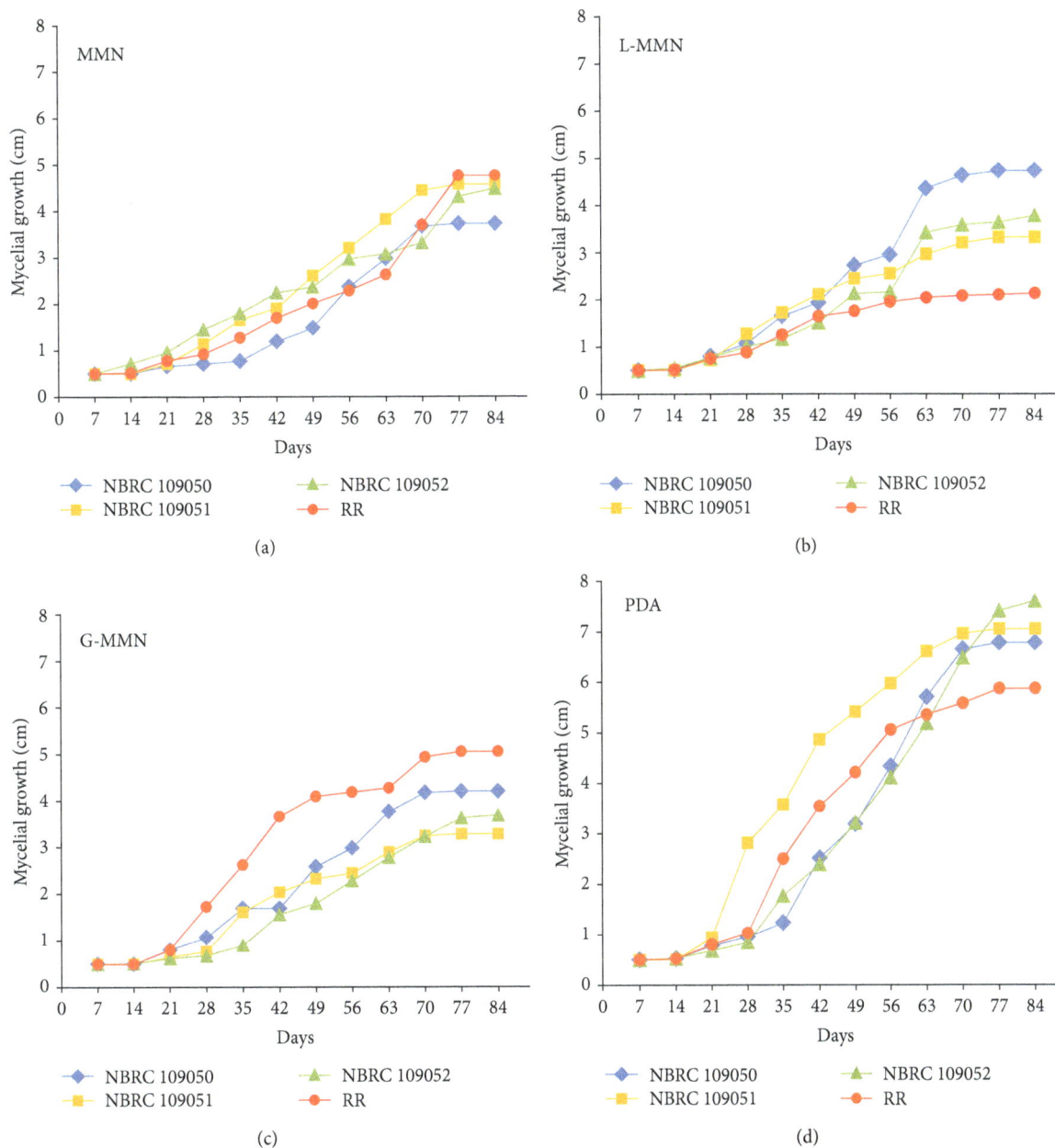

FIGURE 1: Effect of culture media on the mycelial growth rates of *T. matsutake* and *R. roseolus*.

for NBRC 109050, NBRC 109051, NBRC 109052, and RR, respectively (Figure 1(a)). Significant differences were found ($P < 0.01$) on MMN medium for each strain. RR strain had the highest and NBRC 109050 had the lowest mycelial growth on MMN medium. In case of L-MMN medium, NBRC 109050 had the highest mycelial growth rates and RR had the lowest growth rates. The mycelial growth rates showed highly significant ($P < 0.01$) differences on L-MMN medium and the growth rates recorded on 84th day of inoculation were around 4.73, 3.32, 3.78, and 2.13 cm for NBRC 109050, NBRC 109051, NBRC 109052, and RR, respectively (Figure 1(b)). On G-MMN medium all of the mycelial growth rates had highly significant ($P < 0.01$) differences. RR strain had the highest growth rates and NBRC 109051 had the lowest. The mycelial

growth rates recorded on the last day of the experiment were more or less 4.20, 3.28, 3.68, and 5.05 cm for NBRC 109050, NBRC 109051, NBRC 109052, and RR, respectively (Figure 1(c)). For PDA medium, all of the ECM strains had highly significant ($P < 0.01$) differences. The growth rates were approximately 6.78, 7.05, 7.60, and 5.85 cm for NBRC 109050, NBRC 109051, NBRC 109052, and RR, respectively (Figure 1(d)). In case of PDA medium, the highest mycelial growth rates were recorded for NBRC 109052 strain and the lowest growth rates were recorded for RR strain. All of the ECM mycelia increased till the end of the experiment on PDA. Active mycelial growth of NBRC 109050, NBRC 109051, and NBRC 109052 strains was observed within 14 to 21 days of incubation. In case of RR strain active mycelial growth

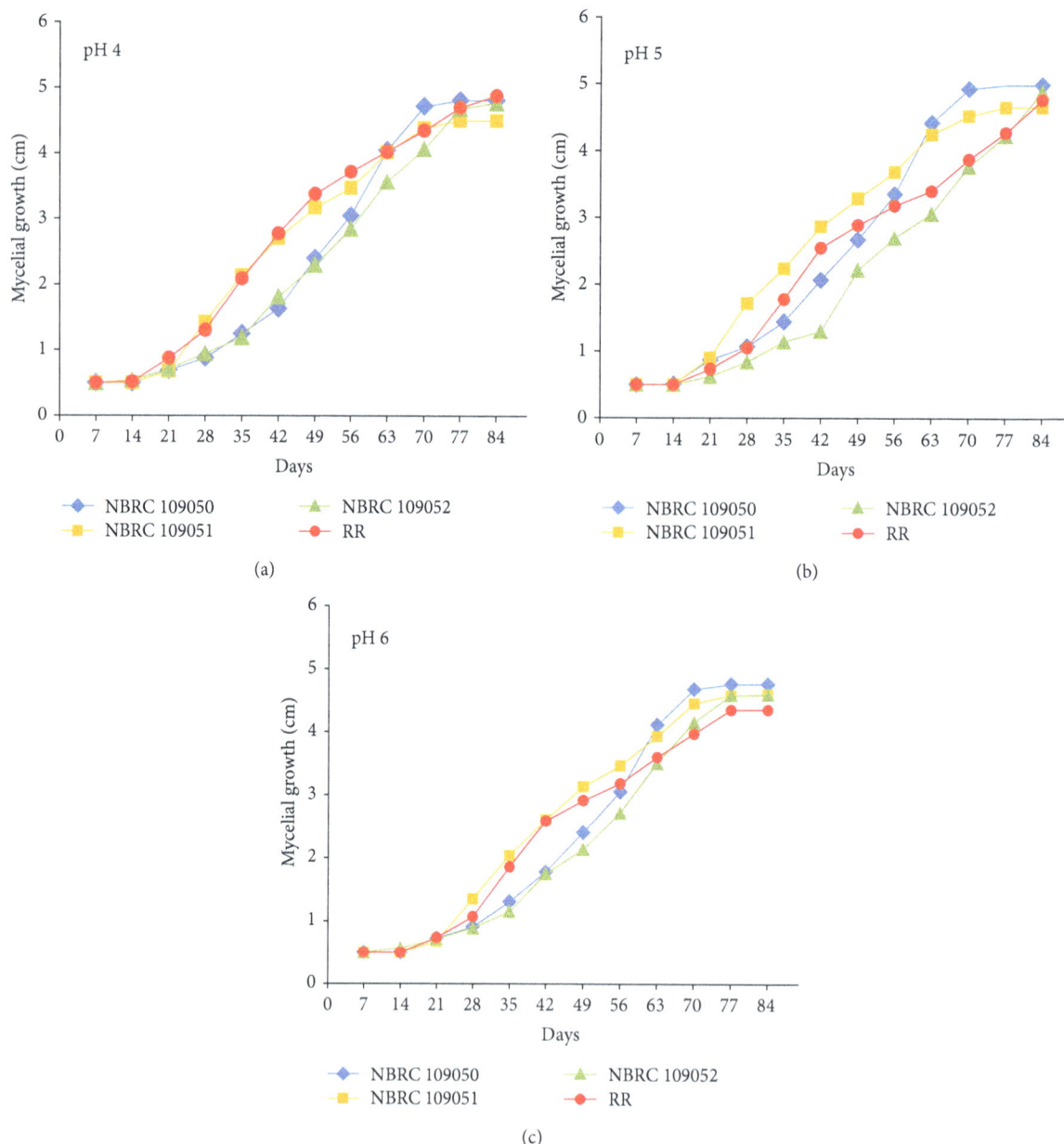

FIGURE 2: Effect of pH on the mycelial growth rates *T. matsutake* and *R. roseolus*.

rates were observed after 7 to 10 days of inoculation on every medium.

3.1.2. The Effect of pH on the Mycelial Growth of Different ECM Strains. On pH 4, NBRC 109050 had the highest growth rates and NBRC 109051 had the lowest growth rates; it was around 4.81, 4.50, 4.77, and 4.77 cm for NBRC 109050, NBRC 109051 NBRC 109052, and RR, respectively, (Figure 2(a)) with highly significant ($P < 0.01$) differences. With pH 5, NBRC 109050 had the highest growth rates and NBRC 109051 had the lowest, and growth rates showed highly significant ($P < 0.01$) differences; the growth rates were 4.99, 4.66, 4.89, and 4.88 cm for NBRC 109050, NBRC 109051, NBRC 109052, and RR, respectively (Figure 2(b)). On pH 6 mycelial

growth rates showed highly significant ($P < 0.01$) differences; the recorded rates were 4.77, 4.61, 4.60, and 4.36 cm for NBRC 109050, NBRC 109051, NBRC 109052, and RR, respectively (Figure 2(c)), on which NBRC 109050 had the highest and RR had the lowest growth rates. In all the tested conditions, interaction with media and pH had highly significant differences ($P < 0.01$) and growth rates were highest on all media with pH 5 conditions. Mycelial growth rates on MMN with pH 5 were close to 3.81, 4.61, 4.61, and 4.93 cm for NBRC 109050, NBRC 109051, NBRC 109052, and RR, respectively. On L-MMN with pH 5 mycelial growth rates were approximately 4.89, 3.41, 4.12, and 2.30 cm for NBRC 109050, NBRC 109051, NBRC 109052, and RR, respectively. In case of G-MMN with pH 5 the growth rates were more

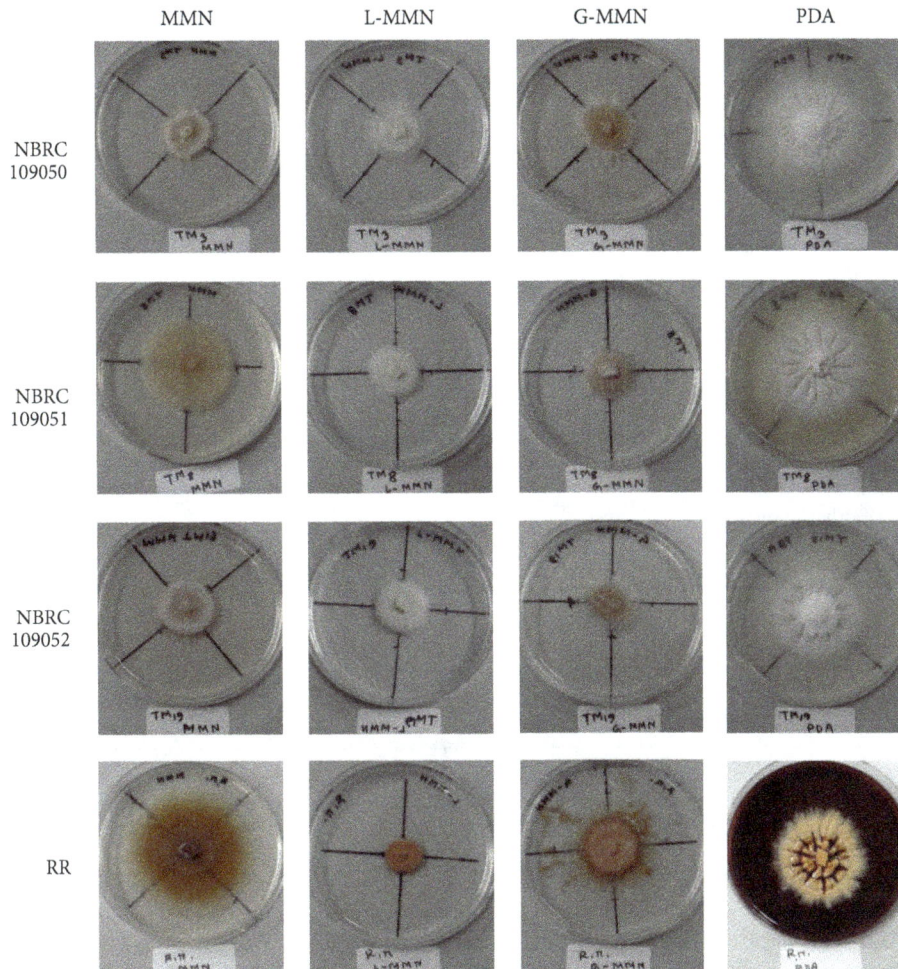

FIGURE 3: Appearance of *T. matsutake* and *R. roseolus* (after 80 days of inoculation) mycelial colonies, on different nutritive culture media at pH 5 level.

or less 4.32, 3.31, 3.92, and 5.31 cm for NBRC 109050, NBRC 109051, NBRC 109052, and RR, respectively. PDA with pH 5, growth rates were about 6.94, 7.31, 8.15, and 6.31 cm for NBRC 109050, NBRC 109051, NBRC 109052, and RR, respectively.

(a) *Morphological Description of NBRC 109050 Mycelial Growth.* NBRC 109050 grown under different media conditions were almost had smooth surface on MMN, L-MMN, and G-MMN, unless on PDA medium in which it became wooly (Table 2) and then after 55 to 60 days the smooth textural surface appeared around the wooly texture. Mycelial colour was bright white on PDA, light brown and light pink (after 65 days) on MMN, white on L-MMN, while it was brown and dull white (after 60 days) on G-MMN (Figure 3). The reverse colours in MMN were light brown and in G-MMN it was brown. On the other hand, the reverse colours for mycelium on PDA and L-MMN were white. Border colours were white on L-MMN, G-MMN, and PDA but light pink on MMN. Borders were clear on MMN and L-MMN but diffuse on PDA and G-MMN. Rifts and lines appeared from centre were present on PDA medium. The numbers of lines

were around 26 for each petri dish, which were originated from the center of the mycelial structure. Aerial growth, media colouration, and exudates were absent.

(b) *Morphological Description of NBRC 109051 Mycelial Growth.* Mycelial textures were wooly on PDA until 50 to 55 days, then around the wooly part the mycelia formed more or less smooth structure, which appeared around wooly structure (Table 2), more or less smooth textures were found on the other medium (Figure 3). The colours of mycelial structures were white on PDA and L-MMN, but they were light brown on MMN and brown on G-MMN. The borders were diffuse on PDA and G-MMN. Clear border was observed on MMN and L-MMN. In all kinds of different media the colour of borders was found white. The reverse colour was light brown on MMN and brown on G-MMN. For PDA and L-MMN the reverse colour was found white. Aerial growth, media colouration, and exudates were absent. On the PDA, rifts and lines were found on the mycelial structure. Lines originated from the center of the mycelial structure for per petri dishes were approximately 14 in number.

TABLE 2: Morphological descriptions of *Tricholoma matsutake* and *Rhizopogon roseolus* on different nutritive culture media at pH 5.

Strains	Culture media	Mycelial texture	Mycelium colour	Border	Border colour	Reverse colour	Aerial growth	Medium colouration	Exudates	Rifts	Lines formed from center (no. of lines per petri dish)
NBRC 109050	MMN	Almost smooth	Light brown and light pink[b]	Clear	Light pink	Light brown	–	–	–	–	–
	L-MMN	Almost smooth	White	Clear	White	White	–	–	–	–	–
	G-MMN	Almost smooth	Brown and dull white[c]	Diffuse	White	Brown	–	–	–	–	–
	PDA	Wooly[a1] and almost smooth	Bright white	Diffuse	White	White	–	–	–	+	+ (around 26 lines)
NBRC 109051	MMN	Almost smooth	Light brown	Clear	White	Light brown	–	–	–	–	–
	L-MMN	Almost smooth	White	Clear	White	White	–	–	–	–	–
	G-MMN	Almost smooth	Brown	Diffuse	White	Brown	–	–	–	–	–
	PDA	Wooly[a2] and almost smooth	Bright white	Diffuse	White	White	–	–	–	+	+ (around 14 lines)
NBRC 109052	MMN	Almost smooth	Light brown and light pink[d]	Diffuse	Light pink	Light brown	–	–	–	–	–
	L-MMN	Almost smooth	White	Diffuse	White	White	–	–	–	–	–
	G-MMN	Almost smooth	Brown and dull white[e]	Diffuse	White	Brown	–	–	–	–	–
	PDA	Wooly[a3] and almost smooth	Bright white	Diffuse	White	White	–	–	–	+	+ (around 11 lines)
RR	MMN	Almost smooth	Dark brown	Diffuse	Dark brown	Dark brown	–	–	–	–	–
	L-MMN	Almost smooth	Dark brown	Clear	Dark brown	Dark brown	–	–	–	–	–
	G-MMN	Almost smooth	Dark brown	Diffuse	Dark brown	Dark brown	–	–	–	–	–
	PDA	Wooly[a4] and almost smooth	Brownish white and chocolate[f]	Diffuse	Chocolate	Chocolate	–	+	+	+	+ (around 16 lines)

[a]a1: Wooly until 55 to 60 days of growth, a2: Wooly until 50 to 55 days of growth, a3: Wooly until 60 to 65 days of growth, a4: Wooly until 40 to 45 days of growth.
[b]Light pink area appeared after 65 days.
[c]Dull white area appeared after 60 days.
[d]Light pink area appeared after 63 days.
[e]Dull white area appeared after 65 days.
[f]Chocolate area appeared after 45 days.

(c) *Morphological Description of NBRC 109052 Mycelial Growth.* NBRC 109052 presented almost the same morphological characteristics like NBRC 109050 on different culture media having small differences (Table 2). The mycelial textures had almost smooth surface in all nutritive media except on PDA. In PDA the textures were wooly (Figure 3) for maximum days and after that almost smooth surface appeared around the wooly structure. Smooth surface was found after 60 to 65 days of inoculation. The colour of the mycelial structure was bright white and white on PDA and L-MMN, respectively. In case of MMN it was light brown and after 63 days light pink mycelial structures were formed around light brown area, but on G-MMN brown colour mycelial structure was developed up to 65 days. After 65 days of inoculation dull white colour mycelial structure was observed to be formed around the brown coloured area on G-MMN. The borders of the mycelial structure were diffuse on all conditions. On L-MMN, G-MMN and PDA the border colour of mycelial structure were white but on MMN it was light pink. The reversed colours were white on both L-MMN and PDA but they were light brown and brown on MMN and G-MMN, respectively. Aerial growth, medium colouration, and exudates were absent in all conditions. Rifts and lines formed from center of the mycelial structure were found only on PDA. The lines which were formed on PDA on mycelial structure were around 11 in number for per petri dish.

(d) *Morphological Description of RR Mycelial Growth.* RR mycelial structures had wooly appearances until 40 to 45 days of inoculation on PDA. After 40 to 45 days of inoculation almost smooth surface of mycelial structure was found to be formed around the wooly structures, on PDA (Table 2). The surface of mycelial growth was almost smooth on other media. Dark brown growths of mycelial structures were found on MMN, L-MMN, and G-MMN. Brownish white and chocolate (after 45 days) coloured mycelial growth structures were observed on PDA (Figure 3). Borders were clear on L-MMN but on other media they were diffuse. In all conditions both border colours and reverse colours were dark brown except on PDA, in which it was chocolate colour. Aerial growth was absent in all states of the media. Media colouration, exudates, rifts, and lines were present only on PDA. Number of lines formed from the center of the mycelial growth was near about 16 in number for each petri dish on PDA.

3.1.3. Host Plant Survival Rates on Improved Media Compositions. Seedlings on L-MMN showed uniform growth with green shoot colour on 80% planlets for both *Pinus* sp. but on PDA both species had the lowest survival rates and were about 10% and 20% for *P. densiflora* and *P. thunbergii*, respectively; they showed severe discoloration and shedding of pine needles. For MMN and G-MMN survival rates were 30% to 50% for *P. densiflora* and 30% to 60% for *P. thunbergii*, respectively. Both *Pinus* sp. have yellowish green shoot colour on MMN and they had light pale green colour on G-MMN.

3.2. Under In Vivo Conditions

3.2.1. Testing of Media Compositions for Mycorrhizal Association with Host Plant in Mixer Soil. After 2 weeks of inoculation, it was found that fungal hyphae bound soil particle together by aggregating soils (Figures 4(a) and 4(b)) during that time necessary precautions were taken not to disturb the soil portion close to the root system of the seedlings. After 7 weeks of inoculation ectomycorrhizal hyphae colonized the root surface and formed aggregated soil around the root surface; elongated lateral roots were noticed lacking root hairs (Figures 4(c) and 4(d)) which were colonized with fine discrete hyphae. Ramified and branched roots were observed in all soil conditions, for all strains the numbers (per plant)

TABLE 3: Survival rates of *Pinus densiflora* and *P. thunbergii* on four different media composition at pH 5 on *in vitro* conditions.

Medium	Number of plants	Plant survival rate % on medium (without ECM)
MMN	10[a]	30
	10[b]	30
L-MMN	10[a]	80
	10[b]	80
G-MMN	10[a]	50
	10[b]	60
PDA	10[a]	10
	10[b]	20

[a] *P. densiflora.*
[b] *P. thunbergii.*
Plant survival rate was specified by the percentage of living plantlets/seedlings.

were maximum on PDB containing soil mixer maximum around 16 to 17 (Table 4) and lower in MMN containing soil mixer which were about 12. For L-MMN soil mixer they were approximately 14 to 15 and on G-MMN they were recorded 12 to 14 in number. Among *T. matsutake* strains NBRC 109052 had always the best mycorrhizal formation with the host plant, and *R. roseolus* strain formed maximum mycorrhizal root around 17 in PDA containing soil mixer and had minimum near about 12 in MMN containing soil mixer.

3.2.2. Host Plant Survival Rates with Ectomycorrhizal Association in Mixed Soil with Media Compositions. Seedlings in L-MMN mixed soil showed uniform growth with green shoot colour on 90% to 100% planlets for both *Pinus* sp. With MMN and PDA mixer soil both species had the lowest survival rates which were around 30% to 60%; they showed discoloration and shedding of pine needles. Both *Pinus* sp. have yellowish green shoot colour on MMN and they had light pale green colour on G-MMN. For all soil mixer conditions ECM association increased the host plant survival rates 10% to 30% (Tables 3 and 4) comparing to the survival rates without ectomycorrhizal inoculation.

4. Discussion

Fungi of ECM group have been widely studied in different parts of the world for their wide range of considerable ecological and economic importance. For this circumstance, these species are often used both for experimental research and applied purposes [22]. This research focused on physical growth rate, morphological characters of mycelia of the ECM fungi, their host plant survival rates, and ECM association with host plant.

We stimulated ECM mycelial growth by different nutritional conditions and found considerably significant differences among the applied *in vitro* conditions without host plant, which speculate that higher concentrations of nutrients may be the reason of mycelial growth rates inhibition. This

(a)   (b)

(c)   (d)

FIGURE 4: (a) Appearance of *T. matsutake* mycelial structure aggregates soil materials adhering to the root surface. Scale bar, 25 mm. (b) Appearance of *R. roseolus* mycelial structure aggregates soil materials adhering to the root surface. Scale bar, 30 mm. (c) Formation of ectomycorrhizal roots on *P. densiflora* (*T. matsutake*). Scale bar, 1.0 mm. (d) Formation of ectomycorrhizal roots on *P. thunbergii* (*R. roseolus*). Scale bar, 1.5 mm.

phenomenon can be explained by two ways: (i) the mycelial catabolism was suppressed by the higher concentration of nutrients because of osmotic pressure as the ectomycorrhizal fungi do not grow under high osmotic pressure and (ii) presence of high concentration of vitamin and hormones increased the activity of some enzymes which could suppress mycelial growth [23, 24]. These kinds of effect were also reported from other scientists with the application of carbohydrate components [25, 26]. It was found that higher concentration of extra glucose had an effect on catabolite suppression in the ECM fungal group. These might be an explanation of our findings for studied strains.

This study shows that media compositions with different pH levels were effective in stimulating mycelial growth rates without the host plant. It has been well known that pH had significant influences on the growth of fungi; in general they grow better between pH 5 and pH 6. Considerable growth rates were also found in some studies in between pH 3.2 and 6.5, but the optimal pH ranges from 4.5 to 5.5. Some scientists stated that optimum pH ranges are mainly related to different species, strains, enzymatic systems, important vitamin entry in the cell, mineral capture, and surface metabolic reactions

[19, 27]. This supports our findings that our studied ECM strains had the highest growth on pH 5.

A relatively close study shows that there are considerable morphological differences among the studied strains of ECM on different media. We were able to distinguish them for different strains but these specific characteristics had no influences on the applied pH levels. For each strain they had similar morphological characteristics on different pH. These might be the results due to the variations among the different strains. Therefore, the results also provide evidence that the mycelial appearance not only varies with the culture media but also changes with the fungal species and strains which suggest considerable intraspecific variation among different strains [19, 23, 24, 27].

The present experiment also showed that plant survival rates were poor on widely used standard MMN formulation and on PDA which may be due to the absence of essential plant nutrients and vitamins, because with the addition of supplemented elements to G-MMN and L-MMN increased the plant survival rates. It suggests that plantlets were not able to produce satisfactory amount of nutrients and vitamins required for normal growth and development. It had been

TABLE 4: Survival rates of *Pinus densiflora* and *P. thunbergii*, appearance of mycelial colony in cultured soil, and formation of mycorrhizal roots (at pH 5 with mixed soil).

Medium	ECM strain	Number of plants	(ECM) Appearance of mycelial colony in cultured soil	ECM association with the plant roots	Plant survival rate % (with ECM)	Formation of ectomycorrhizal roots[c] (approximately for per plant, mean value)
MMN	NBRC 109050	10[a]	+	+	40	12
	NBRC 109051	10[a]	+	+	40	12
	NBRC 109052	10[a]	+	+	50	12
	RR	10[b]	+	+	60	12
L-MMN	NBRC 109050	10[a]	+	+	90	14
	NBRC 109051	10[a]	+	+	90	14
	NBRC 109052	10[a]	+	+	100	15
	RR	10[b]	+	+	100	15
G-MMN	NBRC 109050	10[a]	+	+	70	12
	NBRC 109051	10[a]	+	+	80	13
	NBRC 109052	10[a]	+	+	80	14
	RR	10[b]	+	+	90	14
PDA	NBRC 109050	10[a]	+	+	30	16
	NBRC 109051	10[a]	+	+	30	17
	NBRC 109052	10[a]	+	+	30	17
	RR	10[b]	+	+	40	17

[a] *P. densiflora*.
[b] *P. thunbergii*.
[c] Formations of ectomycorrhizal roots were counted in the plant roots with naked eyes and were determined by morphological root characteristics indicative of mycorrhizal formation.

reported that nutrients especially micronutrients and vitamins play an important role on plant survival rates, growth, development establishment, and resistance capacity to a wide range of variations [28, 29]. This phenomenon supports our findings, which confirms that *Pinus* sp. need sufficient amount of nutrients and vitamins for survival with normal growth and development.

We found that the numbers of ramified and branched roots were higher in case of all strains on PDA mixed soil but plant survival rates were poor. Kusuda et al. [26] suggested that the supply of carbohydrate plays an obligate role for some of mycorrhizal formation; this might be a reason that the strains in our study formed maximum mycorrhizal roots in soil mixer containing PDA. They also stated that exogenous carbohydrates sometimes also reduce the fungal requirement for root carbohydrates which may be a reason for the variations of mycorrhizal colonization of the studied strains. Results showed that the mycorrhizal associations of studied strains increased plant survival rates 100% with L-MMN mixed soil which contained minimum amount of carbohydrate among the used medium compositions. This may prove the definite nutrient demand of the plantlets during mycorrhizal synthesis. Plant survival rates reached 90–100% with ectomycorrhizal inoculation by the addition of vitamin mixture (thiamine, myo-inositol, nicotinic acid and pyridoxine, usually supplemented to plant tissue culture) which also indicated that pine seedlings required symbiotic helper such as the ectomycorrhizal fungi for their highest

survival rates. Normally, the addition of thiamine meets the vitamin demands of plants during mycorrhizal association. Fungal colonization on roots can benefit the tree by forming hyphal network that effectively increases plant nutrient absorptive surface area and benefited each other by symbiotic associations [30]. These phenomena support our results, that culturing pine seedlings with ectomycorrhizal inoculation increased plant survival rates.

The results have important attribution to future research with these strains and might have some important implication for ECM inoculation along with their mycelial production. Observations indicate that NBRC 109050, NBRC 109051, NBRC 109052, and RR are suitable for inoculation to pine seedlings due to their adaptability under a wide range of *in vitro* and *in vivo* conditions, but among them strain NBRC 109052 and RR were the best for *in vitro* and *in vivo* conditions, which indicates the growth variation between and within the species [23, 27]. Trappe [31], Parladé et al. [32] and Marix et al. [33] suggested that intraspecific variability of fungi plays an important role on controlling inoculation and exhibits the physiological capacity to form abundant ECM on the desired host. Habitat differences might be reflected on the *in vitro* growth conditions; also in the colonization patterns they indicate some degree of specialization or host preference among the strains [32, 33].

To summarize, our findings show that formulations of media on *in vitro* and *in vivo* conditions varied the mycelial growth of the studied ECM strains whereas the additional nutrients are essential for plant establishment for its own potentiality. It also confirms that *Pinus* sp. may form successful ECM association *in vivo* with number of special nutrient and vitamin support. A balanced nutrient and the inclusion of different vitamins are vital for plant establishment and successful ectomycorrhizal association. Several new techniques may be complemented with adapted medium composition and may thereby increase the proportion of successful ECM association on *Pinus* sp. The procedure of enriching media can successfully increase the host plant survival rates without ECM and satisfactory ECM associations with host plant. With ECM association the plant survival rate achieved the highest survival rates. The mycelial growth of *T. matsutake* and *R. roseolus* performed best on MMN and PDA media without host plant *in vitro* condition, whereas host plant survival rates were poor without ECM association *in vitro* condition on MMN and PDA. Methods based on enriching media formulations for ECM with new strains can have further scope for future research work.

5. Conclusion

To conclude, formulation of media on *in vitro* and *in vivo* conditions varies the mycelial growth and development of ECM strains, whereas additional nutrients are essential for plant survival. Our findings confirm that *Pinus* sp. forms successful associations with ECM on *in vivo* when supplied with a number of special nutrients and vitamins. A balanced nutrient composition and inclusion of vitamins are essential for successful ectomycorrhizal associations and highest host plant survival. Several new techniques may be complemented with adapted media formulation and may thereby increase the proportion of successful ECM association with *Pinus* sp. The findings of our research may further be employed with different synthesis techniques carried out in plantation and reforestation areas with different ECM fungi. This opens new prospective in enriching plantation forest research where ectomycorrhizal association can flourish to protect the decline of ECM mushroom and pine forest in Japan.

References

[1] S. Smith and D. J. Read, *Mycorrhizal Symbiosis*, vol. 640, Academic Press, London, UK, 1996.

[2] S. E. Smith and D. J. Read, *Mycorrhizal Symbiosis*, Academic Press, London, UK, 2008.

[3] F. Islam and S. Ohga, "The response of fruit body formation on in situ condition *Tricholoma matsutake* by applying electric pulse stimulator," *ISRN Agronomy*, vol. 2012, Article ID 462724, 6 pages, 2012.

[4] S. T. Chang, "Global impact of edible and medicinal mushrooms of human welfare in the 21 century: non-green revolution," *International Journal of Medicinal Mushrooms*, vol. 1, pp. 1–7, 1999.

[5] G. C. Yen and C. Y. Hung, "Effects of alkaline and heat treatment on antioxidative activity and total phenolics of extracts from Hsian-tsao (*Mesona procumbens* Hemsl.)," *Food Research International*, vol. 33, no. 6, pp. 487–492, 2000.

[6] I. C. F. R. Ferreira, P. Baptista, M. Vilas-Boas, and L. Barros, "Free-radical scavenging capacity and reducing power of wild

edible mushrooms from northeast Portugal: individual cap and stipe activity," *Food Chemistry*, vol. 100, no. 4, pp. 1511–1516, 2007.

[7] L. Barros, R. C. Calhelha, J. A. Vaz, I. C. F. R. Ferreira, P. Baptista, and L. M. Estevinho, "Antimicrobial activity and bioactive compounds of Portuguese wild edible mushrooms methanolic extracts," *European Food Research and Technology*, vol. 225, no. 2, pp. 151–156, 2007.

[8] C. C. Weng, *Taste quality of Grifola frondosa, Morcbella esculenta and Termitomyces albuminosus mycelia and their application in food application in food processing [M.S. thesis]*, National Chung-Hsing University, Taichung, Taiwan, 2003.

[9] W. Yun, I. R. Hall, and L. A. Evans, "Ectomycorrhizal fungi with edible fruiting bodies 1. *Tricholoma matsutake* and related fungi," *Economic Botany*, vol. 51, no. 3, pp. 311–327, 1997.

[10] N. Ohnuma, K. Amemiya, R. Kakuda, Y. Yaoita, K. Machida, and M. Kikuchi, "Sterol constituents from two edible mushrooms, *Lentinula edodes* and *Tricholoma matsutake*," *Chemical and Pharmaceutical Bulletin*, vol. 48, no. 5, pp. 749–751, 2000.

[11] H. W. Lim, J. H. Yoon, Y. S. Kim, M. W. Lee, S. Y. Park, and H. K. Choi, "Free radical-scavenging and inhibition of nitric oxide production by four grades of pine mushroom (*Tricholoma matsutake* Sing.)," *Food Chemistry*, vol. 103, no. 4, pp. 1337–1342, 2007.

[12] T. Ebina, T. Kubota, N. Ogama, and K. I. Matsunaga, "Antitumor effect of a peptide-glucan preparation extracted from a mycelium of *Tricholoma matsutake* (S. Ito and Imai) Sing," *Biotherapy*, vol. 16, no. 3, pp. 255–259, 2002.

[13] K. Matsunaga, T. Chiba, and E. Takahashi, "Mass production of Matsutake (*Tricholoma matsutake*) mycelia and its application to functional foods," *Bioindustry*, vol. 20, pp. 37–46, 2003.

[14] V. P. Cirillo, W. A. Hardwick, and R. D. Seeley, "Fermentation process for producing edible mushroom mycelium," USA Patent no. 2, 928, 210; 1960.

[15] A. Guerin-Laguette, L. M. Vaario, W. M. Gill, F. Lapeyrie, N. Matsushita, and K. Suzuki, "Rapid in vitro ectomycorrhizal infection on *Pinus densiflora* roots by *Tricholoma matsutake*," *Mycoscience*, vol. 41, no. 4, pp. 389–393, 2000.

[16] R. Molina and J. G. Palmer, "Isolation, maintenance, and pure culture manipulation of ectomycorrhizal fungi," in *Methods and Priciples of Mycorrhizal Research*, N. C. Schenck, Ed., pp. 115–129, American Phytopathological Socity, St Paul, Minn, USA, 1982.

[17] R. L. Peterson and S. M. Bradbury, "Use of plant mutants, interspecific variants and non-hosts in studying mycorrhiza formation and function," in *Mycorrhiza: Structure, Function, Molecular Biology and Biotechnology*, A. K. Varma and B. Hock, Eds., Springer, Berlin, Germany, 1995.

[18] I. Langer, D. Krpata, U. Peintner, W. W. Wenzel, and P. Schweiger, "Media formulation influences in vitro ectomycorrhizal synthesis on the European aspen *Populus tremula L.*," *Mycorrhiza*, vol. 18, no. 6-7, pp. 297–307, 2008.

[19] L. Barros, P. Baptista, and I. C. F. R. Ferreira, "Influence of the culture medium and pH on the growth of saprobic and ectomycorrhizal mushroom mycelia," *Minerva Biotecnologica*, vol. 18, no. 4, pp. 165–170, 2006.

[20] M. C. Park, S. G. Sim, and W. J. Cheon, "Methods of preparing *Tricholoma matsutake*-infected young pine by coculturing aseptic pine seedlings and .*T. matsutake*," US Patent no. 7,269,923 B2; 2007.

[21] H. C. Chung, D. H. Kim, and S. S. Lee, "Mycorrhizal formation and seedling growth of *Pinus densiflora* by *in vitro* synthesis with

the inoculation of ectomycorrhizal fungi," *Mycobiology*, vol. 30, no. 2, pp. 70–75, 2002, Copyright by The Korean Society of Mycology.

[22] I. R. Hall and Y. Wang, "Methods for cultivating edible mycorrhizal mushroom," in *Mycorrhiza Manual*, A. Varma, Ed., pp. 99–114, Springer, Berlin, Germany, 1998.

[23] T. Hatakeyama and M. Ohmasa, "Mycelial growth characteristics in a split-plate culture of four strains of the genus *Suillus*," *Mycoscience*, vol. 45, no. 3, pp. 188–199, 2004.

[24] T. Lubbehusen, V. González Polo, S. Rossi et al., "Protein kinase A is involved in the control of morphology and branching during aerobic growth of *Mucor circinelloides*," *Microbiology*, vol. 150, no. 1, pp. 143–150, 2004.

[25] T. Hatakeyama and M. Ohmasa, "Mycelial growth of strains of the genera *Suillus* and *Boletinus* in media with a wide range of concentrations of carbon and nitrogen sources," *Mycoscience*, vol. 45, no. 3, pp. 169–176, 2004.

[26] M. Kusuda, M. Ueda, Y. Konishi, K. Yamanaka, T. Terashita, and K. Miyatake, "Effects of carbohydrate substrate on the vegetative mycelial growth of an ectomycorrhizal mushroom, *Tricholoma matsutake*, isolated from Quercus," *Mycoscience*, vol. 48, no. 6, pp. 358–364, 2007.

[27] L. L. Hung and J. M. Trappe, "Growth variation between and within species of ectomycorrhizal fungi in response to pH *in vitro*," *Mycologia*, vol. 75, pp. 234–241, 1983.

[28] E. J. Jokela, W. W. McFee, and E. L. Stone, "Micronutrient deficiency in slash pine: response and persistence of added manganese," *Soil Science Society of America Journal*, vol. 55, no. 2, pp. 492–496, 1991.

[29] A. Shalata and P. M. Neumann, "Exogenous ascorbic acid (vitamin C) increases resistance to salt stress and reduces lipid peroxidation," *Journal of Experimental Botany*, vol. 52, no. 364, pp. 2207–2211, 2001.

[30] J. V. D. Rousseau, D. M. Sylvia, and A. J. Fox, "Contribution of ectomycorrhiza to the potential nutrient-absorbing surface of pine," *New Phytologist*, vol. 128, no. 4, pp. 639–644, 1994.

[31] J. M. Trappe, "Selection of fungi for ectomycorrhizal inoculation in nurseries,"" *Annual Review of Phytopathology*, vol. 15, pp. 203–222, 1977.

[32] J. Parladé, J. Pera, and J. Luque, "Evaluation of mycelial inocula of edible *Lactarius species* for the production of *Pinus pinaster* and *P. sylvestris* mycorrhizal seedlings under greenhouse conditions," *Mycorrhiza*, vol. 14, no. 3, pp. 171–176, 2004.

[33] D. H. Marix, S. B. Maul, and C. E. Cordell, "Application of specific ectomycorrhizal fungi in world forestry,"" in *Frontiers in Industrial Mycology*, G. F. Leatham, Ed., Chapman & Hall, New York, NY, USA; Kluwer Academic Publishers, Dordrecht, The Netherlands, 1992.

9

On-Farm Evaluation of Beans Varieties for Adaptation and Adoption in Kigoma Region in Tanzania

Tulole Lugendo Bucheyeki[1,2] and Tuaeli Emil Mmbaga[3]

[1] University of KwaZulu-Natal, Private Bag X01, Scottsville 3209, South Africa
[2] Tumbi Agricultural Research and Development Institute, P.O. Box 306, Tabora, Tanzania
[3] Selian Agricultural Research and Development Institute, P.O. Box 6024, Arusha, Tanzania

Correspondence should be addressed to Tulole Lugendo Bucheyeki; tlbucheyeki@yahoo.co.uk

Academic Editors: A. Berville, G. M. Dal Bello, and M. Ruiz

On-farm beans research was carried out in Kigoma region, Tanzania. Objectives were to evaluate beans varieties for yield under farmers' management conditions and to assess farmers' preferences on beans varieties. Nine farmers from three villages with three farmers per village participated in beans trials. A randomized complete block design with five plots per replications was used to evaluate five bean varieties: Lyamungo 90, Jesca, Uyole 94, Kablanketi, and Kigoma yellow (control). Beans were planted on a $10 \, m \times 2.5 \, m$ plot at a spacing of $50 \, cm \times 20 \, cm$. Data was subjected to analysis using ANOVA table in GenStat statistical computer software. Three villages × three seasons resulted in nine environments which were used for stability analysis. Farmers developed their criterion to assess the performance and acceptability of beans varieties. Lyamungo 90 and Jesca ranked high and outyielded other varieties with an average yield of 1430.00 and 1325.67 $kg \, ha^{-1}$, respectively. Genotypes sum of squares accounted for the most of the variability (89.12%). Introduction of high yielding bean varieties with the desired farmers' traits is expected to revamp beans production and contribute to the improved food security in Tanzania.

1. Introduction

Common bean (*Phaseolus vulgaris* L.) is an important herbaceous annual grain legume in the world chiefly grown as a cheap source of protein among majority of Sub-Saharan African people [1].

Farmers frequently use it as a vital component in crop rotation for its ability to fix nitrogen [2, 3]. According to FAO-STAT [4] estimate for the year 2006, world beans production was 1235 $kg \, ha^{-1}$ while that of Africa was 799 $kg \, ha^{-1}$. The average beans yield per annum in many African countries is always lower than that of the world. Lack of improved varieties associated with edaphic and biotic factors has been cited as one of the primary sources of lower beans production [5]. In Tanzania, common bean is an important food and cash crop which is mostly grown by small-holder farmers [6]. However, common bean production in Tanzania is low and does not meet the increasing demand. The average yield is 741 $kg \, ha^{-1}$ which is lower than that found in the developed countries [4]. The low beans yield is mostly contributed by the use of unimproved varieties. Farmers use the locally available varieties with low yield potential. The result is low yield per area and reduced beans production (Figure 1).

In addition to the lack of improved varieties and high seed demand during planting seasons, farmers use recycled seeds [7, 8]. This has stalled production of beans in the country and calls for more breeding efforts to curb the problem.

Development of new varieties requires full participation of stakeholders [9]. On-farm trials have been reported by researchers as vital tools for speeding up of breeding processes and enhanced cultivars adoption rates in farming communities [10, 11]. On-farm trial enables the incorporation of farmers' opinions and ensures testing of technologies under farmers' management conditions [12]. There are reports of increased rate of adoption and reduced variety abandonment when farmers' knowledge and experiences are acknowledged [13, 14]. To speed up variety evaluation, testing, and eventually

TABLE 1: Six testing sites description for three seasons (2003–2006) in Kasulu district, Kigoma region.

Farmer	District	Village	Longitude	Latitude	Altitude (masl)	Max. temp (°C)	Max. temp (°C)	Rainfall (mm)
John Bichila	Kasulu	Titye	030 17′281″E	04 40′756″S	1148	28.70	18.70	1014.80
Fabian Ntalumanga	Kasulu	Titye	030 17′596″E	04 40′185″S	1138	29.30	19.10	1026.60
Selina Dulubaye	Kasulu	Titye	030 17′584″E	04 40′163″S	1139	29.10	18.10	1180.90
Geralid Bilaro	Kasulu	Nyenge	030 15′786″E	04 37′080″S	1139	29.80	18.30	1025.90
Elias Kapisi	Kasulu	Nyenge	030 15′782″E	04 37′096″S	1140	29.70	18.00	1138.20
Margerth Mussa	Kasulu	Nyenge	030 15′783″E	04 37′046″S	1145	29.60	16.00	1039.90
Koladi Mussa	Kasulu	Kanazi	030 11′728″E	04 31′497″S	1236	29.30	15.70	1204.40
Mohamed Issa	Kasulu	Kanazi	030 13′371″E	04 33′981″S	1231	30.70	16.70	1126.40
Japhet Kipara	Kasulu	Kanazi	030 13′423″E	04 33′669″S	1257	31.30	18.90	1028.20

TABLE 2: Average yield of beans varieties for three seasons (2003–2006).

Variety	Seasons			
	2003/2004	2004/2005	2005/2006	Mean
Lyamungo 90	1394	1427	1469	1430.00
Jesca	1326	1334	1317	1325.67
Uyole 94	880	870	857	869.00
Kablanketi	924	924	878	908.67
Kigoma yellow	684	732	728	714.67
Grand mean	1041	1058	1050	1049.60
SED	48.2	49.2	50.3	
CV	9.8	9.9	10.2	

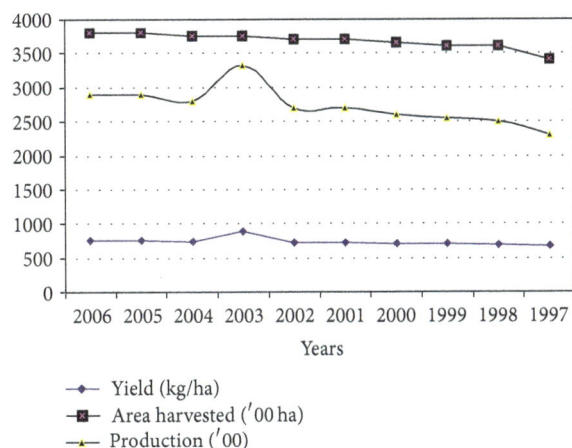

FIGURE 1: Yield of beans and proportion area in Tanzania (1997–2006). Source: (FAOSTAT, 2008).

introducing new bean varieties to farmers, on-farm bean trials were conducted in Kigoma region, Tanzania. Objectives were to evaluate bean varieties for yield under farmers' management conditions and to assess farmers' preferences on bean varieties.

2. Materials and Methods

Common beans on-station trials were conducted in Kigoma region, Tanzania, from 2003 to 2006 seasons. Researcher-managed trials (same managements) were planted in three villages with three farmers per village. Villages included Kanazi, Nyenge, and Titye (Table 1).

A randomized complete block design (RCBD) with five plots per replications was used to evaluate five beans varieties: Lyamungo 90, Jesca, Uyole 94, Kablanketi, and Kigoma yellow (Control). Beans were planted on a 10 m × 2.5 m plot at a spacing of 50 cm × 20 cm. Data was subjected to analysis using ANOVA table in GenStat statistical computer software. Stability analysis was employed to evaluate varieties sensitivity in different farmers' fields. Three villages × three seasons resulted in nine environments which were used for stability analysis. To assess farmers, preferences on beans cultivars, participating farmers gathered together and jointly select-ed ten criteria for ranking the cultivars: large seed size, good taste, short cooking time, early maturity, high market demand, high yielding, disease resistance, insect-pest resistance, cooking quality, and suitability to short rainfall farming system (njera season). A scale of 1–5 was used to assess these traits with the definition as follows: 1 = not preferred, 2 = less preferred, 3 = moderately preferred, 4 = highly preferred, and 5 = excellent. Farmers were given 5 grains and asked to place 1, 2, 3, 4, or 5 grains to score a given trait and cultivar. Seeds were counted, and the largest total count was ranked first.

3. Results

There were high significant yield differences among varieties across three seasons ($P < 0.001$) with Lyamungo 90 and Jesca outyielding other varieties (Table 2). Lyamungo 90 and Jesca recorded an average yield of 1430.00 and 1325.67 kg ha^{-1},

TABLE 3: AMMI analysis of variances of five bean varieties across nine environments.

SOV	DF	SS	MS	F	F pr	SS or GE × SS%
Blocks	18	776278	43127	4.48	<0.001	6.65
Genotypes	4	10409365	2602341	270.17	<0.001	89.12
Environment	8	89684	11211	0.26	0.97661	0.77
G × E	32	404514	12641	1.31	0.17020	3.46
AMMI model						
IPCA1	11	266533	24230	2.52	0.00969	65.89
IPCA2	9	80958	8995	0.93	0.50149	20.01
IPCA3	7	51734	7391	0.77	0.61648	12.79
Residual	5	5289	1058	0.11	0.98983	1.31
Total treat	44	10903563	247808	25.73	<0.001	
Error	72	693518	9632			
Total	134	2373359	92338			

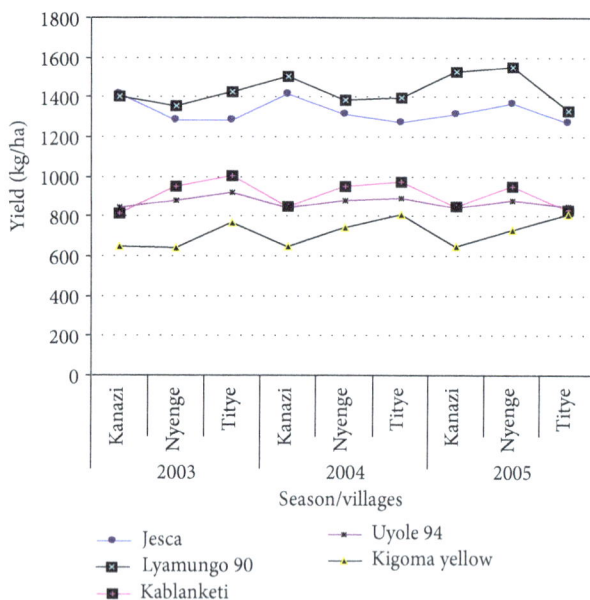

FIGURE 2: On-farm performance of five bean varieties across three villages for three seasons (2003–2006).

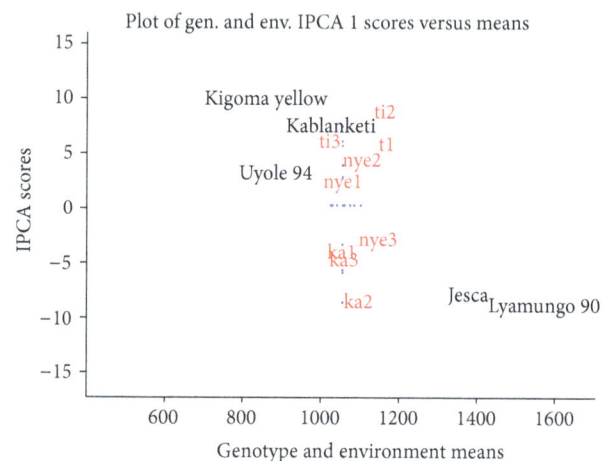

FIGURE 3: IPCA1 scores of five bean varieties, nine environments, and genotype × environment means. Key: t1, 2, and 3 = Titye for seasons 1, 2, and 3; ka1, 2, and 3 = kanazi for seasons 1, 2, and 3; nye1, 2, and 3 = Nyenge for season 1, 2, and 3.

respectively. The overall average yield for all varieties was 1049.60 kg ha^{-1}.

Results also revealed that the local check (Kigoma yellow) recorded the lowest average yield of 714.67 kg ha^{-1}.

Figure 2 elaborates the performance of five bean varieties grown in three villages for three seasons. Similar beans yield trends as that of Table 2 were observed.

When additive main effects and multiplicative interactions (AMMI) analysis of five bean varieties across nine environments was performed (Table 3), the IPCA1 was found to be significant. However, IPCA2 and IPCA3 were not significant.

The genotypes sum of squares (SS) accounted for the most of the variability (89.12%). In addition, results further showed that G × E interactions were superior to environment effects. However, G × E interactions were not significant to denote less importance of its joint effect. Figure 3 elaborates

interaction principal component analysis (IPCA) scores against genotypes and environment means.

Figure 3 showed varieties being more dispersed than environments. From this figure, Lyamungo 90 and Jesca bean varieties were allocated on the high yield environments while Kigoma yellow (local check) was on the lower yield environments. Lyamungo 90 and Jesca bean varieties were placed far from zero which is the indication of variety instability. Uyole 94 showed fairly high G × E stability.

Tables 4 and 5 show farmers' assessment on five beans across three seasons.

By using 10 beans traits developed by them, farmers ranked fourth Kigoma yellow (local check) while Lyamungo 90 and Jesca were ranked first and second, respectively. The developed beans traits for assessment were included.

4. Discussions

On-farm research on beans revealed significant differences ($P < 0.001$) among varieties with Kigoma yellow recording

TABLE 4: Farmers on-farm evaluation of five bean varieties for three seasons (2003–2006).

Variety	Seasons			Ranks
	2003/2004	2004/2005	2005/2006	
Lyamungo 90	1	2	1	1
Jesca	2	3	2	2
Uyole 94	5	3	5	5
Kablanketi	3	1	3	3
Kigoma yellow	4	4	4	4

Key: ranks: 1 = highly preferred, 5 = least preferred.

TABLE 5: Farmers evaluation criteria of five bean varieties at the end of three years project, 2006.

Variety	Criteria										Total	Rank
	Se	Ta	Co	Ma	Mark	Yield	Disease	Pest	C	S		
Lyamu	4	4	5	4	3	4	4	3	2	3	36	1
Jesca	3	3	4	2	3	5	3	3	5	3	34	2
Uyole	1	1	2	3	2	2	2	2	2	2	19	5
Kablan	5	5	3	2	4	1	2	3	3	2	30	3
Kigo	3	2	3	1	2	1	3	3	3	4	25	4

Key: Se = large seed size, Ta = good taste, Co = short cooking time, Ma = early maturity, Mark = high market demand, yield = high yielding, disease = disease resistance, pest = insect-pest resistance, C = cooking quality, and S = suitability to short rainfall farming system. Varieties: Lyamu = Lyamungo 90, Jesca, Uyole = Uyole 94, Kablan = Kablanketi, and Kigo = Kigoma yellow.
Scores: 5 = highly preferred, 1 = least preferred.

the lowest yield. These results clearly indicate the superiority of the introduced varieties over the local variety in the area. The recorded yield was above the national beans yield of 741 kg ha^{-1} [4]. The observed yield was in accordance to Mekbib [15] who recorded the yield range of 1511–2216 kg ha^{-1}. The high yielding varieties which were accompanied by farmers' preferences on the new varieties observed in this study suggest the possibility of increased adoption rate in the area. Farmers tend to adopt new technologies that fit their selection criteria [12]. Additionally, high yielding varieties have an added advantage of improving food security and raising farmers' income through reduction of uncertainties and unexpected crop failures provided they are accepted by the community [16].

Farmers' assessment on varieties revealed farmers' power on varieties evaluation and selection. Farmers ranked high Lyamungo 90 and Jesca bean varieties. The reasons given by farmers were high yielding and short cooking time. To improve beans breeding program in Tanzania, these traits are expected to be incorporated in the breeding processes. Statistical analysis ranked the fifth locally adapted variety (Kigoma yellow).

Farmers differed with researchers by ranking it fourth according to their developed criteria. Kigoma yellow was ranked fourth instead of being last because it fits Kigoma farming systems of long and short cropping seasons. It can be planted in both seasons. This variety had adapted Kigoma conditions and thus became suitable for two production seasons. Farmers plant this variety for provision of an added food security to the families [17]. Varieties evaluation and selection which use farmers criteria to meet specific objectives have been reported by other researchers [18, 19].

Stability analysis revealed that yield variations were mostly contributed by genotypes (89.12%). These findings support observations by Jutsum et al. [20]. However, González et al. [21] recorded high environmental variation contribution than genotypes effects. The plot of the interaction principal components analysis against the genotypes and environment means enabled the visual comparison of environments, genotypes, and their interactions. This plot revealed Lyamungo 90 and Jesca as unstable varieties. In addition, these varieties recorded the highest yields across environments. Uyole 94 showed relatively high G × E interaction stability. The reason could be that Uyole 94 was developed at Uyole Agricultural Centre to suite highland environments as that of Kigoma region. Although this variety obtained the lowest rank from farmers' assessment, its stability and yielding relatively high than Kigoma yellow could be a better variety for risk avoidance against production uncertainties [22, 23].

5. Conclusion

To speed up variety introduction to farmers, on-farm beans experiments were conducted in Kigoma region, Tanzania. Lyamungo 90 and Jesca recorded the highest yield across the environments. Farmers ranked them high which was attributed to high yielding and short cooking time. Introduction of high yielding bean varieties with the desired farmers' traits could revamp beans production and could contribute to the improved food security in the region. Information gathered by this study can be utilized by plant breeders and farmers for incorporation of farmers preferred traits into beans breeding program and bean farming systems.

References

[1] T. Dzudie, J. Scher, and J. Hardy, "Common bean flour as an extender in beef sausages," *Journal of Food Engineering*, vol. 52, no. 2, pp. 143–147, 2002.

[2] B. A. Medvecky, Q. M. Ketterings, and F. M. Vermeylen, "Bean seedling damage by root-feeding grubs (*Schizonycha* spp.) in Kenya as influenced by planting time, variety, and crop residue management," *Applied Soil Ecology*, vol. 34, no. 2-3, pp. 240–249, 2006.

[3] A. Krouma, J. J. Drevon, and C. Abdelly, "Genotypic variation of N_2-fixing common bean (*Phaseolus vulgaris* L.) in response to iron deficiency," *Journal of Plant Physiology*, vol. 163, no. 11, pp. 1094–1100, 2006.

[4] FAO, "Production data," 2008, http://faostat.fao.org/faostat.

[5] P. H. Graham and P. Ranalli, "Common bean (*Phaseolus vulgaris* L.)," *Field Crops Research*, vol. 53, no. 1–3, pp. 131–146, 1997.

[6] BACAS, *Bureau for Agriculture Consultancy and Advisory Services.: Final Report. Baseline Survey on the Agricultural Research System Under the Department of Research and Training, Volume 1 Western Zone. Synthesis of Main Findings and Recommendations*, Sokoine University of Agriculture, Morogoro, Tanzania, 2000.

[7] C. R. Doss, W. Mwangi, H. Verkuijl, and H. deGroote, "Adoption of maize and wheat technologies in eastern Africa: a synthesis of the findings of 22 case studies," Working Paper 03-06, CIMMYT Economics, Sonora, Mexico, 2003.

[8] J. Ouma, H. DeGroote, and M. Gethi, "Focused participatory rural appraisal of farmer's perceptions of maize varieties and production constraints in the moist transitional zone in Eastern Kenya," Economic Working Paper 02-01, CIMMYT and KARI, Nairobi, Kenya, 2002.

[9] J. R. Witcombe, A. Joshi, and S. N. Goyal, "Participatory plant breeding in maize: a case study from Gujarat, India," *Euphytica*, vol. 130, no. 3, pp. 413–422, 2003.

[10] T. Assefa, G. Abebe, C. Fininsa, B. Tesso, and A.-R. M. Al-Tawaha, "Participatory bean breeding with women and small holder farmers in eastern Ethiopia," *World Journal of Agricultural Sciences*, vol. 1, pp. 28–35, 2005.

[11] D. L. Romney, P. Thorne, B. Lukuyu, and P. K. Thornton, "Maize as food and feed in intensive smallholder systems: management options for improved integration in mixed farming systems of east and southern Africa," *Field Crops Research*, vol. 84, no. 1-2, pp. 159–168, 2003.

[12] C. K. Kaizzi, H. Ssali, and P. L. G. Vlek, "Differential use and benefits of Velvet bean (*Mucuna pruriens* var. *utilis*) and N fertilizers in maize production in contrasting agro-ecological zones of E. Uganda," *Agricultural Systems*, vol. 88, no. 1, pp. 44–60, 2006.

[13] C. M. Moser and C. B. Barrett, "The disappointing adoption dynamics of a yield-increasing, low external-input technology: the case of SRI in Madagascar," *Agricultural Systems*, vol. 76, no. 3, pp. 1085–1100, 2003.

[14] J. Gressel, A. Hanafi, G. Head et al., "Major heretofore intractable biotic constraints to African food security that may be amenable to novel biotechnological solutions," *Crop Protection*, vol. 23, no. 8, pp. 661–689, 2004.

[15] F. Mekbib, "Simultaneous selection for high yield and stability in common bean (*Phaseolus vulgaris*) genotypes," *The Journal of Agricultural Science*, vol. 138, no. 3, pp. 249–253, 2002.

[16] Z. R. Khan, D. M. Amudavi, C. A. O. Midega, J. M. Wanyama, and J. A. Pickett, "Farmers' perceptions of a 'push-pull' technology for control of cereal stemborers and Striga weed in western Kenya," *Crop Protection*, vol. 27, no. 6, pp. 976–987, 2008.

[17] R. J. Hillocks, C. S. Madata, R. Chirwa, E. M. Minja, and S. Msolla, "Phaseolus bean improvement in Tanzania, 1959–2005," *Euphytica*, vol. 150, no. 1-2, pp. 215–231, 2006.

[18] G. Abebe, T. Assefa, H. Harrun, T. Mesfine, and A. M. Al-Tawaha, "Participatory selection of drought tolerant maize varieties using mother and baby methodology: a case study in the semi arid zones of the central rift valley of Ethiopia," *World Journal of Agricultural Sciences*, vol. 1, pp. 22–27, 2005.

[19] D. Soleri, S. E. Smith, and D. A. Cleveland, "Evaluating the potential for farmer and plant breeder collaboration: a case study of farmer maize selection in Oaxaca, Mexico," *Euphytica*, vol. 116, no. 1, pp. 41–57, 2000.

[20] A. R. Jutsum, J. M. Franz, J. W. Deacon et al., "Commercial application of biological control: status and prospects [and discussion]," *Philosophical Transactions of the Royal Society of London Series B, Biological Sciences*, vol. 318, pp. 357–373, 1988.

[21] A. M. González, A. B. Monteagudo, P. A. Casquero, A. M. de Ron, and M. Santalla, "Genetic variation and environmental effects on agronomical and commercial quality traits in the main European market classes of dry bean," *Field Crops Research*, vol. 95, no. 2-3, pp. 336–347, 2006.

[22] A. J. McDonald, P. R. Hobbs, and S. J. Riha, "Does the system of rice intensification outperform conventional best management? A synopsis of the empirical record," *Field Crops Research*, vol. 96, no. 1, pp. 31–36, 2006.

[23] J. D. Reece and J. Sumberg, "More clients, less resources: toward a new conceptual framework for agricultural research in marginal areas," *Technovation*, vol. 23, no. 5, pp. 409–421, 2003.

10

Effect of Heat Moisture Treatment Conditions on Swelling Power and Water Soluble Index of Different Cultivars of Sweet Potato (*Ipomea batatas* (L). Lam) Starch

Suraji Senanayake,[1] **Anil Gunaratne,**[2] **KKDS Ranaweera,**[1] **and Arthur Bamunuarachchi**[3]

[1] *Department of Food Science & Technology, University of Sri Jayewardenepura, Sri Lanka*
[2] *Faculty of Agricultural Sciences, Sabaragamuwa University of Sri Lanka, Belihuloya, Sri Lanka*
[3] *"ON—SITE" Consultancy, Training & Trade Systems, 128/22, Poorwarama Road, Kirulapone, 5 Colombo, Sri Lanka*

Correspondence should be addressed to Suraji Senanayake; surajisena@gmail.com

Academic Editors: J. B. Alvarez and C. Ramsey

A study was done to analyse the change in swelling power (SP) and the water soluble index (WSI) of native starches obtained from five different cultivars of sweet potatoes (swp 1 (Wariyapola red), swp 3 (Wariyapola white), swp 4 (Pallepola variety), swp 5 (Malaysian variety), and swp 7 (CARI 273)) commonly consumed in Sri Lanka. Extracted starch from fresh roots, two to three days after harvesting has been modified using 20%, 25%, and 30% moisture levels and heated at 85°C and 120°C for 6 hours and determined the SP and WSI. Results were subjected to general linear model, and analysis of variance (ANOVA) was carried out by using MINITAB version 14. Overall results showed a significantly high level ($P < 0.05$) of SP and WSI in all the cultivars of moisture—temperature treated starches than their native starch. Correlation analysis showed an effect on SP with the variation in the cultivar, temperature, and moisture; temperature combination and moisture alone had no significant effect. Significantly high levels of swelling power ($P > 0.05$) were observed in 20%—85°C, and 30%—120°C and the highest amount of swelling in the modified starch than its native form was observed in swp 7 cultivar. Results revealed a nonlinear relationship in the WSI with the cultivar type, moisture level, and the lower moisture—temperature combinations but higher temperature—moisture combinations had a significant effect. SP and WSI had a slight positive linear relationship according to analysis. Based on the results, a significantly high level of swelling and water solubility of native starches of different cultivars of sweet potatoes can be achieved by changing the moisture content to 30% and heating at 120°C for 6 hours.

1. Introduction

For a wide range of starch applications, native starches cannot be used due to inability to bring out the desired properties. Native starches can be modified to obtain the desired qualities by starch modification methods. Chemical modification of starch molecules is commonly used in achieving the desired properties. Also by using specific moisture and temperature conditions, some physicochemical properties of starch can be altered. There are more trends in the world for physical modification of starch which is used in food industrial applications as there is an increasing difficulty in obtaining regulatory approval of the new chemical reagents and higher levels of treatment as described by BeMiller [1]. Since most physical

modifications involve only water and heat, these hydrothermal treatments are considered to be natural and safe materials by Jacobs and Delcour [2].

Two basic types of heat-moisture treatments are commonly employed in modifying the physicochemical properties of starch [3]. This involves the storage of starch at specific levels of moisture and heat for a specific period of time without causing a significant level of starch gelatinization. Treatment of starch with excess moisture is referred to as "annealing" [4–6], and the term "heat-moisture treatment" (HMT) is used when restricted levels of moisture are applied in the literature [7, 8]. These two types of physical modifications occur at temperatures above the glass transition temperatures of the relevant starch and often below the gelatinization temperatures which depend on the specific moisture contents used for the treatment. Heat-moisture treatment effect on changing the functional properties of wheat, maize, potato, barley, cassava, yam, and legume starches were reported by many researchers [8–11]. It was observed that the HMT has increased the gelatinization temperature, enzymatic susceptibility, solubility, swelling volume, and changes in the X-ray diffraction patterns. The changes in these parameters vary depending on the source of the starch and the HMT conditions.

Temperature and moisture conditions are often selected without considering the exact gelatinization temperature of the starch at that particular moisture level, and the observed results on HMT may have been affected by the partial gelatinization of starch [12]. Not much work was done or reported on the effects of HMT at different combinations of temperature-moisture levels and treatment times on the properties of sweet potato starch. This study was done to observe the effect of different heat-moisture conditions on the swelling power and the water soluble index of five different cultivars of sweet potato (*Ipomea batatas* (L). Lam) starch.

2. Materials and Methods

Matured tubers of sweet potatoes, namely, swp 1 (Wariyapola red), swp 5 (Malaysian variety), swp 7 (CARI 273), swp 3 (Wariyapola white), and swp 4 (Pallepola variety) were collected from different areas in Dambulla, Gokarella, and Horana regions, Sri Lanka. Random samples selected from market areas have been identified at the Horticultural Crop Research and Development Institute, Gannoruwa, Kandy, Sri Lanka. Starch samples were prepared two to three days after harvesting.

2.1. Starch Extraction. Starch separation was carried out according to the method described by Takeda et al. [13] with slight modifications. Fresh tubers were washed, peeled, diced, and dipped in ice water containing 100 ppm sodium metabisulphite to minimize browning. Diced sample was wet milled at low speed in a laboratory scale blender with 1 : 2 w/v of tap water for 2 minutes and filtered through a gauze cloth. Residue was repeatedly wet milled and filtered for thrice, and suspension was kept overnight for settling of starch. The supernatant was decanted, and the settled residue was further purified with repeated suspension in tap water (1 : 2 v/v) followed by the settling for 3 hours. The purified starch was dried at 35°C, sifted through 300 μm sieve, sealed, and packed for analysis.

2.2. Heat-Moisture Treatment (HMT). Starch samples (20 g) were adjusted to 20%, 25%, and 30% moisture levels and placed in tubes with a sealing cap and equilibrated at room temperature for 12 hours. Samples were heated at 85°C and 120°C for 6 hours. Occasional shaking was done to samples within the treatment period for homogeneous distribution of moisture. After treatment, the samples were cooled to room temperature and dried at 40°C to a uniform moisture level of 10% and equilibrated at room temperature for 2 days.

2.3. Swelling Power (SP) and Water Soluble Index (WSI). Swelling power (SP) of the native and the HMT treated starch was determined according to the method of Gunaratne et al. [14]. Starch (100 mg, db) was weighed directly into a screw-cap test tube, and 10 mL distilled water was added. The capped tubes were placed on a vortex mixer for 10 seconds and incubated at 85°C water bath for 30 minutes with frequent mixing. The tubes were cooled to room temperature in an iced water bath and centrifuged at 2000 ×g for 30 minutes; the supernatant was removed, and the remaining sediment in the tube was weighed (W_s). The supernatant was dried to constant weight (W_1) in a drying oven at 100°C. The water swelling power was calculated as follows:

$$SP = \frac{W_s}{[0.1 \times (100\% - WSI)]} \, (g/g), \qquad (1)$$

where WSI = $W_1/0.1 \times 100\%$.

2.4. Statistical Analysis. Results were subjected to general linear model, and analysis of variance (ANOVA) and correlation analysis for this study were carried out by using MINITAB version 14.

3. Results and Discussion

Results showed a significantly high level ($P < 0.05$) of swelling power and water soluble index in all the cultivars of HMT starch than their native starch (Tables 1 and 2). Also the results indicated a significant effect of cultivar, temperature, and moisture-temperature combination on the variation of SP in HMT starch and also showed no effect from the used moisture contents ($P < 0.05$). Correlation analysis showed a significant level of influence on SP by the moisture-temperature combination (Figure 1).

Lowest level of SP was found in the native starch of Swp 7 and comparatively high level of increase in SP was observed after the heat-moisture treatment, than the other cultivars. High level of swelling in all the cultivars of HMT starch than the native starch may be due to the amylose which complex with lipids in starch granules that are in helical form may change its physical form due to HMT and becomes more amorphous so that it can readily combine with moisture and increase swelling. But in most recent studies by researchers [8, 15] revealed a decrease in swelling with varying levels of HMT.

There is no significant difference ($P > 0.05$) in SP in the moisture-temperature combinations, 20%, 85°C, 20%—85°C, 20%—120°C, and 25%—120°C. Combinations of 25%—85°C, and 30%—85°C—had the lowest SP at the level $P > 0.05$, and significantly highest level of SP was observed in the 30%—120°C—combination. From the total correlation analysis of moisture-temperature combination, significantly high levels of SP ($P > 0.05$) were observed in 20%—85°C, and 30%—120°C. Therefore, 20%—85°C—can be used economically in increasing the swelling power of native starch.

Correlation analysis revealed there is no significant effect ($P < 0.05$) on WSI from the cultivar type, moisture level, and the moisture-temperature combination (Figure 2) but temperature affects the WSI significantly. Therefore, all moisture levels with 120°C temperature showed higher WSI than

Effect of Heat Moisture Treatment Conditions on Swelling Power and Water Soluble Index of Different Cultivars
of Sweet Potato (Ipomea batatas (L). Lam) Starch

89

TABLE 1: Changes in swelling power (g/g) of different sweet potato starches after 6 hr HMT.

		Swp 1	Swp 3	Swp 4	Swp 5	Swp 7
Native starch		7.9 ± 0.1^e	8.7 ± 0.2^d	8.7 ± 0.1^d	8.0 ± 0.1^e	5.8 ± 0.1^d
Temperature	Moisture					
	20%	11.5 ± 1.2^d	12.0 ± 0.9^a	12.5 ± 0.8^a	12.1 ± 0.2^c	8.5 ± 0.7^c
85°C	25%	12.9 ± 0.5^d	$10.5 \pm 0.5^{b,c}$	11.4 ± 0.2^b	11.0 ± 0.6^c	5.8 ± 0.3^d
	30%	$13.9 \pm 0.2^{a,c}$	10.1 ± 0.3^b	10.3 ± 0.4^c	9.7 ± 0.7^d	5.4 ± 0.2^d
	20%	15.4 ± 0.3^a	10.5 ± 0.3^b	12.6 ± 0.2^a	11.8 ± 0.1^c	10.0 ± 0.2^b
120°C	25%	$15.7 \pm 0.2^{a,b}$	$11.6 \pm 0.1^{a,b}$	$11.7 \pm 0.2^{a,b}$	13.1 ± 0.1^b	10.3 ± 0.1^b
	30%	$14.3 \pm 0.2^{a,b,c}$	$10.1 \pm 0.1^{b,c}$	12.8 ± 0.1^a	15.1 ± 0.1^a	12.1 ± 0.1^a

Values represented by different superscripts in each column are different at $P > 0.05$ level.

TABLE 2: Variation of water soluble index (%) in different sweet potato starches after 6 hr HMT.

		Swp 1	Swp 3	Swp 4	Swp 5	Swp 7
Native starch		1.6 ± 0.1^d	1.8 ± 0.1^f	1.3 ± 0.1^e	2.1 ± 0.1^e	0.5 ± 0.1^d
Temperature	Moisture					
	20%	3.9 ± 0.2^c	3.2 ± 0.3^e	4.2 ± 0.3^d	4.1 ± 0.3^d	2.8 ± 0.3^c
85°C	25%	3.3 ± 0.4^c	4.1 ± 0.4^d	$5.0 \pm 0.1^{b,c}$	2.3 ± 0.6^e	3.5 ± 0.5^c
	30%	$8.6 \pm 0.6^{a,b}$	5.6 ± 0.2^c	6.1 ± 0.3^f	4.1 ± 0.3^d	2.4 ± 0.2^c
	20%	9.2 ± 0.2^a	8.5 ± 0.5^b	4.7 ± 0.2^c	6.6 ± 0.2^c	7.8 ± 0.1^b
120°C	25%	9.0 ± 0.2^a	9.6 ± 0.1^a	5.5 ± 0.2^b	11.5 ± 0.3^a	8.2 ± 0.1^b
	30%	8.1 ± 0.1^b	8.5 ± 0.1^b	7.7 ± 0.2^a	9.1 ± 0.2^b	13.5 ± 0.3^a

Values represented by different superscripts in each column are different at $P > 0.05$ level.

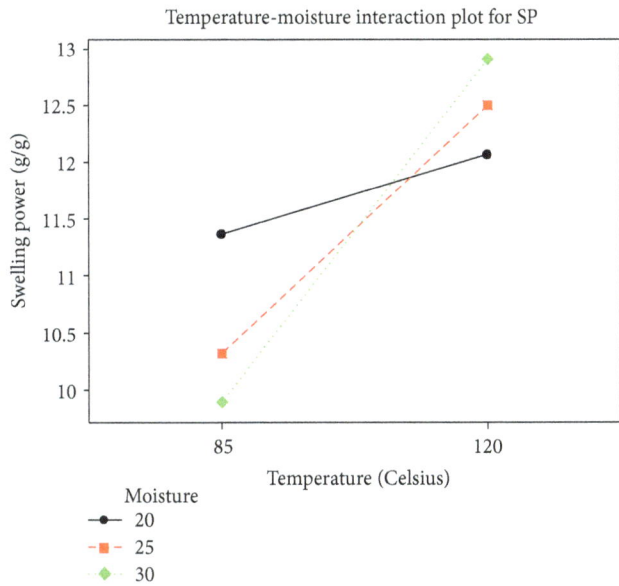

FIGURE 1: Temperature-moisture combined effect on swelling power (SP) of different starches.

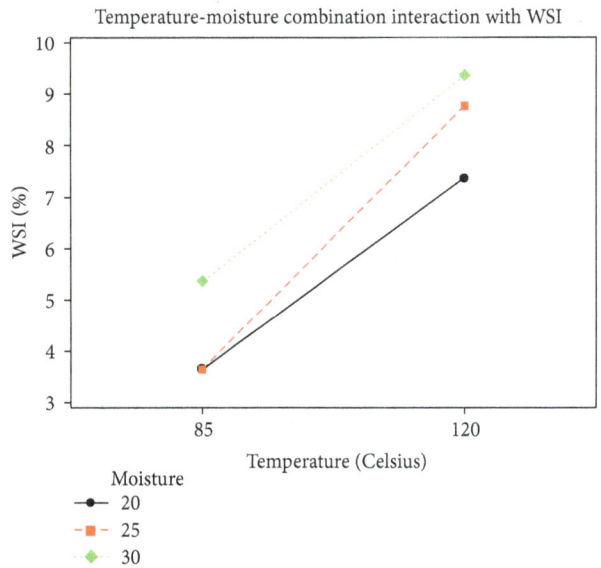

FIGURE 2: Temperature-moisture combined effect on WSI of different starches.

the combinations at 85°C. From all the combinations 30% moisture and 120°C had the highest level of WSI (Table 2).

Previous studies indicate different trends of change in solubility or WSI in various starches due to HMT [8, 15]. This may be also due to the fact that physical change in starch granules due to heat-moisture treatment. Our results indicate that the higher temperature used for HMT had more

WSI than the lower temperature used. Higher solubility of starch may be due to the degradation of starch at higher temperatures. When considering the relationship between SP and WSI, there is a slight positive linear correlation (Pearson correlation coefficient 0.489). Analysis also indicated that there is no significant effect ($P < 0.05$) on the SP and WSI relationship from the cultivar type and moisture but slight effect from the temperature.

Overall results for HMT of native starch for 6 hrs have shown increased levels of SP and WSI than the native starch, and it is recommended to carry out the observations for different time periods with more variations in moisture-temperature combinations to get a more conclusive idea in this area.

4. Conclusion

The application of HMT on native starch showed increased levels of swelling and solubility thus obtained higher values for SP and WSI than their native counterparts. Results indicated a significant level of effect from moisture-temperature combination on SP and increase in WSI due to high level of temperature. Overall results showed the highest level of SP and WSI in moisture-temperature combination of 30%—120°C, and swp 7 cultivar had the highest level of increase in SP and WSI than its native form compare to other cultivars.

References

[1] J. N. BeMiller, "Starch modification: challenges and prospects," Starch, vol. 49, no. 4, pp. 127–131, 1997.

[2] H. Jacobs and J. A. Delcour, "Hydrothermal modification of granular starch, with retention of the granular structure: a review," Journal of Agricultural and Food Chemistry, vol. 46, no. 8, 1998.

[3] R. Stute, "Hydrothermal modification of starches: the difference between annealing and heat—moisture treatment," Starch, vol. 44, no. 6, pp. 205–214, 1992.

[4] H. Jacobs, R. C. Earlingen, S. Clauwaert, and J. A. Delcour, "Influence of annealing on the pasting properties of starches from varying botanical sources," Cereal Chemistry, vol. 72, pp. 480–4487, 1995.

[5] C. A. Knutson, "Annealing of maize starches at elevated temperatures," Cereal Chemistry, vol. 67, pp. 376–384, 1990.

[6] I. Larsson and A. C. Eliasson, "Annealing of starch at intermediate water content," Starch, vol. 43, no. 6, pp. 227–2231, 1991.

[7] L. S. Collado and H. Corke, "Heat-moisture treatment effects on sweet potato," Food Chemistry, vol. 65, no. 3, pp. 339–346, 1999.

[8] R. Hoover, T. Vasanthan, N. J. Senanayake, and A. M. Martin, "The effects of defatting and heat-moisture treatment on the retrogradation of starch gels from wheat, oat, potato, and lentil," Carbohydrate Research, vol. 261, no. 1, pp. 13–24, 1994.

[9] T. E. Abraham, "Stabilization of paste viscosity of cassava by heat moisture treatment," Starch, vol. 45, pp. 131–1135, 1993.

[10] J. W. Donovan, K. Lorenz, and K. Kulp, "Differential scanning calorimetry of heat—moisture treated wheat and potato starches," Cereal Chemistry, vol. 60, pp. 381–3387, 1983.

[11] R. Hoover and T. Vasanthan, "Effect of heat-moisture treatment on the structure and physicochemical properties of cereal, legume, and tuber starches," Carbohydrate Research, vol. 252, pp. 33–53, 1994.

[12] R. C. Eerlingen, H. Jacobs, H. Van Win, and J. A. Delcour, "Effect of hydrothermal treatment on the gelatinisation properties of potato starch as measured by differential scanning calorimetry," Journal of Thermal Analysis, vol. 47, no. 5, pp. 1229–1246, 1996.

[13] Y. Takeda, A. Suzuki, and S. Hizukuri, "Influence of steeping conditions for kernels on some properties of corn starch," Starch, vol. 40, no. 4, pp. 132–1135, 1988.

[14] A. Gunaratne, A. Bentota, Y. Z. Cai, L. Collado, and H. Corke, "Functional, digestibility, and antioxidant properties of brown and polished rice flour from traditional and new-improved varieties grown in Sri Lanka," Starch, vol. 63, no. 8, pp. 485–492, 2011.

[15] L. Sair, "Heat—moisture treatment of starches," Cereal Chemistry, vol. 44, pp. 8–26, 1967.

11

Impact of Time of Weeding on Tobacco (*Nicotiana tabacum*) Growth and Yield

Mashezha Ian,[1] **Rukuni Dzingai,**[2] **Manyangarirwa Walter,**[1] **and Svotwa Ezekia**[2]

[1] *Africa University, P.O. Box 1320, Mutare, Zimbabwe*
[2] *Tobacco Research Board, Kutsaga Farm, P.O. Box 1909, Harare, Zimbabwe*

Correspondence should be addressed to Mashezha Ian; ian.mashezha@gmail.com

Academic Editors: G. M. Dal Bello, C. H. Kao, and M.-J. Simard

An experiment laid in a Randomized Complete Block Design with 4 blocks and 5 treatments, was done at Kutsaga Research Station in the 2012 / 2013 season to study the impact of time of weeding on tobacco production. The treatments comprised of different times of weed control with a weed free treatment as the control. The variables measured were stalk heights at 5, 6 and 7 weeks after planting and, leaf expansion measurements were also recorded at 9, 10 and 11 weeks. Leaf yield was measured at untying using a digital scale. Results showed that Tobacco stalk heights were affected at 5 W.A.P since significant differences ($P < 0.05$) were noted among the treatments. Suppressive effects of weeds were shown at 6 and 7 W.A.P due to effective competition (RCI > 0) in all other treatments excluding the control. There were significant differences ($P < 0.05$) among the treatments on leaf expansion at 9, 10 and 11 W.A.P. The treatment weeded at 4 W.A.P showed leaf yield that was significantly higher (L.S.S = 270.8) than the treatment weeded at 2 W.A.P. Basing on the 3 reaps recorded, time of weeding had an influence tobacco yield.

1. Introduction

Yield of tobacco depends on the implementation of good agronomic practices and good management of insect pests and weeds [1]. Tobacco yield loss due to weed competition is the most important factor that causes yield and quality reduction. Weed infestation of broad-leaved grasses and sedges in tobacco growing areas of Zimbabwe has reduced the primary industry productivity and profitability, and seriously limits the long-term sustainability of the crop [2]. Weeds compete directly with the desired crop for nutrients, water, light, carbon dioxide, and oxygen, and this competition significantly reduces crop yield and quality [3]. Weed suppressive impacts on tobacco are becoming a huge problem for the large commercial famers and small scale farmers. This can be largely attributed to the presence of competitive weeds and lack of capital to improve technologies of controlling them [4].

Wilson [5] estimated a 77% yield reduction and 10% quality reduction due to weed infestation in tobacco at various unknown densities of various weed species. Niell [6] also reported 12% loss due to vigorous weeds at 2-week growth stage and 64% yield loss when weeds were left in the critical weed free period of 2–6 weeks. Tobacco production at low input costs of weed management has been reported by Klingman [7] who observed a significant low gross return per unit area of the land than tobacco at high input cost of weed control. If weed management is not improved, tobacco production will not be realized, and farming in general will continue to be full of drudgery and unattractive to the younger generations [8].

Delay in weed management of aggressive weeds such as *Acathospermum hispidum*, *Amaranthus hybridus*, *Cyperus rotundus*, and *Cyperus esculentus* results in difficulties in controlling them. Their distribution in all is the primary cause of tobacco yield loss in small-holder agriculture [3]. Inevitably, the time of first weeding is spread from one to six weeks after transplanting tobacco, and the weeding operation is done once or twice but rarely three times in most field crop operations [9]. In small-holder farming environments, labor shortages are commonly encountered during the onset of the rainy season. This delay in weed control exposes the tobacco crop to weed competition effects. Effective weed control practices in such situations will require development of appropriate techniques and judicious weeding to cater for labor constraints [9]. High population densities due to delays in weed management will enhance the weed species with higher competitive ability. The inability to accurately predict when weeds first begin to affect yield is creating high levels of risk when relying on scheduled weed control programs [10].

This research work seeks to establish the optimum growth stage at which the tobacco is weeded for maximum growth, development, and yield. It also sought to establish the impact of delayed weeding on the weed densities in tobacco lands. It was hypothesized in this research that an appropriate timing of weed removal for maximum crop growth, development, and yield can be established. The generated information can guide practitioners to minimize the frequency of weeding, thereby reducing the cost of production without compromising yield and quality of flue cured tobacco.

2. Materials and Methods

The study was carried out at Kutsaga Research Station during the 2012/2013 season. The research station is in Natural Region II receiving annual rainfall of 800 mm to 1000 mm per annum. The rainfall occurs during a single rainy season from November to April [11]. The experiment was done in open field. The soils at the Research Station are sandy loams derived from granite [12]. The mean annual temperature is 21°C with insignificant frost occurrence in the months of June and July.

The land was ploughed using a tractor drawn plough to a depth of 38 cm. Ridging and fertilizer applications were done on October 20, 2012. Compound C (6N : 15P : 12K) was applied at a rate of 700 kg/ha, based on soil analysis. The nematicide ethylene dibromide (EDB) was also applied at a rate of 13 litres per hectare, two weeks before planting. Tobacco seedlings of T66 variety were used. This is a slow ripening cultivar with a yield potential of more than 3500 kg/ha cured leaf at Kutsaga. The crop was transplanted on October 25, 2012.

There were a total of 5 treatments in the trial shown in Table 1.

The experiment was laid in a Randomized Complete Block Design with 4 blocks. The design was done to block against varying soil texture, weed species distribution, weed population distribution, and slope. The experimental plots 3.6 m × 17.92 m (3 rows), and the blocks, measured 18 m × 17.92 m (15 rows). Weeding was done using hand hoe method in the plots according to treatment specification. Total weed removal was done at 8 W.A.P. Stalk heights were measured on 30 sampled plants per assessment row of each treatment using 1 meter ruler. These measurements were recorded at 5, 6, and 7 W.A.P.

Leaf expansion measurements were recorded using 1 meter rulers measuring the length and width of the leaves at 9, 10, and 11 W.A.P, and this was done when the crop was topped to remove apical dominance. Yield was measured after flue curing by untying the tobacco leaves and measuring weight of leaves from specified treatments at a digital scale. A general analysis of variance (ANOVA) was used in the stalk height, leaf area, and flue-cured tobacco yield using GENSTAT Ninth Edition (9.2) statistical package.

TABLE 1: Different weed control treatments applied to cultivar T66 at Kutsaga Research Station during 2012/2013 season.

Treatment	Time of weeding	Delay period
1	Standard practice (Control)	No delay
2	2 W.A.P	Delayed by 2 week
3	4 W.A.P	Delayed by 4 weeks
4	6 W.A.P	Delayed by 6 weeks
5	Weeded only at 8 W.A.P	Unweeded

The relative competition intensity (RCI) was quantified by wide set of competition intensity of the effect size of competition [13, 14]. The RCI measures the proportional decrease in plant performance due to weed competition. The RCI values are obtained from the following model:

$$\text{RCI} = \frac{Y_{\text{no weed}} - Y_{\text{weed}}}{Y_{\text{no weed}}}, \tag{1}$$

where Y is the measurement of the crop performance or of crop yield, $Y_{\text{no weed}}$ represents weed free plot and Y_{weed} represents weed infested plot.

3. Results and Discussion

3.1. Impact of Time of Weeding on Tobacco Growth

3.1.1. Effect on Stalk Height. There were no significant differences ($P > 0.05$) among the 5 treatments for stalk height at 5 W.A.P. The control treatment showed the highest mean stalk height (9.88 cm) followed by the treatment that was weeded at 2 W.A.P (9.05 cm). Although there was effective competition in other treatment excluding the control as shown by the RCI values, the competition was less as compared to the competition intensity observed from the RCI values at 6 and 7 W.A.P. As the plant developed, there were still no significant differences ($P > 0.05$) at 6 W.A.P among the 5 treatments. The RCI values observed at this stage had increased of all the treatments showing an increase in competition intensity on all plants exposed to competitive weed density environment. These results observed in this study on the competitiveness of weeds on tobacco growth were similar to those observed by Aguyoh and Masiunas [15]. The increasing of weed density promoted a reduction in the dry bean production, reaching losses sometimes greater than 50%, for the highest densities.

Significant treatment differences ($P < 0.05$) were observed among the treatments at 7 W.A.P which was due to an extensive competitive environment (RCI > 0) as observed in unweeded treatment showing an increase of RCI value from 0.447 to 0.59. The plants that were in the control treatment were exposed to a competitive free environment (RCI = 0) because the RCI was constant at zero from 5 W.A.P to 7 W.A.P. However there were no significant (L.S.D = 7.60) differences between the control (28.66 cm) and the treatment that was weeded at 2 W.A.P (21.59 cm). Significant differences (L.S.D = 7.60) were observed between the standard practice and the treatment weeded at 6 W.A.P. Results are shown in Tables 2 and 3.

TABLE 2: Means for stalk heights (cm) at 5, 6, and 7 weeks after planting at Kutsaga Research Station in the 2012/2013 season.

Time of weed removal (W.A.P)	Weeks after planting		
	5	6	7
Standard practice (Control)	9.88	13.5	28.66
2	9.05	11.18	21.59
4	8.41	10.56	18.85
6	7.66	9.97	15.82
Unweeded	5.21	7.46	11.74
Significance of F	NS	NS	* * *
L.S.D$_{0.05}$	—	—	7.06
CV (%)	31.70	24.80	25.50

$***$ Denote significance at $P < 0.05$, NS: not significant at $P > 0.05$.

TABLE 3: The intensity of the effect of size of competition on stalk heights at 5, 6, and 7 weeks after planting.

Time of weed removal (W.A.P)	RCI and stalk height					
	RCI	SH	RCI	SH	RCI	SH
Standard practice	0	9.88	0	13.5	0	28.66
2	0.084	9.04	0.172	11.18	0.247	21.59
4	0.149	8.41	0.218	10.56	0.342	18.85
6	0.225	7.66	0.262	9.97	0.448	15.82
Unweeded	0.472	5.21	0.447	7.46	0.590	11.74

RCI: relative competition intensity; S.H: stalk height; W.A.P: weeks after planting.

3.1.2. Effect on Leaf Area. Leaf area at 9 W.A.P showed significant treatment differences ($P < 0.05$) which was due to treatment effects at early stages of the crop. The treatment that was weeded at 2 W.A.P (20.25 cm) and weeded at 4 W.A.P (24.32 cm) showed no significant difference (L.S.D = 18.59) from the control (35.20 cm), although the treatment that was weeded at 4 W.A.P showed a higher leaf area which also indicated the appropriate time of weed control in tobacco basing on this study. Maw and Mullinix [16] reported that leaf area influences biomass production and yield of crop plants. Tobacco high yields are mainly attributed to the significant improvement in leaf areas as a result of increased leaf widths and lengths, and most of the broad types tend to produce a balanced proportion of market desirable dark and light colored cured leaf [17].

As the crop developed, a same trend on the results of leaf area was observed from 9 W.A.P to 11 W.A.P where significant differences ($P < 0.05$) among the treatments on leaf area were also observed at 10 W.A.P. The unweeded treatment was the lowest (22.75 cm), and there was no significant difference (L.S.D = 14.75) from treatment that was weeded at 2 W.A.P (34.34 cm) and treatment that was weeded at 6 W.A.P (38.94 cm). This indicated that weeding at 2 W.A.P was too early because of development of a competitive weed population from 2 W.A.P to 8 W.A.P when total weed removal was done. As the experiment advanced to 11 W.A.P, significant treatment differences ($P < 0.05$) among the treatments on leaf area were also observed. The

TABLE 4: Means for leaf area at 9, 10, and 11 weeks after planting at Kutsaga Research Station in the 2012/2013 season.

Time of weed removal (W.A.P)	Weeks after planting		
Standard practice	35.20	45.79	49.87
2	20.25	34.34	39.07
4	24.32	38.94	44.50
6	7.24	27.39	35.63
Unweeded	3.01	22.75	30.32
Significance of F	*	* *	* * *
LSD$_{0.05}$	18.59	14.75	10.74
CV (%)	69.32	8.30	17.4

*, **, and *** denote significance at $P < 0.05$, 0.01, and 0.001, respectively. NS: not significant.

control had a higher leaf expansion measurement compared to the treatment that was weeded at 2 W.A.P (39.87 cm), and treatment weeded at 4 W.A.P (44.50 cm) showed no significant difference ($P > 0.05$) from the control. Results on the effect on leaf area are shown in Table 4.

3.1.3. Impact of Time of Weeding on Yield. There were no significant differences ($P > 0.05$) among the treatments on yield recorded from reap 1. This was mainly due to a few numbers of leaves harvested at this stage, and for treatments weeded at 6 W.A.P. At this stage the unweeded treatment had more harvested leaves due to false ripening. Significant differences ($P < 0.05$) were noted among the treatments at reaps 2 and 3. The treatment weeded at 4 W.A.P was significantly higher (L.S.D = 192.8) than the treatment weeded at 2 W.A.P. The same trend was also observed on the yield recorded for reap 3. This support the suggested appropriate time of weeding in tobacco basing on this study as indicated from the results on leaf area in this study. The stage of yield reducing competition which was observed to be from 4 W.A.P by Cousins [18] also supports the appropriate time of weeding suggested from the yield results observed from this study. This study also indicated the relationship on leaf area and yield of tobacco. High weed densities in treatment weeded at 2 W.A.P and 6 W.A.P and the unweeded treatment contributed to lower yields recorded. Weed density increase resulted in leaf yield reduction. The results are shown in Table 5.

4. Conclusion

Based on results of this study, the most important stage to avoid yield reduction was from 4 W.A.P to the topping stage 9 W.A.P. The critically damaging period of weed competition can be effectively countered by weed control in early stages of weed germination and growth, although subsequent cultivation may be required to ensure continued control of all germinating weeds. In situation where labor shortages are experienced weed control practises should be done at 4 W.A.P.

TABLE 5: Mass at untying first reaping, second reaping, and third reaping (KG/HA).

Time of weed removal (W.A.P)	Reap			
	Reap 1	Reap 2	Reap 3	All groups
Standard practice (control)	153.53	474.86	459.71	1088.1
2	74.08	267.97	220.9	562.95
4	73.67	460.5	364.71	898.88
6	84.15	294.33	289.0	667.48
Unweeded	77.56	234.38	211.29	523.23
Significance of F	NS	*	*	*
L.S.D$_{0.05}$	—	192.8	143.6	270.80
CV (%)	63.3	41.5	28	25

*Denote significance at $P < 0.05$.
NS: not significant; W.A.P: weeks after planting.

Acknowledgments

The authors would like to acknowledge the Tobacco Research Board Zimbabwe (T.R.B) for the assistance rendered during this study, for allowing me the opportunity to use the facilities of the T.R.B, and for funding this study and Dr. Walter Manyangarirwa of Africa University for helpful discussion.

References

[1] J. L. Stocks, "Flue-cured tobacco production in Zimbabwe," CORESTA Zimbabwe, pp. 205–209, 1994.

[2] FAO, Issues in the global tobacco economy, Food and Agricultural Organization of the United Nations of Rome, 2003.

[3] S. Mabasa, A. M. Rambakudzibga, O. Mandiringana, C. Ndebele, and F. Bwakaya, "A survey of maize production practices in three communal areas of Zimbabwe," in *Paper Presented at the Rockefeller Soil Fertility Network Meeting*, p. 28, Kadoma, Zimbabwe, July 1995.

[4] A. S. Adegoroye, O. A. Akinyemiju, and F. Bewaji, "Weeds as a constraint to food production in Africa. Management and the African farmer," in *Proceedings of the ICIPE/World Bank Conference on Integrated Pest Management in Africa*, ICIPE Science Press, 1989.

[5] R. W. Wilson, "Effects of cultivation on growth of tobacco," Tech. Rep. 116, Agricultural Experiment Station, Raleigh, NC, USA, 1995.

[6] A. Niell, *A Windmill Guide to Controlling Losses in Tobacco Production*, Windmill Private, 1996.

[7] G. C. Klingman, "Weed control in flue-cured tobacco," *Tobacco Science*, vol. 11, pp. 115–119, 1967.

[8] I. O. Akobundu, "Weeds in human affairs in sub-Saharan Africa: implications for sustainable food production," *Weed Technology*, vol. 5, pp. 680–690, 1991.

[9] A. M. Rambakudzibga, A. Makanganis, and E. Mangosho, "Competitive influence of *Eleusine indica* and other weeds on the performance of maize grown under controlled and open field conditions," *Zimbabwe Agricultural Journal*, vol. 75, pp. 14–18, 2000.

[10] S. Z. Knezevic, S. P. Evans, E. E. Blankenship, R. C. Van Acker, and J. L. Lindquist, "Critical period for weed control: the concept and data analysis," *Weed Science*, vol. 50, no. 6, pp. 773–776, 2002.

[11] V. Vincent and R. G. Thomas, *An Agricultural Survey of Southern Rhodesia Part I: An Agroecological Survey*, Government Printers, Salisbury, UK, 1960.

[12] K. W. Nyamapfene, *Soils of Zimbabwe*, Nehanda Publishers, Harare, Zimbabwe, 1st edition, 1991.

[13] J. B. Grace, "On the measurement of plant competition intensity," *Ecology*, vol. 76, no. 1, pp. 305–308, 1995.

[14] D. E. Goldberg, T. Rajaniemi, J. Gurevitch, and A. Stewart-Oaten, "Empirical approaches to quantifying interaction intensity: competition and facilitation along productivity gradients," *Ecology*, vol. 80, no. 4, pp. 1118–1131, 1999.

[15] J. N. Aguyoh and J. B. Masiunas, "Interference of redroot pigweed (*Amaranthus retroflexus*) with snap beans," *Weed Science*, vol. 51, no. 2, pp. 202–207, 2003.

[16] B. W. Maw and B. Mullinix, "Comparing six models of various complexities for calculating leaf area from measurements of leaf width and length," *Alteration of Leaf Shape in Field Tobacco Science*, vol. 36, pp. 40–42, 1992.

[17] R. T. Garvin, "Flue cured Tobacco yield estimation in crop research," *Zimbabwe Science News*, vol. 19, no. 5-6, pp. 65–67, 1985.

[18] L. T. V. Cousins, *The effects of weed competition on flue-cured tobacco yield and quality (Nicotiana tabacum) [Ph.D. thesis]*, Department of Agriculture, University of Rhodesia, 1979.

12

Screening for Salt Tolerance in Eight Halophyte Species from Yellow River Delta at the Two Initial Growth Stages

Liu Xianzhao,[1,2,3] Wang Chunzhi,[2] and Su Qing[4]

[1] *College of Architecture and Urban Planning, Hunan University of Science and Technology, Xiangtan 411201, China*
[2] *College of Geography and Planning, Ludong University, Yantai 264025, China*
[3] *Department of Geography, Linyi University, Linyi 264000, China*
[4] *College of Life Science, Hunan University of Science and Technology, Xiangtan 411201, China*

Correspondence should be addressed to Liu Xianzhao; xianzhaoliu@sina.com

Academic Editors: N. Hulugalle, C. H. Kao, and M. Zhou

Screening of available local halophytes for salinity tolerance is of considerable economic value for the utilization of heavy salt-affected lands in coastal tidal-flat areas and other saline areas. In this study, the germination and seedling pot experiments on salt tolerance of eight halophytic species from Yellow River Delta, China, at seven NaCl concentrations (0, 50, 100, 150, 200, 250, and 300 mM), were conducted at both growth stages. Results showed that germination rate and germination index decreased with an increase in NaCl concentration. The higher germination rates were obtained from *Tamarix chinensis* and *Suaeda salsa* seeds exposed to 0~200 mM NaCl. At the seedling stage, the salt tolerances of eight halophytes were also different from each other. *Tamarix chinensis* had significantly greater fresh biomass and plant height in relative terms than the others in all salt treatments. The order of the relative growth yield in seedling was *Tamarix chinensis* > *Suaeda salsa* > *Salicornia europaea* > *Limonium bicolor* > *Atriplex isatidea* > *Apocynum venetum* > *Phragmites australis* > *Sesbania cannabina*. The comprehensive analysis showed that *Tamarix chinensis* had the highest tolerance to salt, followed by *Suaeda salsa*, and the salt tolerance of *Sesbania cannabina* was the lowest.

1. Introduction

Soil salinization is one of the most serious impediments to agricultural production both in China and the other regions of the world [1, 2]. According to statistics, there is about 9.54×10^8 hm^2 of saline soil worldwide, and seven percent of the land surface and five percent of cultivated lands are affected by salinity [3, 4]. China has all kinds of saline soils with a total area of about 0.99×10^8 hm^2. In the Yellow River Delta alone, there is approximately 44.3×10^4 hm^2 of seashore salinized tidal flat, including heavily salinized soil (soil salinity is over 0.6%) and saline-alkaline bare land with an area of 23.63×10^4 hm^2. And this number continues to grow at a rate of 2.2×10^4 hm^2 per year due to a variety of natural and human activities [5–7]. On these saline lands, it is not suitable for growth of traditional crops because

of extreme salinity and other adverse factors. If plant salt tolerance cannot be improved, then vast amounts of soils may be left uncultivated. This will severely threaten the national food security and biomass energy production.

In recent years the development thinking of solonchak agriculture is brought up at home and abroad. Namely, salinized soils can be utilized through the transgenic technology or the breeding of salt-tolerant plants, which may successfully maintain a relatively reasonable yield and an increased growth on salt-affected soils. And then some approaches to overcome the salinity problem have been put forward by various plant scientists [1, 2, 8]. For example, large scale soil was ameliorated to meet the needs of plant growth by either altering farming practices to prevent soil salinization or implementing schemes to remediate salinized soils. One of the most feasible and economic paths, which is of prime importance, is the screening of available local halophytes for salinity tolerance, despite its limitation of time required and environment dependence.

Halophytes are plants, which are able to grow or survive in saline conditions and have considerable importance as food, forage, and material source for biofuel [9]. Cultivation of halophytes or salt-tolerant plants on the salt soil in some cases would spare arable land and fresh water for conventional agriculture. Despite strong adaptations to saline environments, halophytes are sensitive to salt stress, like many other traditional crops, and may not sustain good growth or yield when they are grown in soils with high salt concentrations [10, 11]. In recent years, extensive research on plant screening for salt tolerance has been conducted, with the aim of providing a relatively tolerant cultivar, but these researches mainly focused on conventional crops, screening criteria, and methods for plants salt tolerance [1, 12–15]. Unfortunately, there are few investigations about the screening of available halophyte species and their responses to heavy saline conditions.

Keeping this in mind, the work presented here was carried out to examine the salinity tolerance of eight local halophytes by germination and pot experiments when exposed to different NaCl concentrations. The aim of this study was to determine how far the ecological amplitude, in relation to salt, of a range of halophyte species may be restricted by their inability to evolve salt tolerance and finally screen the most adaptive salt-tolerant halophytes which can grow well on the heavy salt soils.

2. Materials and Methods

The 8 halophyte species (*Tamarix chinensis, Suaeda salsa, Atriplex isatidea, Apocynum venetum, Sesbania cannabina, Salicornia europaea, Phragmites australis,* and *Limonium bicolor*) were used in this study. The seed material was collected during October 2010 from Halophytes Garden of Dongying in the Yellow River Delta, China, and stored for four months at room temperature (22°C) and 50% relative humidity. Prior to experimentation, seed samples were surface sterilized using 5% sodium hypochlorite solution for five minutes and thoroughly rinsed with distilled water.

2.1. Germination Experiment. The germination experiment was conducted in a growth chamber at $26 \pm 2°C$, with 12 h day length, at a light intensity of 36 Wm^{-2} and relative humidity of 76%. 30 surface sterilized seeds of each halophyte were placed on moistened filter paper in a 9 cm plastic Petri dish. The plastic Petri dishes were arranged in a completely randomized design, with five replicates, seven salt treatments (0, 50, 100, 150, 200, 250, and 300 mM NaCl), and the 8 halophyte species mentioned above. 5 mL of appropriate treatment solution was applied on alternate days to each Petri dish after rinsing out the previous solution. A seed was considered to have germinated when both plumule and radicle had emerged ≥ 2 mm. The number of germinated seeds was counted daily for 7 days. The rate of germination was expressed a percentage of the number of germination seeds divided by the number of

tested seeds for each treatment. The germination index (GI) was estimated using a modified Timson index [16] of germination velocity, $GI = \sum G/t$ (where G is the number of seed germination every day, and t is the corresponding number of days for germination test). The relative germination index (RGI) is expressed as a ratio of germination index under salt stress to germination index in control treatment.

2.2. Seedling Experiment. Seedling Experiments for salt tolerance were conducted to evaluate the effects of salinity on halophyte growth in a glasshouse at $24 \pm 3°C$ day temperature and $12 \pm 1°C$ night temperature. For each halophyte, twenty to thirty seeds, depending upon 8 halophytic species, were sown in a 3 L PVC (polyvinyl chloride) pot with a mixture containing 70% perlite and 30% dry sand. After two weeks, seedlings were thinned to leave seven uniform and healthy seedlings in each pot. All the pots were irrigated for one week with Hoagland solution every alternate day. The concentrations of NaCl used were exactly the same as we used in the germination experiment, namely, 0, 50, 100, 150, 200, 250, and 300 mM. The experiment was laid out in a completely randomized design with five replications in each treatment. NaCl treatment was applied to 3-week-old plants and lasted for 5 weeks. The salt concentration was increased in aliquots of 25 mM on alternate days until the appropriate salt treatment was reached. Plants were watered twice a day with about 50 mL solution applied each time per pot. The symptoms of plant injury at salt stress were observed at any time during the course of the experiment. After 35 days of NaCl treatment, two measurements were undertaken in each treatment for both plant height and fresh weight of above-ground biomass. Salt tolerance of plant has been assessed by measuring the relative growth rate and the salt damage rate, which were expressed as follows:

relative growth rate

$$= \frac{\text{plant growth at salt concentration}}{\text{plant growth at control treatment}} \times 100\%,$$

salt-injury rate

$$= \frac{\text{number of plants with salt-injury symptoms}}{\text{total number of plants}}$$

$$\times 100\%.$$

(1)

2.3. Statistical Analysis. Data were analyzed statistically using a one-way analysis of variance (ANOVA), linear regression, and correlation analysis, and the means of each treatment were analyzed by Duncan's multiple range test. All statistical methods were performed with SAS software [17].

3. Results

3.1. Effect of Different Concentrations of NaCl on the Seed Germination. One-way ANOVA of the data for germination rate of the 8 halophyte seeds indicated that salt treatment

TABLE 1: Results of one-way ANOVA for germination rate of the eight species at different NaCl treatments.

Species	df	Mean square	F	P
Sesbania cannabina	6	16.63	21.48	<0.01
Limonium bicolor	6	19.32	20.04	<0.01
Suaeda salsa	6	31.32	27.08	<0.01
Tamarix chinensis	6	38.14	19.16	<0.01
Apocynum venetum	6	21.54	14.28	<0.01
Salicornia europaea	6	14.45	33.87	<0.01
Phragmites australis	6	15.76	18.93	<0.01
Atriplex isatidea	6	41.46	30.37	<0.01

TABLE 2: Results of one-way ANOVA for germination index of the eight species at different NaCl treatments.

Species	df	Mean square	F-value	P value
S. cannabina	6	15.78	41.25	<0.001
L. bicolor	6	13.32	84.16	<0.001
S. salsa	6	16.89	89.56	<0.001
T. chinensis	6	18.34	59.51	<0.001
A. venetum	6	17.52	71.30	<0.001
S. europaea	6	16.95	46.27	<0.001
P. australis	6	14.35	21.31	<0.001
A. isatidea	6	11.64	12.33	<0.05

had a significantly adverse effect on total germination rate (Table 1). As the NaCl concentration increased, the eight species all exhibited a decreasing trend of germination rate (Figure 1). For example, compared with the control, the total germination rates of Suaeda salsa at 50, 100, 150, 200, 250, and 300 mM NaCl treatments were 94.5%, 87.2%, 81.4%, 78.4%, 69.3%, and 56.2% of the control, respectively. Although the germination of halophyte seeds was strongly inhibited when they were subjected to salt stress, the degree of inhibition differed markedly. The eight halophytic species tested in this study could be classified into three groups depending on their ability to germinate in saline medium. First group was the most salt tolerant and had the smallest degree of salt inhibition. These plants (Suaeda salsa, Tamarix chinensis, Apocynum venetum, and Salicornia europaea) in this group had higher final germination rate (>80%) when the salinity was less than 100 mM NaCl and still had more than 40% of the germination rate even on the highest salinity condition (300 mM NaCl) (Figures 1(c)~1(d)). The second was the moderate tolerant halophytes, including Sesbania cannabina and Limonium bicolor. Their seeds also retained high germination rate at low saline (≤50 mM NaCl) and nonsalt treatments, whereas the germination rate decreased significantly at 150 mM NaCl (Figures 1(a) and 1(b)). The third was the least tolerant at the germination stage. The halophytes, Phragmites australis and Atriplex isatidea, each had the lowest germination rate as compared to the other six halophytes even if the NaCl concentration was very low, or under the without salinity conditions (Figures 1(g) and 1(h)). This may be related to the lower viability of the two species.

Variance analysis of the data for GI also showed that salt treatment had a significant effect on germination of the eight species (Table 2). The GI had the same change trend with that of germination rate. As the salt concentration increased, the GI of the eight halophyte seeds gradually decreased (Table 3). In the eight species Tamarix chinensis under all NaCl treatments had the highest GI, followed by Suaeda salsa and Apocynum venetum. And when the salinity concentration was less than 150 mM, the GI of Tamarix chinensis was more than 84%. However, the other five halophytes, Sesbania cannabina, Salicornia europaea, Phragmites australis, Atriplex isatidea, and Limonium bicolor, each showed a very poor GI (lower than 40.0%) both in nonsalt and all salt treatments (Table 3). Among them, the GI of Phragmites australis and

Atriplex isatidea rapidly decreased to 0.1% and 1.7%, respectively, at 250 mM NaCl, and when the concentration of NaCl was over 200 mM, the germination of Sesbania cannabina and Limonium bicolor was completely inhibited (Table 3). The results indicated that there was tremendous difference in ability of halophytes to germinate in the presence of NaCl salinity.

From Table 4, it could be seen that the relative germination index had a significantly linear negative relationship with NaCl concentration (all $r < -0.9$, all $P \le 0.01$), which again showed that salt treatment had a significantly adverse effect on germination of halophytic seeds. According to the linear regression equations listed in Table 4 and the method proposed by Bai et al. [18] and Zeng et al. [11], let the relative germination index be equal to 75%, 50%, and 25% respectively; we could obtain the appropriate, critical, and limited values of salt concentration for germination of the eight halophyte seeds under salt stress conditions. Data in Table 4 demonstrated that at the germination stage, Tamarix chinensis was the most salt tolerant of all the tested halophyte seeds, which was consistent with the results observed in Figure 1.

3.2. Effect of Different NaCl Treatments on the Growth of Halophytes at the Seedling Stage. Analyses of variance of the data for relative growth for the eight halophytes showed that five weeks of salt treatment had a significant influence on growth characteristics measured (plant height $P \le 0.01$; above-ground biomass $P \le 0.01$) at the seedling stage. As the salt concentration increases, the degrees of growth inhibition and salt-injury rate increase. High NaCl treatment (≥200 mM) had an obviously adverse inhibition on all tested halophytes, with the relative growth rate (fresh weight of above-ground biomass and plant height) being fast reduced, and the salt-injury rate increased rapidly as the NaCl concentration increased above 150 mM (Figures 2 and 3). For example, the seedlings of Sesbania cannabina and Phragmites australis were dying at 300 mM NaCl (Figure 2), and their salt-injury rates were drastically increased from 5.1% (Sesbania cannabina) and 4.5% (Phragmites australis) at 150 mM NaCl to nearly 100% at 300 mM NaCl (Figure 3), accompanied by death of individual whole plant.

However, when tested plants were under hyposaline, there were some differences in relative growth and salt-injury

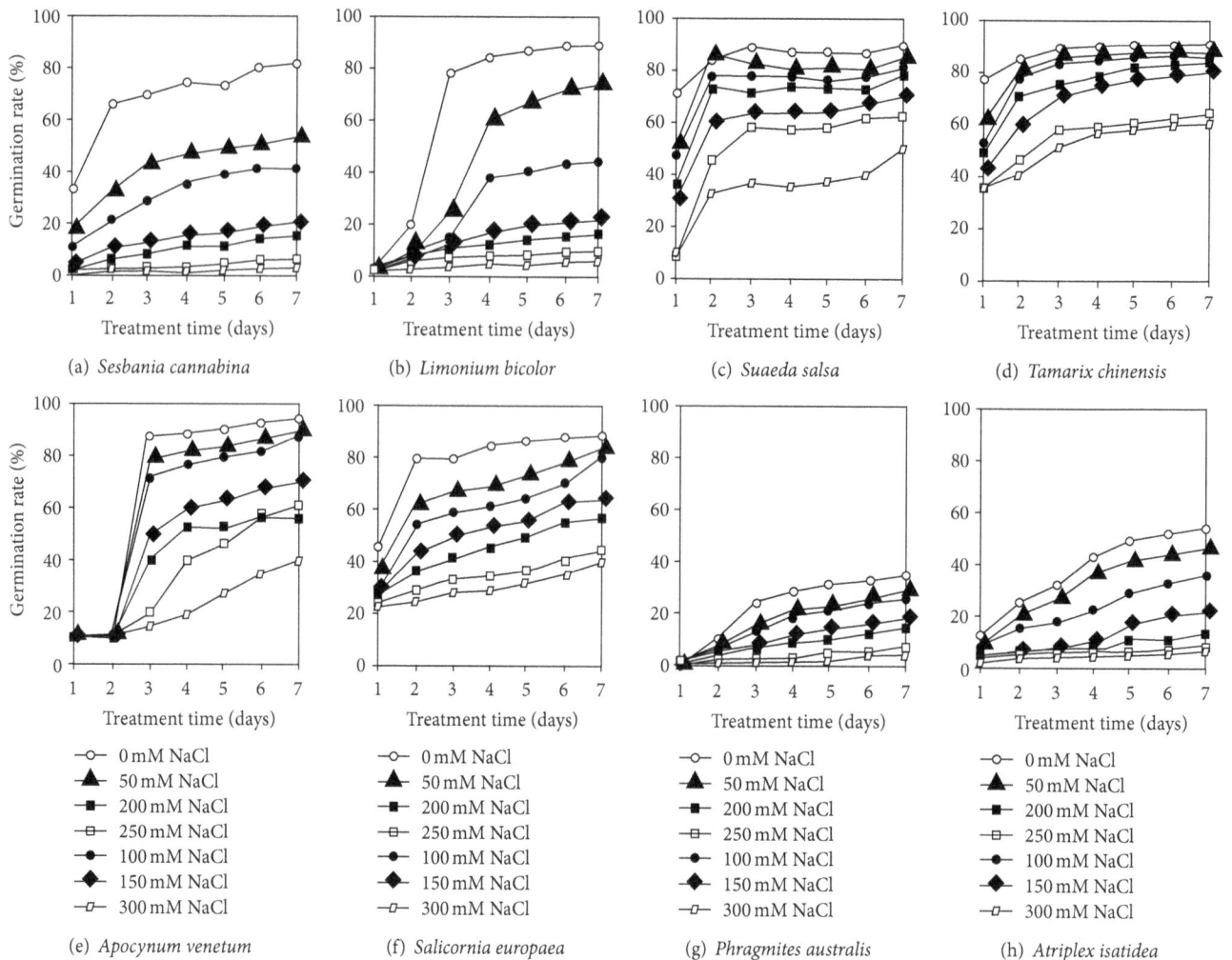

FIGURE 1: Changes in germination rates with treatment days at different NaCl concentrations.

TABLE 3: Germination indexes of 8 halophyte seeds at different NaCl concentrations.

Species	Germination index						
	0 mM	50 mM	100 mM	150 mM	200 mM	250 mM	300 mM
S. cannabina	28.7 ± 4.10	13.3 ± 2.71	11.5 ± 2.35	9.6 ± 1.60	7.8 ± 1.64	0.0 ± 0.00	0.0 ± 0.00
L. bicolor	36.4 ± 3.51	32.0 ± 3.68	18.4 ± 3.80	6.7 ± 1.09	1.4 ± 0.12	0.0 ± 0.00	0.0 ± 0.00
S. salsa	89.2 ± 4.25	84.8 ± 4.60	81.6 ± 3.69	77.1 ± 2.94	50.3 ± 4.24	42.4 ± 3.14	21.5 ± 2.65
T. chinensis	98.3 ± 0.87	95.6 ± 1.15	94.1 ± 2.00	84.7 ± 5.31	55.4 ± 7.51	49.4 ± 2.40	36.7 ± 3.78
A. venetum	65.2 ± 1.93	61.7 ± 3.77	59.5 ± 1.71	45.0 ± 5.13	38.5 ± 4.67	33.5 ± 3.07	21.1 ± 4.25
S. europaea	34.0 ± 5.33	28.4 ± 4.15	30.9 ± 4.72	26.8 ± 3.02	25.4 ± 2.34	21.8 ± 4.37	18.8 ± 3.64
P. australis	19.7 ± 4.55	17.9 ± 4.62	13.6 ± 3.68	4.4 ± 3.07	1.5 ± 1.09	0.1 ± 0.21	0.0 ± 0.00
A. isatidea	11.8 ± 2.11	8.1 ± 1.17	9.0 ± 1.82	5.5 ± 1.05	4.2 ± 0.30	1.7 ± 0.34	0.4 ± 0.09

Values are means ± SE ($n = 5$).

rate for different halophytic species. At 50~100 mM NaCl the relative growth rates of *Tamarix chinensis* and *Salicornia europaea* increased gradually with an increase in salinity and were greater than those of the control plants (Figure 2).

There was also a certain increase in relative growth for *Suaeda salsa* at less than 200 mM NaCl as well as for *Limonium bicolor* exposed to 50 mM NaCl compared with

the control treatment, which showed that low salt stress could promote the seedling growth of some halophytes. And within the scope of this treatment, all the halophytes examined in this study did not show apparent symptoms of salt injury for their salt-tolerant ability at the seedling stage, except for the fact that two halophytes, *Sesbania cannabina* and *Phragmites australis*, presented some symptoms of salt

TABLE 4: Correlation between relative germination index of eight halophyte seeds and salt concentration.

Species	Regression equation	Regression coefficient (R^2)	Correlated coefficient (r)	Salt tolerance (mM NaCl)		
				A	B	C
S. cannabina	$y = -0.2576x + 78.15$	0.8492	-0.9215^{**}	12.2	109.3	206.3
L. bicolor	$y = -0.2210x + 89.63$	0.8316	-0.9119^{**}	66.2	179.3	292.4
S. salsa	$y = -0.2557x + 109.91$	0.9078	-0.9528^{**}	136.5	234.3	332.1
T. chinensis	$y = -0.2297x + 109.18$	0.9099	-0.9539^{**}	148.8	257.6	366.5
A. venetum	$y = -0.2287x + 104.31$	0.9633	-0.9815^{**}	128.2	237.5	346.8
S. europaea	$y = -0.2349x + 98.52$	0.9141	-0.9561^{**}	100.1	206.6	312.9
P. australis	$y = -0.3875x + 99.48$	0.9107	-0.9543^{**}	63.2	127.7	192.2
A. isatidea	$y = -0.3332x + 97.57$	0.9518	-0.9756^{**}	67.7	142.7	217.8

Note: ** denotes significant at ≤0.01 level. In the regression equation, y is relative germination index, and x is salt concentration. A: appropriate value of salt concentration for seed germination; B: critical value of salt concentration for seed germination; C: limited value of salt concentration for seed germination.

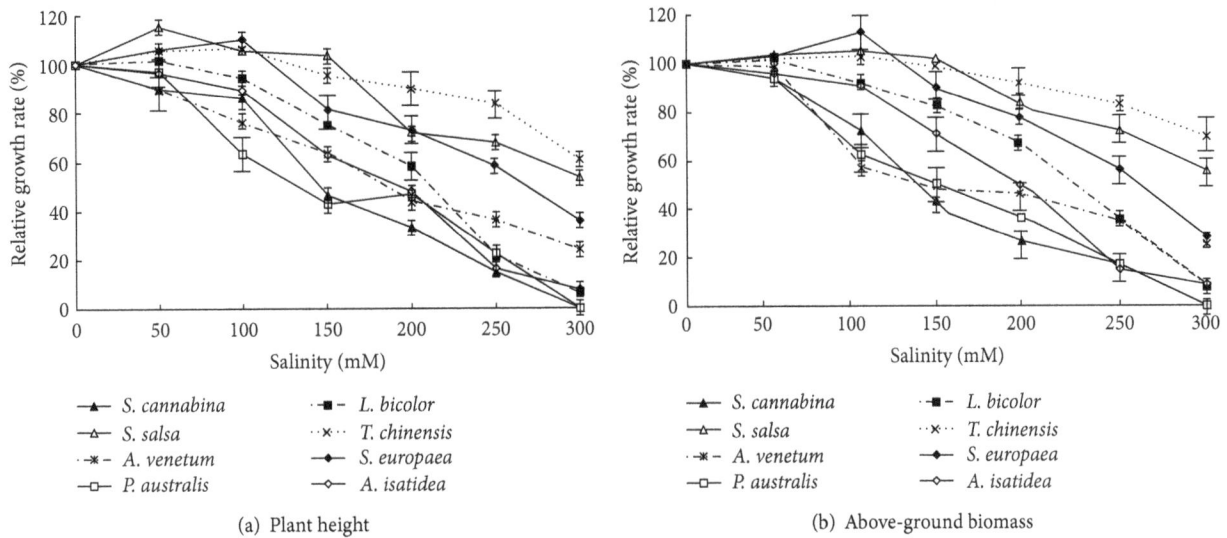

(a) Plant height

(b) Above-ground biomass

FIGURE 2: Relative growth rate of the eight halophytes after 5-week exposure to different salt concentrations. Each point represents the mean value ± SD.

injury (leaf margin scorch, leaf turning yellow, and falling off) and a more growth inhibition in relative terms under 150 mM NaCl treatment (Figure 3). This again indicated that different halophytes responded differently to NaCl treatment.

The above results showed that the salt tolerance of eight halophyte seedlings were different from each other. Among them, Sesbania cannabina had the lowest salt tolerance, whose average relative growth rate was just 49.1% with the average salt-injury rate of 41.9%, while Tamarix chinensis had the highest salt tolerance as compared to the other seven species in salt treatments, whose average relative growth increment was as high as 91.1%, only accompanied by the average salt-injury rate of 5.7% (Table 5). On the basis of their performance in relative growth and salt-injury rate (Figures 2 and 3, Table 5), in all the tested halophytes, the sequence of salt tolerance from strong to weak was as follows: Tamarix chinensis, Suaeda salsa, Salicornia europaea, Limonium bicolor, Atriplex isatidea, Apocynum venetum, Phragmites australis, and Sesbania cannabina.

4. Discussion

Screening of available halophytes for salinity tolerance is of considerable value for the economic utilization of salt-affected soils in coastal tidal-flat areas. To explore salt tolerance of 8 halophyte species at the germination and seedling stages was examined in our study. In the germination experiment, salt stress markedly induced lower germination rate and germination index of the eight halophyte seeds (Tables 1~3, Figure 1). For instance, for halophytes Atriplex isatidea, Phragmites australis, Sesbania cannabina, and Limonium bicolor, at 300 mM NaCl, only less than 5% seeds germinated, and their germination indexes were also nearly 0% (Table 3). This showed that germination of some halophyte seeds might be completely inhibited under high salt conditions. Similar results were also determined in some species of Reaumuria trigyna [19], Aeluropus lagopoides [20], and Salsola vermiculata [8].

A significant and negative correlation was found between data for relative germination index and NaCl concentration

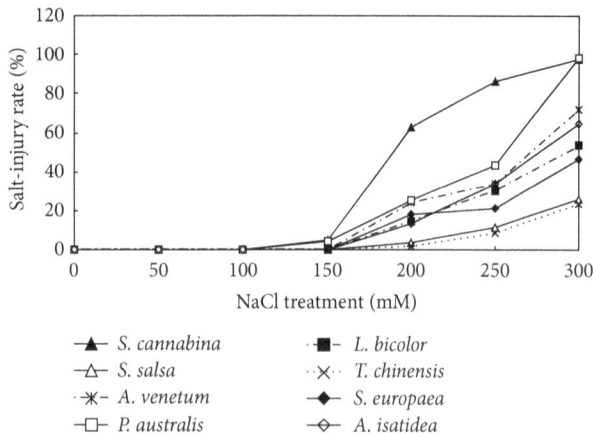

FIGURE 3: Salt-injury rate (%) of the eight halophytes after 5-week growth in different salt concentrations.

TABLE 5: Mean salt-injury rate and mean relative growth rate of the eight halophytes after 5-week exposure to different salt concentrations.

Species	Mean salt-injury rate (%)	Mean relative growth rate (%)
S. cannabina	41.9	49.1
L. bicolor	16.3	63.7
S. salsa	6.8	86.4
T. chinensis	5.7	91.1
A. venetum	21.6	52.3
S. europaea	14.3	77.1
P. australis	28.6	51.3
A. isatidea	18.7	54.1

Note: the mean relative growth rate here was above-ground biomass.

(Table 4). The higher the NaCl concentration, the lower the germination index of halophytes (Table 3). The decrease in ability of halophytes to germinate under salinity conditions was probably due to a reversible osmotic effect that induced dormancy by saline stress [21]. The germinability of halophytes at salinity stress was reported by Chen et al. [22], who found that the germination rate of *Apocynum venetum* seeds increased with NaCl concentrations below 150 mM, implicating that low salt stress could promote the seed germination of halophytic species. In this experiment, however, the germination of tested halophyte seeds decreased even if the NaCl concentration was low (Figure 1, Table 3), except for the fact that 50 mM NaCl had a slight promotion to germination of *Suaeda salsa* on the second day of the salt treatment (Figure 1(c)). The reasons for the discrepancy between our result and result reported by Chen et al. [22] could be attributed to two aspects. Firstly, low salt stress might inhibit seed germination of halophytes as well as break the seed dormancy, eventually promoting seed germination. Secondly, Chen et al. [22] only considered germination rate under salinity stress regardless of germination index reflecting both the number of germinated seeds and the speed of seed germination while assessing the salt tolerance at germination. Thus, it was likely to overestimate the salt tolerance of halophyte seeds, as there was an evidence to show that, according to regression equation between seed germination and salt concentration, the appropriate, critical, and ultimate salinity for seeds germination calculated using germination index was less than the corresponding values calculated by relative germination rate [23]. Therefore, although germination rate/relative germination rate has been widely used as an index to determine effects of salt stress on the seed germination, it is not likely that this measurement is appropriate to screen halophytes for salt tolerance.

It has been reported that, regardless of the salt concentration used, salt stress had different degrees of inhibition on the growth of plants [15, 24, 25]. The typical symptom of salinity injury to the plant is the growth retardation, leaf shrink with yellow, and shedding or death, due to the inhibition of cell elongation [26]. In the present study, at the seedling stage,

low concentrations of NaCl promoted the seedlings growth of four species, *Tamarix chinensis*, *Suaeda salsa*, *Salicornia europaea*, and *Limonium bicolor* (Figure 2), and the eight halophytes, except for *Sesbania cannabina* and *Phragmites australis*, showed no apparent symptoms of salt injury within 50~150 mM NaCl (Figure 3), indicating no significant effects of low salt stress on the growth of halophytes mentioned above. Similar performances at low salinity were also found in *Puccinellia tenuiflora* [27] and *Atriplex centralasiatica* [28]. For example, Qi et al. [29] and Li et al. [30] found that the growth of *Suaeda salsa* increased significantly with NaCl concentrations when exposed to hyposaline environment (<150 mM NaCl). NaCl solution with high concentration (>150 mM NaCl) caused a reduction in growth and an increase in salt-injury rate of all tested halophytes due to salt stress (Figures 2 and 3). However, the degree of growth inhibition of halophyte species varied at high NaCl concentration. For example, the salt tolerance of *Sesbania cannabina* was the worst in all the tested species, and its individual plant was almost dead after 35 days of 300 mM NaCl treatment. In contrast, *Tamarix chinensis* had the highest salt tolerance; regardless of severity of the salt stress, the relative growth increment remained above 65% (Figure 2) with the average salt-injury rate of 5.7% (Table 5). The inconsistencies existed in salt tolerance among halophytes, which were possibly due to differences in interspecies metabolic rates and sensitivities of different halophytes to salt stress.

Why could low salt stress promote the growth of plant? There were two possible reasons for this. Firstly, halophytes have a nutritional requirement for sodium and an optimal salt concentration during the process of growth (halophyte cannot complete its life cycle for the lack of Na). When the salt concentration in the external solution of plants was less than the optimal salt concentration of halophytes, the increasing of external salinity would probably reduce cell osmotic potential of most halophytes, resulting in the enhancement of water-absorption ability of plant and thereby stimulating the growth of seedlings. Secondly, this may be an adaptation of halophytes to salt stress by accelerating growth in order to reduce the salt concentration [31].

As for the inhibition of high salt stress on growth of halophytes, it could be attributed to decreases in cell metabolism and the toxicity of Na^+ that caused irreversible damage due to prolonged exposure to high concentrations of NaCl [32]. It is generally accepted that, under high salinity conditions, high sodium concentrations in the external solution of plant cells will produce a variety of negative consequences [33]. As only NaCl was used in this study, high salinity might very well lead to ionic imbalance, with excess Na^+ and Cl^- ions having a continual damage on function and structure of cell membrane and leading to membrane dysfunction and cell death [34, 35]. On the other hand, due to the competition for absorptions of Na^+, K^+, Cl^-, and Ca^{2+} to outer membrane, excess Na^+ and Cl^- ions interfered with plants absorption of potassium and calcium and resulted in deficiencies of nutritional elements such as K^+ and Ca^{2+} in plants [36, 37], affecting the growth and development of plants. High salinity also inflicted hyperosmotic shock on plants, as chemical activity of water was decreased, causing a loss of cell turgor. In addition, salt stress generated an increase in reactive oxygen species (ROS), which leaded to decreases in plant photosynthetic capacity [38]. Despite the absence of data of salt stress effects on leaf photochemistry in this study, the fact that NaCl treatment caused a very significant reduction in net CO_2 assimilation, being closely correlated with biomass and plant height, has been proven [39]. Therefore, it appeared that stomatal limitation of photosynthesis was an important factor reducing halophytes growth in saline conditions. Owing to such factors, tested halophytes in this study demonstrated reduced growth at high salinity (Figure 2).

5. Conclusion

Perfectly, results in this study indicated that the eight halophytes had different responses to salt stress at the two initial growth stages. *Tamarix chinensis* and *Suaeda salsa* produced significantly greater above-ground biomass in relative terms and had a higher germination capacity and a lower salt-injury rate than all the other halophytes. Therefore, *Tamarix chinensis* and *Suaeda salsa* were categorized as the most salt tolerant among the eight halophytes appraised. Although the tolerance observed in the present study may or may not be conferred at the adult stage, the performance of seedlings under saline conditions has been considered highly predictive of the response of adult plants to salinity. Thus, the highly salt-tolerant halophytes found in this study would be of considerable economic value for improvement of the heavily saline soils and increasing yield on salt-affected soils in coastal areas. However, the tolerance of halophytes to NaCl single salt only at the germination and seedling stages was investigated in this study. Thus, it is obvious that further field experiment is required to identify exactly the salt tolerance of the eight halophytes at the adult stage.

Acknowledgments

The authors wish to acknowledge the support of the research work by the Science and Technology Development Plan of Yantai (2010163), Natural Science Foundation of Shandong Province (ZR2011DM007), and Key Laboratory Foundation of Soil and Water Conservation and Environmental Protection in Shandong Province (STKF201004).

References

[1] Z. Chen, I. Newman, M. Zhou, N. Mendham, G. Zhang, and S. Shabala, "Screening plants for salt tolerance by measuring K^+ flux: a case study for barley," *Plant, Cell and Environment*, vol. 28, no. 10, pp. 1230–1246, 2005.

[2] K. C. Ravindran, K. Venkatesan, V. Balakrishnan, K. P. Chellappan, and T. Balasubramanian, "Restoration of saline land by halophytes for Indian soils," *Soil Biology and Biochemistry*, vol. 39, no. 10, pp. 2661–2664, 2007.

[3] T. J. Flowers, A. Garcia, M. Koyama, and A. R. Yeo, "Breeding for salt tolerance in crop plants—the role of molecular biology," *Acta Physiologiae Plantarum*, vol. 19, no. 4, pp. 427–433, 1997.

[4] S. C. Praxedes, C. de Lacerda, F. M. DaMatta, J. T. Prisco, and E. Gomes-Filho, "Salt tolerance is associated with differences in ion accumulation, biomass allocation and photosynthesis in cowpea cultivars," *Journal of Agronomy and Crop Science*, vol. 196, no. 3, pp. 193–204, 2010.

[5] T. Zhang, S. Zeng, Y. Gao et al., "Assessing impact of land uses on land salinization in the Yellow River Delta, China using an integrated and spatial statistical model," *Land Use Policy*, vol. 28, no. 4, pp. 857–866, 2011.

[6] H. Fang, G. Liu, and M. Kearney, "Georelational analysis of soil type, soil salt content, landform, and land use in the Yellow River Delta, China," *Environmental Management*, vol. 35, no. 1, pp. 72–83, 2005.

[7] Q. Ye, G. Liu, G. Tian et al., "Geospatial-temporal analysis of land-use changes in the Yellow River Delta during the last 40 years," *Science in China D*, vol. 47, no. 11, pp. 1008–1024, 2004.

[8] I. R. Guma, M. A. Padrón-Mederos, A. Santos-Guerra, and J. A. Reyes-Betancort, "Effect of temperature and salinity on germination of *Salsola vermiculata* L. (Chenopodiaceae) from Canary Islands," *Journal of Arid Environments*, vol. 74, no. 6, pp. 708–711, 2010.

[9] J. C. Dagar, "Ecology, management and utilization of halophytes," *Bulletin of the National Institute of Ecology*, vol. 15, no. 1, pp. 81–89, 2005.

[10] M. A. Khan and K. H. Sheith, "Effects of different levels of salinity on seed germination and growth of *Capsicum annuum*," *Biology Journal*, vol. 22, no. 1, pp. 15–16, 1996.

[11] Y. Zeng, Z. Cai, J. Ma, F. Zhang, and B. Wang, "Effects of salt and water stress on seed germination of halophytes *Kalidium foliatum* and *Halostachys caspica*," *Chinese Journal of Ecology*, vol. 25, no. 9, pp. 1014–1018, 2006.

[12] M. C. Shannon and C. L. Noble, "Genetic approaches for developing economic salt tolerant crops," in *Agricultural Salinity Assessment and Management*, K. K. Tanji, Ed., ACSE Manuals and Reports on Engineering Practice No. 71, p. 161, ASCE, New York, USA, 1990.

[13] T. J. Flowers and A. R. Yeo, "Breeding for salinity resistance in crop plants: where next?" *Australian Journal of Plant Physiology*, vol. 22, no. 6, pp. 875–884, 1995.

[14] S. Krishnaraj, B. T. Mawson, E. C. Yeung, and T. A. Thorpe, "Utilization of induction and quenching kinetics of chlorophylla fluorescence for in vivo salinity screening studies in wheat (*Triticum aestivum* vars Kharchia 65 and Fielder)," *Canadian Journal of Botany*, vol. 71, no. 1, pp. 87–92, 1993.

[15] S. Sevengor, F. Yasar, S. Kusvuran, and S. Ellialtioglu, "The effect of salt stress on growth, chlorophyll content, lipid peroxidation and antioxidative enzymes of pumpkin seedling," *African Journal of Agricultural Research*, vol. 6, no. 21, pp. 4920–4924, 2011.

[16] M. A. Khan and I. A. Ungar, "Seed polymorphism and germination responses to salinity stress in *Atriplex triangularis* Willd," *American Journal of Botany*, vol. 71, pp. 481–489, 1984.

[17] SAS Institute Inc, *SAS/STAT 9. 2 User's Guide*, SAS Institute, Cary, NC, USA, 2008.

[18] Y. E. Bai, J. Yi, A. L. Gu, and Z. J. Guo, "Studies on salt tolerance of seeds of 8 rhizomatous grasses," *Grassland of China*, vol. 21, no. 2, pp. 55–59, 2005 (Chinese).

[19] Y. Xue and Y. Wang, "Influence of light, temperature and salinity on seed germination of *Reaumuria trigyna* Maxim," *Plant Physiology Communications*, vol. 43, no. 4, pp. 708–710, 2007.

[20] S. Gulzar and M. A. Khan, "Seed germination of a halophytic grass *Aeluropus lagopoides*," *Annals of Botany*, vol. 87, no. 3, pp. 319–324, 2001.

[21] M. Mehrun-Nisa, M. A. Khan, and D. J. Weber, "Dormancy, germination and viability of *Salsola imbricata* seeds in relation to light, temperature and salinity," *Seed Science and Technology*, vol. 35, no. 3, pp. 595–606, 2007.

[22] Y. Y. Chen, G. Q. Li, J. Meng, and J. M. Cao, "Effect of sodium chloride stress to the seed germination and seedling growth of *Apocynum venetam* L.," *Chinese Wild Plant Resources*, vol. 26, no. 2, pp. 49–51, 2007 (Chinese).

[23] D. H. Yu, "Effects of NaCl stress on the seed germination and seedling growth of Limonium Bicolor," *Journal of Huazhong Normal University*, vol. 42, no. 3, pp. 435–439, 2008 (Chinese).

[24] S. Kusvuran, S. Ellialtioglu, F. Yasar, and K. Abak, "Effects of salt stress on ion accumulation and activity of some antioxidant enzymes in melon (*Cucumis melo* L.)," *Journal of Food, Agriculture and Environment*, vol. 2, no. 5, pp. 351–354, 2007.

[25] H. Y. Dasgan and S. Koc, "Evaluation of salt tolerance in common bean genotypes by ion regulation and searching for screening parameters," *Journal of Food, Agriculture and Environment*, vol. 7, no. 2, pp. 363–372, 2009.

[26] F. Yasar, S. Ellialtioglu, and K. Yildiz, "Effect of salt stress on antioxidant defense systems, lipid peroxidation, and chlorophyll content in green bean," *Russian Journal of Plant Physiology*, vol. 55, no. 6, pp. 782–786, 2008.

[27] C. W, Yang, D. C. Shi, and Y. Zhang, "Effects of complex salt and alkali conditions on the germination of seeds of *Puccinellia tenuiflora*," *Prataculturae Sinica*, vol. 15, no. 5, pp. 45–51, 2006.

[28] C. Z. Li, X. J. Liu, W. Huang, and H. L. Qiao, "Effect of salt stress on seed germination and its recovery of Atriplex centralasiatica," *Journal of Agricultural University of Hebei*, vol. 28, no. 6, pp. 1–4, 2005 (Chinese).

[29] C. H. Qi, N. Han, and B. S. Wang, "Effect of different salt treatments on succulence of Suaeda salsa seedlings," *Chinese Bulletin of Botany*, vol. 22, no. 2, pp. 175–182, 2005 (Chinese).

[30] W. Q. Li, X. J. Liu, and K. F. Zhao, "Growth, development and ions distribution of three halophytes under salt stress," *Chinese Journal of Eco-Agriculture*, vol. 14, no. 2, pp. 49–52, 2006 (Chinese).

[31] B. Mandák and P. Pyšek, "Effects of plant density and nutrient levels on fruit polymorphism in *Atriplex sagittata*," *Oecologia*, vol. 119, no. 1, pp. 63–72, 1999.

[32] M. A. Khan and I. A. Ungar, "Effects of light, salinity, and thermoperiod on the seed germination of halophytes," *Canadian Journal of Botany*, vol. 75, no. 5, pp. 835–841, 1997.

[33] J. Zhu, P. M. Hasegawa, and R. A. Bressan, "Molecular aspects of osmotic stress in plants," *Critical Reviews in Plant Sciences*, vol. 16, no. 3, pp. 253–277, 1997.

[34] R. Serrano, J. M. Mulet, G. Rios et al., "A glimpse of the mechanisms of ion homeostasis during salt stress," *Journal of Experimental Botany*, vol. 50, pp. 1023–1036, 1999.

[35] S. Chookhampaeng, "The effect of salt stress on growth, chlorophyll content proline content and antioxidative enzymes of pepper (*Capsicum annuum* L.) seedling," *European Journal of Scientific Research*, vol. 49, no. 1, pp. 103–109, 2011.

[36] A. Polle, "Defence against photo-oxidative damage in plants," in *Oxidative Stress and Molecular Biology of Antioxidant Defences*, J. G. Scandalios, Ed., pp. 623–666, Cold Spring Harbour Laboratories, Cold Spring Harbor, NY, USA, 1997.

[37] O. Borsani, V. Valpuesta, and M. A. Botella, "Evidence for a role of salicylic acid in the oxidative damage generated by NaCl and osmotic stress in *Arabidopsis* seedlings," *Plant Physiology*, vol. 126, no. 3, pp. 1024–1030, 2001.

[38] A. Price and G. Hendry, "Iron catalysed oxygen radical formation and its possible contribution to drought damage in nine native grasses and three cereals," *Plant, Cell & Environment*, vol. 14, no. 5, pp. 477–484, 1991.

[39] L. Shabala, T. McMeekin, and S. Shabala, "Osmotic adjustment and requirement for sodium in marine protist thraustochytrid," *Environmental Microbiology*, vol. 11, no. 7, pp. 1835–1843, 2009.

13

Canopy Light Signals and Crop Yield in Sickness and in Health

Jorge J. Casal[1,2]

[1] IFEVA, Facultad de Agronomía, Universidad de Buenos Aires and CONICET, Avenida San Martín 4453, 1417 Buenos Aires, Argentina
[2] Fundación Instituto Leloir, Instituto de Investigaciones Bioquímicas de Buenos Aires, CONICET, 1405 Buenos Aires, Argentina

Correspondence should be addressed to Jorge J. Casal; casal@ifeva.edu.ar

Academic Editors: G. M. Dal Bello and E. Perez-Artes

Crop management decisions such as sowing density, row distance and orientation, choice of cultivar, and weed control define the architecture of the canopy, which in turn affects the light environment experienced by crop plants. Phytochromes, cryptochromes, phototropins, and the UV-B photoreceptor UVR8 are sensory photoreceptors able to perceive specific light signals that provide information about the dynamic status of canopy architecture. These signals include the low irradiance (indicating that not all the effects of irradiance occur via photosynthesis) and low red/far-red ratio typical of dense stands. The simulation of selected signals of canopy shade light and/or the analysis of photoreceptor mutants have revealed that canopy light signals exert significant influence on plant performance. The main effects of the photoreceptors include the control of (a) the number and position of the leaves and their consequent capacity to intercept light, via changes in stem height, leaf orientation, and branching; (b) the photosynthetic capacity of green tissues, via stomatic and nonstomatic actions; (c) the investment of captured resources into harvestable organs; and (d) the plant defences against herbivores and pathogens. Several of the effects of canopy shade-light signals appear to be negative for yield and pose the question of whether breeding and selection have optimised the magnitude of these responses in crops.

1. Light Signals in Crops

1.1. Light as a Source of Energy and Light as a Signal. The biomass produced by a crop can be accounted for by the product of three variables: the incident radiation, the efficiency to intercept the incident radiation, and radiation use efficiency (the relationship between plant dry matter and radiation intercepted), integrated for the duration of the growth cycle [1]. In turn, the yield of grain crops can be accounted for by the product of the biomass by the harvest index. Light has fundamental importance in crop yield due to its function in photosynthesis. The aim of this paper is to present an overview of the experimental evidence that supports the often neglected contribution of light as a signal (i.e., as a source of information) in crop yield. A light signal is a variable aspect of the light environment, perceived by specific sensory receptors, which affects selected plant traits. In this sense, while the action of light as a source of energy is explicit among the aforementioned components involved in biomass and yield generation, implicit in the other components there are effects of light as a source of information about the environment.

Grain crops experience two major groups of light signals: (a) light signals related to season and (b) light signals related to the status of the canopy. The first category is relatively simple; it includes photoperiod. Photoperiod changes with time of the year and also with latitude. Therefore, the photoperiods at which a crop is exposed will depend on sowing date and location, but they will change during the course of the growth period. The second category is somewhat more complex, as it includes changes in irradiance and spectral composition both in time and space. In this paper, we will focus on canopy light signals.

1.2. Plant Sensory Photoreceptors. We know the major photoreceptors present in plants, and therefore we can use the photoreceptors to unequivocally define signals; that is, there is no point in enumerating all the aspects of the light affected by the canopy, only those perceived by photoreceptors are justified. Plants have a diverse array of sensory photoreceptors that are involved (or at least predicted to be involved) in the perception of canopy light signals [2, 3]: phytochromes [4], cryptochromes [5], phototropins [6], and UVR8 [7] (Figure 1). The discovery of the UV-B photoreceptor UVR8

FIGURE 1: Phytochrome, cryptochrome, phototropin and the UV-B photoreceptor UVR8 perceive signals of the canopy light environment. The light incident on the canopy has high irradiance, high red/far-red ratio and high blue/green ratio. As light penetrates within the canopy, these values become reduced and the activity of the photoreceptors is also reduced.

is recent [8], and its potential role in plant adjustment to the degree of canopy shade is largely speculative. The function of a photoreceptor is to connect specific light signals to selected physiological responses via a signal transduction network. Phytochrome, cryptochrome, and phototropin molecules have an apoprotein and one or two chromophores, which are involved in light capture. In the case of UVR8, the UV-B signal is captured by selected amino acids.

Some of the photoreceptors listed above are not just a single photoreceptor but a family with several members encoded by different genes. The phytochrome apoproteins are encoded by a small family of genes involving three main clades: *PHYA, PHYB,* and *PHYC* [9]. In some species, the *PHYB* lineage includes different members. For instance, in *Arabidopsis thaliana,* the *PHYB* lineage includes *PHYB, PHYD,* and *PHYE,* and therefore, this species has five phytochromes (phytochromes A, B, C, D, and E). In tomato, the *PHYB* clade includes *PHYB, PHYB2,* and *PHYE,* and this species has five phytochromes (phytochromes A, B1, B2, E, and F or C) [10]. In grasses, the phytochrome gene family contains three members—*PHYA, PHYB,* and *PHYC*—[11, 12]. However, in maize (e.g., inbred B73), the *PHYA, PHYB,* and *PHYC* genes are duplicated, indicating the presence of six potentially functional phytochrome genes in this species: *PHYA1, PHYB1,* and *PHYC1* genes on chromosome 1, PHYA2 and PHYC2 on chromosome 5S, and *PHYB2* on chromosome 9L [13]. Maize phytochrome duplicate genes (homeologs) map to syntenic regions of the genome suggesting that these gene duplications were generated as a consequence of an ancient tetraploid event [13]. Plants have two types of cryptochromes: cryptochrome 1 and cryptochrome 2. In *Arabidopsis thaliana,* there are two cryptochrome genes: CRY1 and CRY2. However, in soybean, there are six cryptochromes: four cryptochrome 1 (GmCRY1a to GmCRY1d) and two cryptochrome 2 (GmCRY2a and GmCRY2b) [14].

This pattern can be accounted for by the paleotetraploid nature of soybean [14]. These are only some examples of the wider available knowledge concerning different photoreceptor families, which should help those who are not experts in the field to understand the literature on the species of their own interest.

If we consider the occurrence of different types (families) of photoreceptors and of different members within a photoreceptor family, it is clear that plants have a surprising multiplicity of photoreceptors. There is some degree of redundancy but also functional specificity. Photoreceptors connect specific light signals to selected physiological responses via signal transduction networks; different photoreceptors perceive different signals and/or control different processes. For instance, the action of phytochromes, cryptochromes, phototropins, and UVR8 is irradiance dependent; therefore, in principle, all these photoreceptors could perceive changes in irradiance associated to the degree of canopy shade (Figure 1). In some cases, whether the range of irradiance dependence of the photoreceptor matches the range of canopy-induced changes in irradiance remains to be elucidated, but given the daily variations in incoming radiation it is likely that at some point the different degrees of canopy shade will affect irradiance in the range of photoreceptor sensitivity. Plants respond to the range of canopy red/far-red ratios (see below), with fine sensitivity to small drops beneath the ratios provided by unfiltered sunlight in open places [15]. Based on the analysis of mutants in *Brassica rapa* [16], *Arabidopsis thaliana* [17], maize [18], sorghum [19], rice [20], and barley [21], it is possible to conclude that the responses to the red/far-red ratio is mediated mainly by phytochrome B. In some cases, various members of the phytochrome B clade contribute to mediate responses to the red/far-red ratio [22]. Phytochrome A also participates in the control of plant responses to canopy shade but apparently by perceiving irradiance, rather than red/far-red ratio [3, 23]. Phytochrome can perceive the red/far-red ratio because it has two forms: Pr, with maximum absorbance at 660 nm (i.e., red light), and Pfr, with maximum absorbance at 730 nm (i.e., far-red light). Upon excitation, Pr is photo-transformed to Pfr and Pfr is photo-transformed to Pr. Therefore, under natural radiation, the two forms reach a steady state that favours Pfr if the red/far-red ratio is high (open places) and Pr if the ratio is low (dense canopies). Only the Pfr form is biologically active, and therefore, phytochrome activity depends on the ratio between red and far-red light. These features correspond to all phytochromes, but in the case of phytochrome A, there are other processes [24] that make of it a good sensor of irradiance but not of red/far-red ratio.

At the other end of the function of photoreceptors are the target processes that they control, and at this point there is also redundancy and specificity. For instance, phytochrome B is the most important in shade-avoidance responses, cryptochromes are important to regulate the investment in photoprotective mechanisms, phototropins are crucial for the rapid and reversible positional adjustments of chloroplasts and leaves, and UVR8 is important in the control of UV-B screens [3]. In maize, the analysis of mutants of *PHYB1* or *PHYB2* genes has revealed some degree of subfunctionalization of these phytochromes because they show some differences

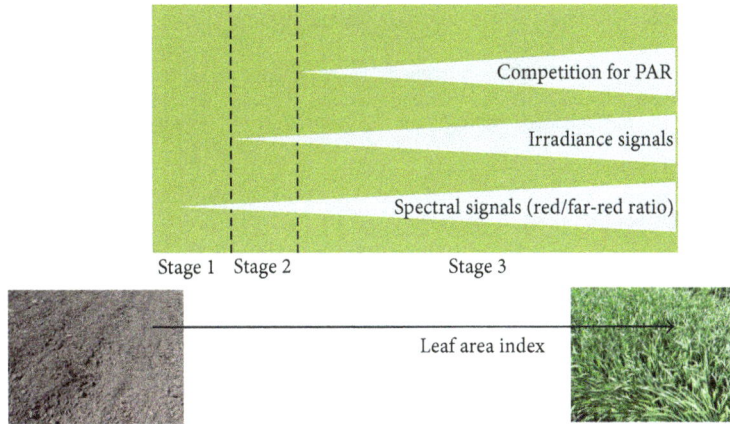

FIGURE 2: Intensity of the canopy shade-light signals and competition for PAR as affected by leaf area index. Three stages are distinguished: at low leaf area indexes (typically lower than 1), stage 1 is characterised by the presence of light signals involving spectral changes (mainly a reduction in the red/far-red ratio). At stage 2, the leaves shade the stem of neighbours, which receive low red/far-red ratios and low irradiance, but leaves do not shade mutually. At stage 3, mutual shading among leaves is established, and therefore the plants receive light signals (both spectral and irradiance signals) and compete for PAR.

in their respective contribution to different light responses [18].

1.3. Canopy Light Signals Are Dynamic. Canopy light signals depend on light attenuation patterns within the canopy and on the optical properties of the leaves. Canopy light attenuation depends on canopy architecture, which can be defined as the size, shape, and orientation of shoot components [25]. The leaf-area index, the arrangement of the plants within the crop (i.e., the relative distance between plants within the rows and between rows, the orientation of rows), and the more erectophile or planophile growth habit of the plants affect light attenuation within canopies and the degree of mutual shading among plants. The photosynthetic pigments (chlorophyll, carotenoids) present in leaves and other green organs absorb a large proportion of the photosynthetically active radiation (PAR, i.e., the radiation between 400 and 700 nm) that they intercept, whereas a small proportion, enriched in the green waveband, is transmitted and reflected. Wavebands out of the 400–700 nm range are also important in the generation of canopy light signals and green leaves absorb strongly in the UV range but much weakly in the far-red range (700–800 nm). Canopy signals of increasing intensity result from two types of changes of the light environment that can be perceived by sensory photoreceptors: (a) the attenuation of irradiance and (b) the change in spectral composition. The latter involves a decrease in the red/far-red ratio and a decrease of blue/green ratio. Phytochrome, cryptochrome, and UVR8 perceive changes in irradiance levels (mainly in the red plus far-red, blue plus UV-A, and UV-B regions of the spectrum, resp.). Phytochrome B perceives the red/far-red ratio [16–21]. Cryptochromes apparently also perceive the blue/green ratio [23, 26, 27]. Very often, plant responses to irradiance (in particular plant responses to PAR) are considered to be mediated by photosynthesis, and only the response to red/far-red is conceptually assigned the category of informational signals. This idea is not entirely correct

because the effects of irradiance can in part be mediated by photosensory receptors.

Figure 2 shows a diagrammatic representation of the dynamics of light signals and competition for PAR in a growing canopy, where three main stages can be distinguished. Canopy light signals anticipate competition for PAR as, at the first stage, before mutual shading among plants is established, selective reflection on green organs alters the spectral distribution of the light (mainly the red/far-red ratio) of the vertically oriented organs (e.g., stems). At this stage, plants surrounded by sparse neighbours receive more far-red light than fully isolated plans [15, 28–30]. At the second stage, the upper leaves project their shade on the stems of neighbours, and therefore they do not seriously affect the ability of neighbour leaves to capture PAR but canopy light signals become more intense. At the third stage, competition for PAR is established when there is mutual shading among leaves. The red/far-red ratio is in itself a very reliable signal of the status of a canopy [31]. In wheat, for instance, the red/far-red ratio measured at the base of the canopy decreases almost linearly with the leaf area index between 0 and 10 [32].

1.4. Canopy Light Signals in the Context of Other Effects of Neighbours. Canopy architecture affects different aspects of the environment, which go beyond changes in the light environment itself to include, for instance, the impact of wind (Figure 3). Actually, differences in canopy architecture can also be associated to differences in the availability of soil resources (water, nutrients). In turn, each one of these aspects of the environment can affect diverse physiological processes (Figure 3). For instance, if we focus on the light environment, increasing mutual shading in more developed or more densely sown crop canopies or as a result of the presence of weeds increases the magnitude of shade-light signals, enhances the competition for PAR, reduces the chances of damage caused either by excessive light absorption by photosynthetic pigments or by UV radiation, lowers plant

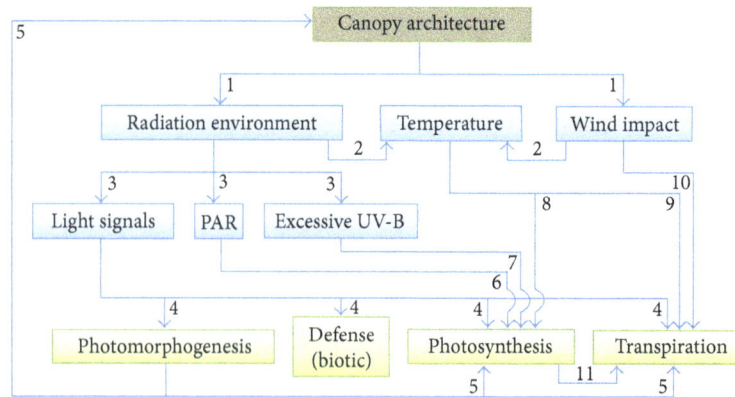

FIGURE 3: Canopy architecture affects multiple aspects of the environment experienced by the shoot, which in turn impact on multiple aspects of plant physiology. Environmental features are shown in blue, and plant processes are shown in green. Arrowheads indicate the direction of the effect. The model represents a simplification of the real world. Canopy architecture affects the sunlight radiation and wind impact received by each plant (1); these features of the environment in turn affect tissue temperature (2). The radiation environment involves three functional categories: the light signals (perceived by sensory receptors), PAR, and damaging UV-B (3). These categories overlap in terms of spectral wavebands (e.g., some wavebands are active both photosynthetically and as a signal). Light signals control plant form (photo-morphogenesis), the status of defences against pathogens and herbivores, and stomatal aperture, which modulates the rates of carbon and water vapour exchange (photosynthesis, respiration) (4). Photosynthesis and transpiration are also affected by photomorphogenesis (e.g., via changes in in stomatal density) (5). The altered photomorphogenesis also modifies canopy architecture (e.g., by changes in stem height, leaf position, etc.) (5). Both PAR (as the source of energy) (6) and UV-B (via its damaging effect) (7) affect photosynthesis, which also depends on temperature (8). Temperature (9), wind impact (10), and photosynthesis (via effects on carbon dioxide concentration and stomatal conductance) (11) affect transpiration rates.

tissue temperature by lowering the radiation load, and so forth (Figure 3).

Given the complex modification of the environment caused by increasing canopy density, estimating the exact quantitative contribution of the light signals perceived by photoreceptors can be cumbersome. Ideally, achieving this goal would require independent simulation of all the changes and their combinations, which is impossible at least with current techniques. However, the literature offers a very wide list of examples, where even suboptimal light signals cause significant effects on plant traits, demonstrating that light signals play a key role in crop performance. For a more detailed analysis of the methods to simulate canopy shade signals, we refer to previous publications [2, 33]. However, some discussion is useful here to illustrate the point. Very often, the low red/far-red ratios of canopy shade are simulated by adding varying amounts of far-red light to a common source of PAR. This procedure prevents confounding effects via photosynthesis, but it does not simulate the lower irra-diance signal also perceived by photoreceptors under dense canopies. Rather, increasing far-red lowers the red/far-red ratio, but it increases irradiance, and this can generate a signal typical of open, rather than shaded places, partially counteracting the impact of lower red/far-red.

1.5. Plant Responses to Canopy Light Signals. Canopy light signals perceived by sensory photoreceptors control several aspects of plant morphology and physiology. Shade sig-nals initiate the so-called shade-avoidance responses (e.g., increased stem growth) that tend to alleviate the degree of shade. In addition, open-canopy versus shade signals perceived by sensory photoreceptors, also help to acclimate

plants to the conditions that they cannot avoid. These responses are often context dependent. There is inter- and intraspecific genetic variation and other aspects of the envi-ronment (e.g., temperature and PAR) can strongly interact with the signals perceived by photoreceptors.

2. Stem Morphology and Physiology in Response to Canopy Light Signals

2.1. Canopy Shade-Light Signals Promote Stem Growth, but Not Always. One of the most obvious and widespread responses to canopy shade-light signals is the promotion of stem growth. In pioneer phytochrome studies, Downs et al. [34] used brief pulses of red or far-red light given at the end of the white light photoperiod (the so-called end-of-day light pulses) to, respectively, establish high or low levels of the active form of phytochrome (Pfr) and observed stem promo-tion in end-of-day far-red light-treated beans. The same methodology reported similar results in numerous species including tobacco (*Nicotiana tabacum*) [35] and tomato (*Solanum lycopersicon*) [36]. Adding far-red light to a source of white light to lower the daytime red/far-red ratio keeping similar levels of PAR has also been shown to promote the stem in many species, including mustard (*Sinapis alba*) [37], *Chenopodium album* [38], and cucumber (*Cucumis sativus*) [39]. Lowering only the red/far-red ratio reaching the stem (i.e., keeping the leaves at a high red/far-red ratio) either by adding far-red light or by reducing red light is effective to promote stem growth in *Sinapis alba* [37], *Datura ferox* [40], and sunflower (*Helianthus annuus*) [41], but the response is more persistent if also the leaves receive a low red/far-red ratio [42]. Stem growth also responds to selective reductions in irradiance in the blue light or red plus far-red light regions of the spectrum [40, 43, 44], which are signals, respectively, perceived by cryptochromes and phytochromes.

Although the promotion of stem growth by shade-light signals is well documented, not all the species have this response. In wheat, low red/far-red ratios may promote the growth of the basal internodes [45]. Although this effect can be significant in relative terms, these internodes remain short and make little contribution to overall plant height. However, the extension of the uppermost internode (peduncle) is delayed (rather than accelerated) by low red/far-red ratios [45, 46].

Stem growth can in principle affect crop yield in different ways: the stem may compete for resources with other parts of the plant, including harvestable organs [47], shorter stems can be more resistant to lodging [48], shorter stems can impair light interception [49], and so forth.

2.2. Stem Length and Stem Dry Matter Accumulation.

In addition to increase stem length, low red/far-red ratios often increase stem dry matter accumulation [35, 50, 51], but this is not necessarily always the case. In mustard (*Sinapis alba* L.), low red/far-red ratios reaching only the stem (a signal typical of sparse canopies) increase stem length but not its dry weight [51]. When both the leaves and the stem of mustard plants are exposed to low red/far-red ratios, the stem does increase its dry weight [52], but this in part is caused by an increased capacity of the leaves to export carbon thanks to the higher activity of the sucrose-phosphate synthase [53]. The case is different in sunflower, where lowering the stem red/far-red ratio is enough to increase its dry weight proportionally to its length increment [51]. In the latter species, if stem extension is physically blocked, the stem recovery of labelled carbon fed to the leaves is also reduced, indicating that increased dry weight would be the consequence of increased stem extension growth [51].

Compared to the wildtype, in the *lh* mutant of cucumber (*Cucumis sativus*), deficient in phytochrome B, stem length is more strongly enhanced than stem dry matter accumulation, and therefore, the *lh* mutant shows a lower dry matter/length ratio and a lower diameter [54]. Anatomical inspection revealed reduced cell diameter, reduced area occupied by load bearing tissues, and reduced number and diameter of xylem vessels in the stem of the *lh* mutant [54]. In turn, these anatomical differences showed functional implications for the *lh* mutant, including reduced field survival due to stem susceptibility to wind impact and reduced stem water conductivity [54].

2.3. Stem Growth Direction.

Stem phototropism is a response that has been extensively investigated under controlled conditions but less considered in field experiments. The blue-light gradient is perceived by phototropin [55]. *Arabidopsis* mutants deficient in phototropin show reduced survival under dense canopies [56]. The red/far-red ratio of the canopy, perceived by phytochrome B, can condition the stem phototropic response [57]. The phototropic response would help to colonise patchy canopies by directing the foliage to less crowded areas [58].

2.4. The Stem Metabolome in Response to Canopy Light Signals.

In sunflower, many metabolites (including reducing sugars and cell-wall carbohydrates) conserve their stem concentration despite the growth promotion of this organ caused by low red/far-red ratios selectively applied to the stem. However, some metabolites do change their concentration in response to low red/far-red ratios [51]. The levels of sucrose, tetradecanoic acid, pentadecanoic acid, and octadecanoic acid decrease in the upper and lower sections of the first internode, while the levels of asparagine 3 and octadecanol decrease only in the upper section. Conversely, the levels of galacturonate, glutarate, saccharate, fructose, and inositol increase in the upper and lower sections of the internode in response to low red/far-red ratios, and the levels of glutamate, pyroglutamate, hexadecanol, and campesterol increase only in the lower section [51].

2.5. Mechanisms Involved in the Control of Stem Growth by Light.

The molecular and cellular mechanisms involved in the promotion of stem growth by shade-light signals have been reviewed recently [3] and are beyond the scope of this paper. However, a brief summary of the current models is informative. The proteins that bind DNA and modify the rate of transcription of the nearby genes are called transcription factors. Many signals modify different aspects of plant physiology by altering the activity of selected transcription factors. The active form (Pfr) of phytochrome B is able to bind different members of the bHLH transcription factors called PHYTOCHROME INTERACTING FACTORS (PIF) in the nucleus. Binding by phytochrome B causes PIF phosphorylation, reduces the ability of PIFs to bind their DNA targets, and causes the labelling for destruction of some members of the PIF family. The low red/far-red signal indicative of shade shifts the steady state of phytochrome B from the active to the inactive (Pr) form, which migrates to the cytoplasm. Released from the negative regulation imposed by Pfr, PIFs bind the promoter region of their target genes, which include auxin synthesis genes. The levels of auxin are increased and stem growth becomes promoted. Connected to this short and simple set of signalling events, there is a complex set of regulatory networks linking light signalling to other hormones such as gibberellins and brassinosteroids.

Despite recent advances in *Arabidopsis thaliana*, the knowledge is more scant in other species and there might be differences. For instance, rice PIF-like protein OsPIL1, which is phylogenetically related to *Arabidopsis* PIF4 and PIF5, promotes growth in rice as PIF4 and PIF5 do in *Arabidopsis*. However, OsPIL1 does not interact with rice phytochrome B and might not be involved in the response to canopy signals [59]. Conversely, OsPIL1 expression is inhibited under drought stress leading to reduced stature [59].

One of the key elements of the "Green Revolution" was the introduction of dwarfing genes. Reduced stem growth reduced the risk of lodging and allowed the incorporation of Nitrogen fertilisation to wheat plants. In addition, these plants with reduced stem growth divert a higher proportion of their photoassimilates to the spike, rather than to the stem. Wheat dwarfing genes were shown many years later to encode DELLA proteins [60]. DELLA proteins are present in the nucleus, where they bind PIFs impeding their binding to DNA and therefore their ability to control transcription [61, 62]. In the presence of the growth-promoting hormone gibberellins, DELLAs are bound by a protein complex involving the activated receptor of gibberellins, and as a result of this, DELLAs become targeted for degradation [60]. DELLA degradation releases PIFs, which then can activate transcription of growth-promoting genes. The dwarfing genes introduced into elite wheat cultivars were mutant alleles of DELLA genes, which produce a mutant protein able to bind PIF but not recognised by the active receptor of gibberellins (the mutation specifically affects the domain of the DELLA proteins that is recognised by the receptor complex). Therefore, these mutant DELLA proteins arrest PIFs and growth even in the presence of gibberellins, causing dwarfism. Dwarfing alleles of DELLA genes reduce the responses to canopy shade signals in *Arabidopsis* [63].

2.6. The Energetic Cost of Enhanced Stem Growth in Response to Canopy Shade-Light Signals. The analysis of the stem-specific changes in the transcriptome of tomato plants transferred from white light to white light plus supplementary far-red light under controlled conditions revealed both rapid and persistent responses [64]. Not unexpectedly, given the strong promotion of stem growth by low red/far-red ratios, the treatment enhanced the expression of genes involved in auxin signalling and cell-wall carbohydrate metabolism. Noteworthy, low red/far red also reduced the expression of genes involved in flavonoid synthesis, isoprenoid metabolism, and dark reactions of photosynthesis. These changes in gene expression were reflected on stem-specific reductions in the levels of flavonoids (anthocyanin, quercetin, and kaempferol) and selected isoprenoid derivatives (chlorophyll and carotenoids) and in photosynthetic capacity. Changes in the levels of jasmonic acid could be involved in these responses. The rate of stem respiration was also strongly reduced in low red/far-red-treated plants. Therefore, by downsizing the stem photosynthetic apparatus and the levels of photoprotective pigments in response to shade-light signals, tomato plants reduce the energetic cost of shade avoidance responses [64]. This is important because shade-avoidance responses can coexist with limited availability of PAR due to mutual shading.

3. Branching in Response to Canopy Light Signals

3.1. Canopy Light Signals Reduce Branching. Low red/far-red ratios reduce tillering in cereals such as wheat [46, 65–67], barley [68], sorghum (*Sorghum bicolor*) [69], teosinte [70], and maize [71]. In maize, the effect is dependent on the cultivar [71]. A mutant of sorghum deficient in phytochrome B shows reduced bud outgrowth even under high red/far-red ratios [69]. Based on the analysis of spring wheat plants grown at three plant population densities with or without neutral shading, Evers et al. [67] have proposed that cessation of tillering is induced when the proportion of PAR intercepted by the canopy exceeds 40–45% and the red/far-red ratio is below 0.35–0.40. Similar responses to

low red/far-red ratios have been observed in forage grasses such as *Lolium multiflorum* [72]. The enrichment of red light beneath the canopy of a natural grasslands promoted tillering in *Paspalum dilatatum* and *Sporobolus indicus* plants [73]. Reduced branching in response to low red/far-red has also been reported for many eudicots, including tobacco [35], tomato [74], and *Trifolium repens* [75]. Tillering is important for grain yield in many crop conditions [76].

3.2. Mechanisms Involved in the Control of Tillering by Canopy Shade-Light Signals. Differences in tillering between maize (*Zea mays* sp. *mays*) and teosinte (*Zea mays* sp. *parviglumis*) can largely be accounted for by the higher (two-fold higher) expression of the *teosinte branched1* (*TB1*) gene in maize than in teosinte [77]. *TB1* encodes a putative basic helix-loop-helix transcription factor that represses the growth of axillary buds and enables the formation of female inflorescences [77]. In sorghum (*Sorghum bicolor*), supplementary far-red light represses bud outgrowth and promotes the expression of *TB1* in the buds [69]. The mutation of the gene encoding phytochrome B has the same consequence. These observations suggest a sequence of signalling events where low red/far-red ratios reduce phytochrome B activity and represses bud outgrowth by promoting the expression of *TB1*.

Another genetic variant in maize with defects in floral development and increased tiller number (six to seven compared to no tillers in the wildtype of the same genetic background) is the *grassy tillers1* (*gt1*) mutant [70]. The *GT1* gene encodes a transcription factor of the type named class I homeodomain leucine zipper (HD-Zip). This gene is expressed in shoot axillary buds, among other organs, and the protein can be found in the nucleus, as expected for a transcription factor [70]. Therefore, the GT1 protein is expected to act in the buds to repress their outgrowth; the *gt1* mutation releases this inhibition, increasing tillering. Teosinte and sorghum plants branch prolifically when grown without signals of neighbours. Lowering the red/far-red ratio promotes the expression of the *GT1* gene and lowers tillering in teosinte. A mutant of phytochrome B in sorghum shows enhanced expression of *GT1* and reduced tillering. Both, *TB1* and *GT1* are involved in the local control bud outgrowth and respond to red/far red. To investigate their functional relationship, the expression of each one of the two genes was investigated in plants mutant for the other. The expression of *TB1* was unaffected by the *gt1* mutation, but the expression of *GT1* was reduced in a *tbl* mutant compared to the wildtype [70]. Taken together, these observations suggest a model where low red/far-red ratios reduce the activity of phytochrome B favouring enhanced expression of *TB1*, which would lead to enhanced *GT1* expression and repression of bud outgrowth [70]. How phytochrome B controls *TB1* expression remains to be elucidated.

3.3. Tiller Death. In field experiments with wheat plants grown at different densities and with different Nitrogen availabilities, the start of tiller death was closely related to the red/far-red ratio reaching the base of the canopy, not to the PAR available per shoot or to Nitrogen levels. This result has been interpreted in terms of a critical red/far-red ratio for the initiation of tiller death in winter wheat [78].

4. Leaf Morphology and Physiology in Response to Canopy Light Signals

4.1. Leaf Position in Response to Canopy Signals. Many crops are sown in rows with a short distance between contiguous plants within the row and wider distances between adjacent rows. This rectangular pattern generates a heterogeneous red/far-red ratio environment, with high values towards the interrow spaces and low values within the row. In maize, these differences appear early in the ontogeny of the crop and persist to flowering [71]. Isolated maize plants grown in the field next to filters reflecting far-red light placed their leaves mainly perpendicular to the direction of the incoming reflected far red (i.e., parallel to the filters). Control plants with filters that did not lower the red/far-red ratio randomly oriented their leaves on the horizontal plane [71]. The ability to reorient leaf growth according to the light signals is cultivar specific and correlates with the ability to reorient leaf growth in response to rectangular sowing arrangements in the field. Maize leaves did not change the position of origin in the meristem. Rather, when far-red light was directed by means of fibre optics to the position where a leaf was expected to appear, the leaf growth turned away from the predicted direction towards any of the sides [71]. The ability to respond to the red/far-red signals reduced mutual shading among leaves, increasing their efficiency to intercept PAR [71].

The vertical angle of the leaves can also respond to canopy signals. Under low red/far-red ratios, the leaves often adopt a more erect position, particularly in rosette plants such as *Arabidopsis* (see [2], for references) but also in other species. Solar tracking by the upper leaves of sunflower cultivars is reduced when the plants are grown under low red/far-red ratios [79]. In some wheat cultivars, low red/far-red ratios induce a more erect position of the tillers [80].

4.2. Leaf Expansion in Response to Canopy Light Signals. The red/far-red ratio has normally large effects on stem growth but weak effects on leaf growth in eudicots. In sunflower, for instance, low red/far-red ratios reduce the early rate of leaf growth and promote leaf growth at later stages, largely not affecting the final size of the leaf [79]. There are, however, cases where leaf area is increased or decreased in response to low red/far-red ratios [81]. In grasses, including *Lolium multiflorum*, *Paspalum dilatatum*, and barley (*Hordeum vulgare*), low red/far-red [68, 82] or low blue light [83] promotes leaf sheath growth. This is a stem-like response that helps to place leaf lamina at higher strata within the canopy.

4.3. Leaf Senescence in Response to Canopy Shade-Light Signals. In dense sunflower crops, the senescence of basal leaves can anticipate anthesis. The beginning of senescence of target leaves of sunflower plants grown in the field at a very low density was advanced both by lowering PAR with the aid of neutral filters and by lowering the red/far-red ratio with the aid of mirrors placed beneath the leaves to selectively reflect far-red light [84]. Conversely, increasing the red/far-red ratio

received by basal leaves of plants grown at high density by means of red-light emitting diodes delayed senescence compared to non-irradiated controls. The effect of red-light enrichment was not photosynthetic as it increased the daily PAR integral in approximately 4%, and a 5% enrichment of PAR with green light had no significant effects [85].

4.4. Stomatal Conductance in Photoreceptor Mutants. The blue-light photoreceptors phototropins and cryptochromes affect stomatal conductance although via different pathways. Phototropins mediate the well-established promotion of stomatal aperture induced by blue light [86]. Contrary to the expectations, mutants of phototropins in *Arabidopsis* have demonstrated that these photoreceptors are important for full stomatal opening even at midday [87]. The effect of cryptochromes appears to be indirect; rather than perceiving the light stimulus causing immediate stomatal opening, cryptochromes would perceive the general status of the blue-light environment, reduce the levels of abscisic acid, and hence condition the subsequent response of stomatal opening to either blue or red light [87]. As for phototropins, the effects of mutations at cryptochrome genes reduces stomatal conductance more strongly at the high irradiances of midday [87]. As a result of these stomatal responses, both phototropins and cryptochromes increase transpiration at the time of the day when atmospheric water demand is maximal. The idea is that the action of these photoreceptors rather than saving water would help to reduce eventual midday limitations of photosynthesis by carbon dioxide availability, with their concomitant risk of diverting the energy captured by photosynthetic pigments to the generation of reactive oxygen species. The irradiance dependency of these responses argues in favour of their role in response to canopy shade [87].

The low red/far red [88] or low irradiances [89] perceived by phytochrome B reduce stomatal density and stomata index in *Arabidopsis*. This reduces stomatal conductance and transpiration and increases long-term water-use efficiency estimated by the analysis of isotopic discrimination against $^{13}CO_2$ [88]. In agreement with the results in *Arabidopsis*, rice mutants deficient in phytochrome B exhibit reduced stomatal density [90]. The enhanced transpiration caused by phytochrome B in open places would be detrimental for the plant water status, but this effect can be compensated by a reduction in leaf area [88] and a higher sensitivity of stomatal conductance to abscisic acid [91]. As discussed above for the effects of phototropin and cryptochrome, enhanced stomatal conductance would reduce the risk of generation of reactive oxygen species. In the case of rice, the phytochrome B mutants exhibited reduced total leaf area per plant, contributing to the lower rates of transpiration, and the root system showed no obvious differences [90].

4.5. Leaf Photosynthesis in Photoreceptor Mutants. Phototropins mediate the adjustment of chloroplast position to maximise the efficient use of PAR. Under high irradiance, the chloroplasts move to the anticlinal wall of palisade cells [92–94], decreasing light absorption and the risk of damage of the photosynthetic by excess excitation [95]. Under low irradiance, the chloroplasts accumulate at the periclinal wall of palisade cells [94], increasing efficient light capture [96].

The mutants lacking phototropins or cryptochromes have reduced photosynthesis in the field [87]. This is caused mainly by nonstomatic effects. The phytochrome B mutant also has reduced photosynthesis but these effects result from a combination of stomatic and nonstomatic limitations [88]. Rice phytochrome mutants also show reduced photosynthesis [90].

5. Root Responses to Canopy Light Signals

5.1. Root Growth. Root growth can be affected by canopy light signals. For instance, low red/far-red ratios reduce root growth in tobacco [35] and *Vigna unguiculata* [97], the number of rooted phytomers in *Trifolium repens* [75], and lateral root formation in *Arabidopsis thaliana* [98]. Some mutants deficient in phytochrome B show reduced root growth as it is the case in cucumber [54], *Lotus japonicus* [99], and *Arabidopsis thaliana* [98]. In *Arabidopsis*, shoot-localised phytochrome B is able to affect the flux of auxin to the root and control the growth of this organ. There are many cases, however, where red/far-red treatments [72] or phytochrome B mutations [90, 91] did not show obvious effects on root biomass accumulation.

5.2. Nodulation. Kasperbauer and Hunt [97] inoculated seeds of southern pea (*Vigna unguiculata* (L.) Walp.) with *Bradyrhizobium japonicum* and grew the plants in pots under photoperiods of 12 h of white light terminated with five minutes of red or far-red light to establish either high or low levels of active phytochrome at the beginning of the night. The far-red light treatment promoted stem elongation, but it reduced root growth and nodule number. Suzuki et al. [99] compared two mutants of *Lotus japonicus* deficient in phytochrome B to the wildtype MG20 grown under white light. The mutants showed approximately the same stature as the wildtype, but they produced less shoot biomass, root biomass, chlorophyll, and number of nodules. When the plants were grown under low red/far-red ratios, the number of nodules decreased in the wildtype and remained low in the mutants (at the same levels observed under high red/far-red ratios). In grafting experiments, the use of the phytochrome B mutant as the scion reduced the number of nodules regardless of the wildtype or mutant rootstocks, whereas the use of MG20 as a scion resulted in wildtype nodule number regardless of the rootstock genotype. These results indicate that phytochrome B in the shoot either stimulates the production of a signal that migrates to the root to promote nodule formation or reduces the production of a signal that migrates to the root to repress nodule formation. The endogenous levels of jasmonoyl isoleucine (the active jasmonic acid derivate) were reduced in the roots of the phytochrome mutant, compared to the MG20 wildtype of *Lotus japonicus* [99]. As expected, adding jasmonic acid at appropriate concentrations increased nodule number in the wildtype under low red/far-red ratios and in the phytochrome mutant grown under high red far-red ratios. Taken together, these observations indicate that plants of *Lotus japonicus* use phytochrome B

to monitor the red/far-red ratio reaching its shoot, and this signal controls the levels of jasmonoyl isoleucine in the root, which in turn controls nodulation. The suppression of nodule development in response to shade-light signals would be part of the mechanisms of autoregulation of nodulation, in this case favouring shade-avoidance over nitrogen fixation under increasing competition for PAR.

6. Reproductive Responses to Canopy Light Signals

6.1. Flowering Time. Low red/far-red accelerate flowering in barley [100] and *Lolium multiflorum* [101]. In soybean, low red/far-red ratios have been shown to correlate with delayed flowering, but since the experimental setting compared incandescent versus fluorescent lighting, other aspects of the light environments were also affected by the treatments [102]. In wheat, the red/far-red ratio has normally no effect on the final number of leaves on the main shoot [45, 46]. In *Arabidopsis*, low red/far-red ratios accelerate flowering by enhancing the expression of *FLOWERING LOCUS T (FT)* [103, 104], that is, the gene encoding the "florigen" involved in the induction of flowering by long days [104]. Consistently with this observation, in the long-day plant *Arabidopsis*, the acceleration of flowering under simulated shade light is maximal when the days are short [105].

6.2. Grain Number per Shoot. Libenson et al. [41] placed selective plastic filters around the stem of sunflower plants grown in large pots arranged at low densities in the field, to reduce the red/far-red ratio reaching the stem, without affecting PAR reaching the leaves. Compared to the controls bearing clear plastic filters, lowering the red/far-red ratio reaching the stem promoted the growth of this organ (both in terms of length and dry matter accumulation) and reduced grain number and grain yield per plant [41]. The promotion of stem growth by the light environment typical of dense commercial stands could reduce the resources available for grain yield in sunflower crops [41].

Heindl and Brun [106] used fluorescent tubes to enrich the red-light environment of the lower part of a soybean canopy during three weeks late in flowering. The treatment had no effects on the flowers produced per node, but it reduced flower abscission and increased seed yield per node [106]. Since field-grown plants were grown at high PAR levels, the effect of supplementary red light could be mediated by increased red/far-red ratios perceived by phytochrome, rather than by increased photosynthesis.

In wheat, the growth of the spike and its developmental progression are delayed by low red/far-red ratios achieved by supplementing sunlight with far-red light during the final hours of the photoperiod [46]. This treatment also delays the growth of the peduncle. Increasing plant densities reduces the total number of floret primordial initiated, increases the rate of floret abortion, and reduces the number of kernels per spike in wheat crops [107]. Similarly, low red/far-red ratios also reduce the number of florets at anthesis and the subsequent number of grains (and grain yield) as a result of reduced floret initiation and increased rate of floret decay [46]. This light signal accelerates the developmental progression of the florets but it causes the subsequent interruption of this progression, which is predicted to result in floret death before anthesis [46]. In these experiments, low red/far-red ratios did not increase wheat plant stature, and therefore it is not possible to account for the reduction of grain number on the basis of resources diverted to the growth of other organs. A more direct action of red/far-red ratio on the development of the spike has been proposed, consistently with the observed associated effects on the expression of developmental genes in the ear [46]. The analysis of the kinetic of spike and stem growth also suggests that floret death involves a developmental decision and is not just the consequence of scarcity of photoassimilates [108]. In rice, the triple *phyA phyB phyC* mutant shows reduced seed production, but this is mainly caused by impaired dehiscence of the anther wall and poor pollination [109].

7. Defence Status in Response to Canopy Light Signals

It is becoming clear that the light signals of dense canopies reduce the defences against biotic agents. These responses are not just the consequence of the morphological responses to canopy light signals. Rather, they are caused mainly by the action of photoreceptors on key points of the defence signalling networks (see [110] for a recent review).

7.1. Defence against Herbivores. Caterpillars of *Manduca sexta* show a higher mass increment when fed on plants of wild tobacco (*Nicotiana longiflora*) exposed to sunlight plus supplementary far-red light than on leaves grown under sunlight with a high red/far-red ratio [111]. Caterpillars of *Spodoptera eridania* grow better on mutant tomato plants lacking phytochromes B1 and B2 than in the isogenic wildtype [111]. The phytochrome B1 and B2 mutant of tomato is also more susceptible to damage by thrips (*Caliothrips phaseoli*), which show preference for the mutant leaves. Plants of *Arabidopsis thaliana* grown in pots at high density or at low density but with supplemental far-red light to provide a signal of neighbours produced leaf tissue that favours weight gain of *Spodoptera frugiperda* caterpillars [112]. In wild tobacco and *Arabidopsis thaliana*, low red/far-red ratio alters the expression of defence-related genes and inhibits the accumulation of herbivore-induced phenolic compounds [111, 112]. A priori, the reduced antiherbivore defences as a result of reduced phytochrome B activity caused by neighbour signals could be either the indirect consequence of shade-avoidance responses or the result of a more direct control of the defence signalling network by phytochrome. In favour of the second interpretation, the *sav3* mutant of *Arabidopsis thaliana*, deficient in auxin synthesis, shows impaired shade avoidance but retains downregulation of defences in response to low red/far-red ratios [112]. Plants grown under low red/far-red ratios show reduced sensitivity to jasmonate, a key hormone in the induction of antiherbivore defences [112]. The *lh* mutant of cucumber, deficient in phytochrome B, showed stronger damage by herbivores in the field, but actual shade had no effect [113].

There are several examples showing the effects of UV-B radiation on plant-herbivore interactions. The intensity of leaf tissue damage caused by phytophagous insects in young seedlings of *Datura ferox* grown in the field decreased with increasing UV-B between 0% and 100% of sunlight values [114]. Comparable differences were observed if UV-B was restricted at the time of insect exposure indicating that the UV-B effect is at least partially on the plant itself. However, in addition to plant responses, phytophagous insects can perceive and avoid UV-B [115]. Similar results were obtained in soybean crops [116] and natural ecosystems [114]. UV-B also reduced the growth of *Manduca sexta* caterpillars on plants of *Nicotiana longiflora* or *N. attenuate* [117]. Plant perception of UV-B causing enhanced resistance to insect herbivores could be mediated by UVR8, but testing this idea awaits evaluation of mutant plants lacking this photoreceptor. Meanwhile, it is not strictly possible to exclude that enhanced resistance derives from UV-B damage.

7.2. Defence against Pathogens. When sprayed with a conidia suspension of the blast fungus *Magnaporthe grisea*, wildtype seedlings of rice developed symptoms in the youngest but not in the oldest leaves. When exposed to the latter treatment, the *phyA phyB phyC* mutant of rice (lacking all phytochromes) shows wildtype density of lesions in young leaves, but it exhibits more than two lesions per cm^2 even in old leaves [118]. When inoculated with *Magnaporthe grisea* or when treated with the defence-related hormones jasmonate or salicylic acid, the wildtype shows enhanced accumulation of the pathogenesis-related class 1 (PR1) protein in old leaves 20 or 24 h later [118]. In young leaves, PR1 accumulation is observed to some degree 48 h after inoculation and reaches the level observed in old leaves at 72 h. In the *phyA phyB phyC* triple mutant, PR1 accumulation is observed only in old leaves but 72 h after inoculation. These observations indicate that in wildtype rice, the leaves develop an age- and phytochrome-dependent ability to respond to fungal infection (and to downstream signals), which is related to the levels of PR1. In addition to the reduced induction of PR1, the triple mutant also had reduced basal levels of PR1 [118].

Wildtype plants of *Arabidopsis* exposed to low red/far-red ratios or mutant for phytochrome B grown at high red/far-red ratios develop more intense disease symptoms than the wildtype grown under high red/far-red ratios when inoculated with a spore suspension of gray mould pathogen *Botrytis cinerea* [110]. The increased susceptibility to *B. cinerea* as a result of reduced phytochrome B activity (either by the light condition or the mutation) is apparently not the consequence of the morphological responses induced by lowering phytochrome B activity. In fact, a mutant deficient in auxin synthesis is severely impaired in morphological responses to low red/far-red ratios while it retains normal defence responses to low red/far-red ratios [110].

Low levels of UV-B enhance the resistance of glasshouse-grown plants of *Arabidopsis* to *B. cinerea*. The level of lesions observed in the mutant lacking UVR8 was similar to that of wildtype plants in the minus UV-B controls, but this mutant failed to respond to UV-B [110]. These observations indicate that UVR8-mediated perception of UV-B enhances resistance to *B. cinerea*, which is therefore not an indirect effect of UV-B-induced damage. The mechanism of action could involve the control of the levels of sinapates by UVR8 [110]. Canopy shade could therefore reduce plant defences by reducing UVR8 activity [110].

Plants of *Arabidopsis* lacking phytochrome B show enhanced growth of incompatible strains of *Pseudomonas syringae* [119, 120]. A screening for plants with defects in shade-avoidance responses helped to identify the *constitutive shade avoidance* (csa) mutant affected in the TOLL/INTERLEUKIN1 RECEPTOR-NUCLEOTIDE BINDING SITE-LEUCINE-RICH REPEAT (TIR-NBS-LRR) gene [120]. TIR-NBS-LRR proteins had previously been implicated in defence responses, indicating an intimate link between the control of plant growth and development by light and plant defences.

8. Early Responses to Canopy Light Signals Impact Subsequent Crop Performance

In growing canopies, phytochrome perception of low red/far-red ratios caused by far-red light reflected on neighbours anticipates mutual shading among plants [15] (Figure 2). These reductions in red/far-red ratio are small but enough to cause plant responses. For instance, seedlings of *Datura ferox, Sinapis alba,* or *Chenopodium album* showed enhanced stem growth (i.e., a typical shade-avoidance response) when exposed to far-red light reflected either on selective mirrors or on green neighbours placed opposite to the side of incoming sunlight (to avoid shading), compared to the controls with senescent (nongreen) neighbours or red plus far-red reflecting mirrors [15].

Markham and Stoltenberg [121] have proposed that the early season red/far-red ratio is important for subsequent grain yield under field conditions. They conducted field experiments, where equidistantly spaced corn plants were grown at 107,600, 53,800, and 3000 plants ha^{-1} from emergence to canopy closure (V7), when all the treatments were thinned to 3000 plants ha^{-1}. The red/far-red ratio decreased with plant density and apparently caused a reduction in the number of tillers per plant, which was associated with a lower per-plant grain yield at a later stage. The availability of PAR, the gravimetric soil water content, and the soil nitrate-nitrogen, phosphorus, or potassium contents were not affected by the early plant density treatments, and therefore cannot account for the observed differences in yield.

The critical period of weed control defines the number of weeks after emergence during which the crop must be weed-free in order to prevent crop yield losses beyond an acceptable amount [122]. This empirical concept is useful in weed management, but in the absence of a deeper understanding of the underlying processes controlling the susceptibility of the crop to the presence of weeds, the values are difficult to extrapolate [122]. One of the obvious mechanisms by which weeds can reduce crop yield is by capturing resources that then become scant for the crop plant. Consistently with the latter interpretation, the presence of weeds in maize crops can anticipate the development of water-deficit or Nitrogen-deficiency symptoms, as well as reduce the interception of PAR [122].

However, the effects of weeds on the crop often appear more strongly linked to a reduced capacity of the crop plant to increase root and leaf area surface to capture resources than to an actual scarcity of resources [122]. Therefore, signals and not just resources, could be important in the interactions between weed and crop plants. The impact of red/far-red light signals produced by weeds on the growth of maize plants was investigated in a series of experiments, where maize plants were grown in a field fertigation system at a low plant density, with or without the presence of neighbours weeds (a mix of *Lolium perenne* L. and *Poa pratensis* L., *L. peremne* alone, or *Amaranthus retroflexus* plants). The weeds were grown in separate pots, fed by separate fertigation lines (ensuring no water movement between root systems) and maintained by manual clipping to prevent direct shading of maize seedlings. Therefore, the weeds did not compete for PAR, water, or nutrients with the crop, but they reduced the red/far-red ratio received by maize plants due to reflected far-red light. Maize plants developing in the weedy environment displayed increased plant height, reduced leaf area, and a transiently increased shoot/root ratio caused by a reduced number of nodal roots and reduced root biomass [123, 124]. Compared with maize growing in the weed-free treatment, the presence of weeds until silking caused a 20% reduction in ear dry weight and most of the effect was caused by the presence of weeds beyond the 8-leaf tip stage [125], but even when weed neighbours were removed 30 days after emergence of the maize crop, reduced kernel number per plant and harvest index were observed in weedy compared to weed-free plants at maturity [123]. The experimental support for the idea of long-term consequences of early weed signals is not limited to maize. A comparable experimental setting demonstrated increased soybean internode elongation, reduced branching, and decreased yield per plant in response to the upwards reflection of far-red light by neighbouring weeds, compared to weed-free controls with high red/far-red ratios [126].

The early low red/far-red ratio signals caused by the presence of weeds can also affect the ability of crop plants to cope with abiotic stress. In some of the maize experiments described in the previous paragraph, after the neighbours were eliminated, fertigation was interrupted 3–5 d in half of the plants of each previous neighbour condition to create a moderate abiotic stress. Maize grain yield and kernel number per plant at maturity were influenced by interactions of early neighbour treatments and stress. The apparent synergy of stresses indicates that early shade avoidance can reduce the tolerance of maize plants to subsequent stressors [123].

9. Genetic Modification of the Impact of Canopy Light Signals

Some (but not all) of the responses to the light signals of dense canopies appear detrimental for yield, but it is not possible to reduce these signals in most commercial crops. However, it would be feasible to search for genotypes with optimised responses to canopy light signals. In this section, we address three issues related to the genetic variability in plant responses to the light signals of dense canopies. First, we will analyse the use of transgenic plants with increased levels of photoreceptors. Second, we will consider whether breeding and selection for yield are leading to cultivars less responsive to the light signals of dense canopies. Third, we will argue the possibility of nonadverted choice of strong or weak plant responses to canopy light signals with the conscious choice of cycle duration.

9.1. Transgenic Plants. Increased interference among plant shoots reduces the activity of photoreceptors (Figure 1). If, due to transgenic expression, plants have elevated levels of photoreceptor molecules, a higher absolute amount would remain active even under shade, and this could result in the reduction of plant responses to canopy density. This idea has been tested in several crops.

Robson et al. [127] cultivated transgenic tobacco plants expressing the gene of phytochrome A (*PHYA*) from *Avena sativa* in the field. High levels of phytochrome A impaired the promotion of stem growth normally caused by the canopy light signals of increasing plant densities. Actually, the presence of neighbours reduced plant stature instead of the normal growth promotion, likely because the transgenic plants responded to far-red light reflected on neighbours as more light rather than as a lower red/far-red ratio. In wildtype crops of tobacco, increasing plant densities have only a moderate influence on plant inequalities because shorter plants become more intensely shaded and induce a stronger shade-avoidance response that tends to compensate the differences [128]. This is not the case in transgenic crops with elevated levels of phytochrome A, unable to respond normally to canopy shade signals, where small plants become rapidly suppressed by their taller neighbours [128]. Therefore, shade-avoidance reactions can contribute to crop yield by reducing the chances of occurrence of very small plants that capture some resources but do not generate yield [129].

Expression of the *Arabidopsis* phytochrome B (*PHYB*) gene in transgenic potato plants increased tuber number and tuber yield in field crops planted at high densities [130]. High phytochrome B levels reduced the decay of maximum photosynthesis normally observed in older leaves, which are more intensively shaded due to their basal position [130]. The higher rates of maximum photosynthesis were related to increased stomatal conductance, largely due to increased aperture of the stomatal pore, rather than increased stomatal density. In glasshouse experiments, these transgenics showed delayed leaf senescence, lower carbohydrate, and higher Nitrogen levels in leaf and stem tissue, leaf photosynthesis, and conductance [131]. However, in contrast to the field experiments, increased conductance was caused by a higher stomatal density. Furthermore, measurements of intercellular leaf CO_2 partial pressure point to nonstomatic limitations of photosynthesis [131]. Although the transgenics bearing high levels of phytochrome B had higher photosynthesis per unit leaf area, they showed reduced ability to cover the soil, underscoring the importance of adequate sensing the canopy light signals to optimally accommodate the leaves, reducing mutual interference [130]. The simplest interpretation of these results is that in dense canopies, plants with higher levels of photoreceptors due to transgenic expression maintain a higher absolute number of active photoreceptor molecules

than the wildtype. As a result of this, some of the effects of increased canopy densities are attenuated. In addition, it is important to note that transgenic modification of photoreceptor levels not only alters the perception of the light environment, but it also modifies the environment itself [130]. In effect, increased photoreceptor levels lead to a shorter, bushier morphology and reduced shading among plants in different rows. In addition, since these plants had a reduced ability to project their foliage towards shade-free areas, the degree of shading within the row was increased [130].

Transgenic japonica rice cv. Nakdong plants expressing *Arabidopsis* phytochrome A grown under sunlight showed reductions in the length of the culm, panicle and leaves, and increments in grain weight and size, which depended on the strength of phytochrome expression [132]. However, transgenic plants showed smaller tiller number and low grain fertility compared to wildtype plants causing yield reductions. Transgenic expression of phytochrome in the indica rice variety Pusa Basmati-1 also reduced plant stature and slightly increased grain weight, but it increased panicle number and grain yield per plant [133]. The different results might reflect differences between the cultivars.

9.2. Breeding and Selection. We have presented examples from different crops where canopy shade-light signals tend to reduce yield potential or yield-related traits. One might therefore expect breeding and selection for yield to reduce the impact of these signals on yield per plant. To test this hypothesis, Ugarte et al. [46] compared the response of tillering, grain yield, grain number, and weight of 1000 grains to red/far-red ratio in ten cultivars released to the Argentinean market at different times of the 20th century. Against the expectations, the most modern cultivars of the series did not respond less than the oldest. Actually, some traits showed the opposite pattern. Thus, breeding and selection for yield are not reducing the impact of the negative control of yield by low red/far-red ratios. One of the scenarios that could account for this pattern is that in commercial crops the low red/far-red ratio could initiate an early adjustment of yield potential that the plant would otherwise experience afterwards, when more intense mutual plant shading occurs. In the field, the red/far-red ratio and the subsequent availability of resources would be correlated and the signal could not seriously reduce yield below the potential. In the low density plants used by Ugarte et al. [46], the low red/far-red treatments would reduce the generation of yield components below the potential because the plants did not mutually shade each other at a later stage; that is, under these conditions, the low red/far-red signal would not correlate with a subsequent scarcity of resources due to mutual shading.

9.3. Does Selection for Cycle Length Affect the Responses to Canopy Shade-Light Signals? The photoperiodic regulation of flowering helps to adjust the duration of crop cycle to the ecological conditions and agricultural needs. The perception of daylength requires the action of photoreceptors to distinguish between light and darkness, and the action of the circadian clock to restrict the sensitivity to light to a given portion of the day [134]. The photoreceptors involved

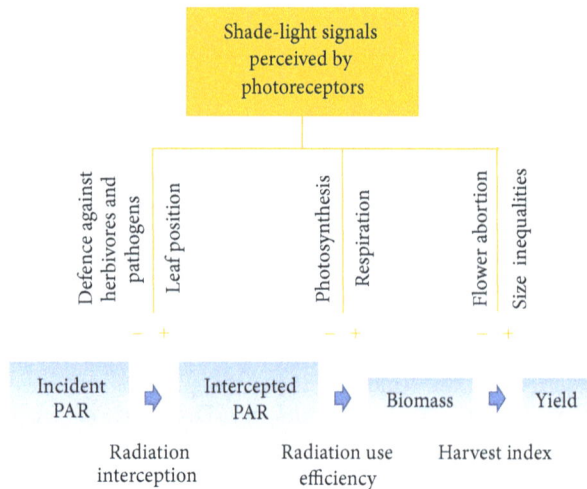

FIGURE 4: Canopy light signals impact on crop yield. Incident radiation is intercepted by the canopy with an efficiency that depends on canopy architecture; intercepted radiation is transformed into biomass with an efficiency that depends on the processes that fix (photosynthesis) and release (mitochondrial respiration, photorespiration) carbon dioxide; part of the biomass is allocated to harvestable organs. The light signals of dense canopies have both positive and negative effects on these processes, and selected examples of the processes that impact on PAR interception, radiation use efficiency, and harvest index are included.

in daylength perception often control other aspects of plant growth and development such as the stem growth response to shade signals. In soybean, the abundance of cryptochrome 1 shows a circadian rhythm controlled by photoperiod and correlates with photoperiodic flowering and latitudinal distribution of soybean cultivars [14]. When expressed in transgenic *Arabidopsis* seedlings under the control of a constitutive promoter, soybean cryptochrome 1 inhibited hypocotyl growth in response to blue light and rescued the long hypocotyl phenotype of the *Arabidopsis* mutant deficient in cryptochrome 1 [14]. In soybean, nine maturity loci (E1 to E8 and J) have been identified, and the E3 and E4 genes encode copies of phytochrome A genes (*GmPHYA3* and *GmPHYA2*) [135]. These observations suggest that selection for different daylength responses and cycle lengths could be driving selection for different degrees of response to shade.

10. Concluding Remarks

Coming back to the beginning of this paper, crop yield is the integral for the growth period of the product of the incident PAR, PAR interception, radiation use efficiency, and harvest index [1], and canopy light signals can affect these terms (Figure 4). Canopy light signals can increase PAR interception, for instance, by guiding leaf and stem growth direction towards the gaps within the canopy. However, canopy light signals can reduce PAR interception by reducing the number and/or area (at least the photosynthetically effective area) of the leaves as a consequence of reduced branching and defences against herbivores and pathogens. Canopy signals could reduce radiation use efficiency by reducing photosynthesis per unit leaf

area via stomatic and/or nonstomatic responses. However, canopy light signals could increase radiation use efficiency by lowering the rate of respiration of selected organs. Canopy signals can reduce harvest index, for instance, by enhancing floret abortion. However, canopy signals can reduce size inequalities and the chance of occurrence of small plants that capture resources but fail to efficiently translate them into yield. Finally, in some crops, canopy light signals can shorten the duration of the cycle by accelerating flowering. The available evidence clearly support supports the strong influence of canopy light signals on crop yield, but we are far from predicting the balance among the positive and negative forces.

Acknowledgments

The author thanks *Agencia Nacional de Promoción Científica y Tecnológica* (ANPCYT, Argentina), University of Buenos Aires, The International Centre for Genetic Engineering and Technology, and *Consejo Nacional de Investigaciones Científicas y Técnicas* (CONICET, Argentina) for their financial support.

References

[1] J. L. Monteith and C. J. Moss, "Climate and the efficiency of crop production in Britain," *Phylosophical Transactions of the Royal Society of London B*, vol. 281, pp. 277–294, 1977.

[2] J. J. Casal, "Shade avoidance," *The Arabidopsis Book*, vol. 10, Article ID e0157, 2012.

[3] J. J. Casal, "Photoreceptor signaling networks in plant responses to shade," *Annual Review of Plant Biology*, vol. 64, 2013.

[4] J. Li, G. Lib, H. Wang et al., "Phytochrome signaling mechanisms," *The Arabidopsis Book*, vol. 9, Article ID e0148, 2011.

[5] X. Yu, H. Liu, J. Klejnot et al., "The cryptochrome blue light receptors," *The Arabidopsis Book*, vol. 8, Article ID e0135, 2010.

[6] W. R. Briggs and J. M. Christie, "Phototropins 1 and 2: versatile plant blue-light receptors," *Trends in Plant Science*, vol. 7, no. 5, pp. 204–210, 2002.

[7] M. Heijde and R. Ulm, "UV-B photoreceptor-mediated signalling in plants," *Trends in Plant Science*, vol. 17, pp. 230–237, 2012.

[8] L. Rizzini, J. J. Favory, C. Cloix et al., "Perception of UV-B by the arabidopsis UVR8 protein," *Science*, vol. 332, no. 6025, pp. 103–106, 2011.

[9] S. Mathews, "Phytochrome-mediated development in land plants: red light sensing evolves to meet the challenges of changing light environments," *Molecular Ecology*, vol. 15, no. 12, pp. 3483–3503, 2006.

[10] R. Alba, P. M. Kelmenson, M. M. Cordonnier-Pratt, and L. H. Pratt, "The phytochrome gene family in tomato and the rapid differential evolution of this family in angiosperms," *Molecular Biology and Evolution*, vol. 17, no. 3, pp. 362–373, 2000.

[11] S. Mathews and R. A. Sharrock, "The phytochrome gene family in grasses (Poaceae): a phylogeny and evidence that grasses have a subset of the loci found in dicot angiosperms," *Molecular Biology and Evolution*, vol. 13, no. 8, pp. 1141–1150, 1996.

[12] R. Kulshreshtha, N. Kumar, H. S. Balyan et al., "Structural characterization, expression analysis and evolution of the red/far-red sensing photoreceptor gene, phytochrome C (*PHYC*), localized on the 'B' genome of hexaploid wheat (*Triticum aestivum* L.)," *Planta*, vol. 221, no. 5, pp. 675–689, 2005.

[13] M. J. Sheehan, P. R. Farmer, and T. P. Brutnell, "Structure and expression of maize phytochrome family homeologs," *Genetics*, vol. 167, no. 3, pp. 1395–1405, 2004.

[14] Q. Zhang, H. Li, R. Li et al., "Association of the circadian rhythmic expression of GmCRY1a with a latitudinal cline in photoperiodic flowering of soybean," *Proceedings of the National Academy of Sciences of the United States of America*, vol. 105, no. 52, pp. 21028–21033, 2008.

[15] C. L. Ballaré, R. A. Sánchez, A. L. Scopel, J. J. Casal, and C. M. Ghersa, "Early detection of neighbour plants by phytochrome perception of spectral changes in reflected sunlight," *Plant, Cell and Environment*, vol. 10, no. 7, pp. 551–557, 1987.

[16] P. F. Devlin, S. B. Rood, D. E. Somers, P. H. Quail, and G. C. Whitelam, "Photophysiology of the elongated internode (*ein*) mutant of *Brassica rapa*: *ein* mutant Lacks a detectable phytochrome B-like polypeptide," *Plant Physiology*, vol. 100, no. 3, pp. 1442–1447, 1992.

[17] P. R. H. Robson, G. C. Whitelam, and H. Smith, "Selected components of the shade-avoidance syndrome are displayed in a normal manner in mutants of *Arabidopsis thaliana* and *Brassica rapa* deficient in phytochrome B," *Plant Physiology*, vol. 102, no. 4, pp. 1179–1184, 1993.

[18] M. J. Sheehan, L. M. Kennedy, D. E. Costich, and T. P. Brutnell, "Subfunctionalization of *PhyB1* and *PhyB2* in the control of seedling and mature plant traits in maize," *Plant Journal*, vol. 49, no. 2, pp. 338–353, 2007.

[19] K. L. Childs, F. R. Miller, M. M. Cordonnier-Pratt, L. H. Pratt, P. W. Morgan, and J. E. Mullet, "The sorghum photoperiod sensitivity gene, *Ma3*, encodes a phytochrome B," *Plant Physiology*, vol. 113, no. 2, pp. 611–619, 1997.

[20] M. Takano, N. Inagaki, X. Xie et al., "Distinct and cooperative functions of phytochromes A, B, and C in the control of deetiolation and flowering in rice," *Plant Cell*, vol. 17, no. 12, pp. 3311–3325, 2005.

[21] M. Hanumappa, L. H. Pratt, M. M. Cordonnier-Pratt, and G. F. Deitzer, "A photoperiod-insensitive barley line contains a light-labile phytochrome B," *Plant Physiology*, vol. 119, no. 3, pp. 1033–1040, 1999.

[22] K. A. Franklin, U. Praekelt, W. M. Stoddart, O. E. Billingham, K. J. Halliday, and G. C. Whitelam, "Phytochromes B, D, and E act redundantly to control multiple physiological responses in Arabidopsis," *Plant Physiology*, vol. 131, no. 3, pp. 1340–1346, 2003.

[23] R. Sellaro, M. Crepy, S. A. Trupkin et al., "Cryptochrome as a sensor of the blue/green ratio of natural radiation in Arabidopsis," *Plant Physiology*, vol. 154, no. 1, pp. 401–409, 2010.

[24] J. Rausenberger, A. Hussong, S. Kircher et al., "An integrative model for phytochrome B mediated photomorphogenesis: from protein dynamics to physiology," *PloS ONE*, vol. 5, no. 5, Article ID e10721, 2010.

[25] G. A. Maddonni, M. E. Otegui, and A. G. Cirilo, "Plant population density, row spacing and hybrid effects on maize canopy architecture and light attenuation," *Field Crops Research*, vol. 71, no. 3, pp. 183–193, 2001.

[26] R. Banerjee, E. Schleicher, S. Meier et al., "The signaling state of Arabidopsis cryptochrome 2 contains flavin semiquinone," *The Journal of Biological Chemistry*, vol. 282, no. 20, pp. 14916–14922, 2007.

[27] J. P. Bouly, E. Schleicher, M. Dionisio-Sese et al., "Cryptochrome blue light photoreceptors are activated through interconversion of flavin redox states," *The Journal of Biological Chemistry*, vol. 282, no. 13, pp. 9383–9391, 2007.

[28] C. L. Ballaré, A. L. Scopel, and R. A. Sánchez, "Photomodulation of axis extension in sparse canopies," *Plant Physiology*, vol. 89, pp. 1324–1330, 1989.

[29] C. L. Ballaré, A. L. Scopel, and R. A. Sánchez, "Far-red radiation reflected from adjacent leaves: an early signal of competition in plant canopies," *Science*, vol. 247, no. 4940, pp. 329–332, 1990.

[30] H. Smith, J. J. Casal, and G. M. Jackson, "Reflection signals and the perception by phytochrome of the proximity of neighbouring vegetation," *Plant, Cell and Environment*, vol. 13, no. 1, pp. 73–78, 1990.

[31] H. Smith, "Light quality, photoperception and plant strategy," *Annual Review of Plant Physiology*, vol. 33, pp. 481–518, 1982.

[32] M. Chelle, J. B. Evers, D. Combes, C. Varlet-Grancher, J. Vos, and B. Andrieu, "Simulation of the three-dimensional distribution of the red:far-red ratio within crop canopies," *The New Phytologist*, vol. 176, no. 1, pp. 223–234, 2007.

[33] C. Fankhauser and J. J. Casal, "Phenotypic characterization of a photomorphogenic mutant," *Plant Journal*, vol. 39, no. 5, pp. 747–760, 2004.

[34] R. J. Downs, S. B. Hendricks, and H. A. Borthwick, "Photoreversible control of elongation of pinto beans and other plants under normal conditions of growth," *Botanical Gazette*, vol. 118, pp. 199–208, 1957.

[35] M. J. Kasperbauer, "Spectral distribution of light in a tobacco canopy and effects of end-of-day light quality on growth and development," *Plant Physiology*, vol. 47, pp. 775–778, 1971.

[36] I. W. Selman and E. O. S. Ahmed, "Some effects of far-red irradiation and gibberellic acid on the growth of tomato plants," *Annals of Applied Biology*, vol. 50, pp. 479–485, 1962.

[37] D. C. Morgan, T. O'Brien, and H. Smith, "Rapid photomodulation of stem extension in light-grown *Sinapis alba* L.—studies on kinetics, site of perception and photoreceptor," *Planta*, vol. 150, no. 2, pp. 95–101, 1980.

[38] D. C. Morgan and H. Smith, "Linear relationship between phytochrome photoequilibrium and growth in plants under simulated natural radiation," *Nature*, vol. 262, no. 5565, pp. 210–212, 1976.

[39] C. L. Ballaré, J. J. Casal, and R. E. Kendrick, "Responses of light-grown wild-type and long-hypocotyl mutant cucumber seedlings to natural and simulated shade light," *Photochemistry and Photobiology*, vol. 54, pp. 819–826, 1991.

[40] C. L. Ballaré, A. L. Scopel, and R. A. Sánchez, "Photocontrol of stem elongation in plant neighbourhoods: effects of photon fluence rate under natural conditions of radiation," *Plant, Cell and Environment*, vol. 14, no. 1, pp. 57–65, 1991.

[41] S. Libenson, V. Rodriguez, R. A. Sánchez et al., "Low red to far-red ratios reaching the stem reduce grain yield in sunflower," *Crop Science*, vol. 42, pp. 1180–1185, 2002.

[42] J. J. Casal and H. Smith, "The loci of perception for phytochrome control of internode growth in light-grown mustard: promotion by low phytochrome photoequilibria in the internode is enhanced by blue light perceived by the leaves," *Planta*, vol. 176, no. 2, pp. 277–282, 1988.

[43] J. J. Casal and R. A. Sánchez, "Impaired stem-growth responses to blue light irradiance in light-grown transgenic tobacco seedlings overexpressing *Avena* phytochrome A," *Physiologia Plantarum*, vol. 91, pp. 268–272, 1994.

[44] J. J. Casal, "Stem extension-growth responses to blue light require Pfr in tomato seedlings but are not reduced by the low phytochrome levels of the aurea mutant," *Physiologia Plantarum*, vol. 91, pp. 263–267, 1994.

[45] J. J. Casal, "Novel effects of phytochrome status on reproductive shoot growth in *Triticum aestivum* L.," *The New Phytologist*, vol. 123, pp. 45–51, 1993.

[46] C. C. Ugarte, S. A. Trupkin, H. Ghiglione, G. Slafer, and J. J. Casal, "Low red/far-red ratios delay spike and stem growth in wheat," *Journal of Experimental Botany*, vol. 61, no. 11, pp. 3151–3162, 2010.

[47] V. O. Sadras and R. F. Denison, "Do plant parts compete for resources? An evolutionary viewpoint," *The New Phytologist*, vol. 183, no. 3, pp. 565–574, 2009.

[48] M. J. Crook and A. R. Ennos, "Stem and root characteristics associated with lodging resistance in four winter wheat cultivars," *Journal of Agricultural Science*, vol. 123, no. 2, pp. 167–174, 1994.

[49] R. Wells, J. W. Burton, and T. C. Kilen, "Soybean growth and light interception: response to differing leaf and stem morphology," *Crop Science*, vol. 33, pp. 520–524, 1993.

[50] C. L. Ballaré, A. L. Scopel, and R. A. Sánchez, "On the opportunity cost of the photosynthate invested in stem elongation reactions mediated by phytochrome," *Oecologia*, vol. 86, no. 4, pp. 561–567, 1991.

[51] M. A. Mazzella, M. I. Zanor, A. R. Fernie, and J. J. Casal, "Metabolic responses to red/far-red ratio and ontogeny show poor correlation with the growth rate of sunflower stems," *Journal of Experimental Botany*, vol. 59, no. 9, pp. 2469–2477, 2008.

[52] J. J. Casal, R. A. Sanchez, A. R. Paganelli-Blau, and M. Izaguirre, "Phytochrome effects on stem carbon gain in light-grown mustard seedlings are not simply the result of stem extension-growth responses," *Physiologia Plantarum*, vol. 94, no. 2, pp. 187–196, 1995.

[53] M. J. Yanovsky, J. J. Casal, G. L. Salerno, and R. A. Sanchez, "Are phytochrome-mediated effects on leaf growth, carbon partitioning and extractable sucrose-phosphate synthase activity the mere consequence of stem-growth responses in light-grown mustard?" *Journal of Experimental Botany*, vol. 46, no. 288, pp. 753–757, 1995.

[54] J. J. Casal, C. L. Ballare, M. Tourn, and R. A. Sanchez, "Anatomy, growth and survival of a long-hypocotyl mutant of *Cucumis sativus* deficient in phytochrome B," *Annals of Botany*, vol. 73, no. 6, pp. 569–575, 1994.

[55] J. M. Christie, "Phototropin blue-light receptors," *Annual Review of Plant Biology*, vol. 58, pp. 21–45, 2007.

[56] C. Galen, J. J. Rabenold, and E. Liscum, "Functional ecology of a blue light photoreceptor: effects of phototropin-1 on root growth enhance drought tolerance in *Arabidopsis thaliana*," *The New Phytologist*, vol. 173, no. 1, pp. 91–99, 2006.

[57] C. L. Ballaré, A. L. Scopel, S. R. Radosevich, and R. E. Kendrick, "Phytochrome-mediated phototropism in de-etiolated seedlings: occurrence and ecological significance," *Plant Physiology*, vol. 100, no. 1, pp. 170–177, 1992.

[58] C. L. Ballare, A. L. Scopel, M. L. Roush, and S. R. Radosevich, "How plants find light in patchy canopies. A comparison between wild-type and phytochrome-B-deficient mutant plants of cucumber," *Functional Ecology*, vol. 9, no. 6, pp. 859–868, 1996.

[59] D. Todaka, K. Nakashima, K. Maruyama et al., "Rice phytochrome-interacting factor-like protein OsPIL1 functions as a key regulator of internode elongation and induces a morphological response to drought stress," *Proceedings of the National Academy of Sciences of the United States of America*, vol. 109, pp. 15947–15952, 2012.

[60] T. Sun, "Gibberellin metabolism, perception and signaling pathways in *Arabidopsis*," *The Arabidopsis Book*, vol. 6, Article ID e0103, 2008.

[61] M. de Lucas, J. M. Daviere, M. Rodríguez-Falcón et al., "A molecular framework for light and gibberellin control of cell elongation," *Nature*, vol. 451, pp. 480–484, 2008.

[62] S. Feng, C. Martinez, G. Gusmaroli et al., "Coordinated regulation of *Arabidopsis thaliana* development by light and gibberellins," *Nature*, vol. 451, no. 7177, pp. 475–479, 2008.

[63] T. Djakovic-Petrovic, M. D. Wit, L. A. C. J. Voesenek, and R. Pierik, "DELLA protein function in growth responses to canopy signals," *Plant Journal*, vol. 51, no. 1, pp. 117–126, 2007.

[64] J. I. Cagnola, E. Ploschuk, T. Benech-Arnold et al., "Stem transcriptome reveals mechanisms to reduce the energetic cost of shade-avoidance responses in tomato," *Plant Physiology*, vol. 160, no. 2, pp. 1110–1119, 2012.

[65] M. J. Kasperbauer and D. L. Karlen, "Light-mediated bioregulation and photosynthate partitioning in wheat," *Physiologia Plantarum*, vol. 66, pp. 159–163, 1986.

[66] J. J. Casal, "Light quality effects on the appearance of tillers of different order in wheat (*Triticum aestivum*)," *Annals of Applied Biology*, vol. 112, pp. 167–173, 1988.

[67] J. B. Evers, J. Vos, B. Andrieu, and P. C. Struik, "Cessation of tillering in spring wheat in relation to light interception and red:far-red ratio," *Annals of Botany*, vol. 97, no. 4, pp. 649–658, 2006.

[68] R. H. Skinner and S. R. Simmons, "Modulation of leaf elongation, tiller appearance and tiller senescence in spring barley by far-red light," *Plant, Cell and Environment*, vol. 16, pp. 555–562, 1993.

[69] T. H. Kebrom, B. L. Burson, and S. A. Finlayson, "Phytochrome B represses *Teosinte Branched1* expression and induces sorghum axillary bud outgrowth in response to light signals," *Plant Physiology*, vol. 140, no. 3, pp. 1109–1117, 2006.

[70] C. J. Whipple, T. H. Kebrom, A. L. Weber et al., "*Grassy tillers1* promotes apical dominance in maize and responds to shade signals in the grasses," *Proceedings of the National Academy of Sciences of the United States of America*, vol. 108, pp. E506–E512, 2011.

[71] G. A. Maddonni, M. E. Otegui, B. Andrieu, M. Chelle, and J. J. Casal, "Maize leaves turn away from neighbors," *Plant Physiology*, vol. 130, no. 3, pp. 1181–1189, 2002.

[72] V. A. Deregibus, R. A. Sánchez, and J. J. Casal, "Effects of light quality on tiller production in *Lolium* spp.," *Plant Physiology*, vol. 72, pp. 900–902, 1983.

[73] V. A. Deregibus, R. A. Sanchez, J. J. Casal, and M. J. Trlica, "Tillering responses to enrichment of red light beneath the canopy in a humid natural grassland," *Journal of Applied Ecology*, vol. 22, no. 1, pp. 199–206, 1985.

[74] D. J. Tucker, "Far-red light as a suppressor of side shoot growth in the tomato," *Plant Science Letters*, vol. 5, no. 2, pp. 127–130, 1975.

[75] M. Lötscher and J. Nösberger, "Branch and root formation in *Trifolium repens* is influenced by the light environment of unfolded leaves," *Oecologia*, vol. 111, no. 4, pp. 499–504, 1997.

[76] P. Bulman and L. A. Hunt, "Relationships among tillering, spike number and grain yield in winter wheat (*Triticum aestivum* L.) in Ontario," *Canadian Journal of Plant Sience*, vol. 68, pp. 583–596, 1988.

[77] J. Doebley, A. Stec, and L. Hubbard, "The evolution of apical dominance in maize," *Nature*, vol. 386, no. 6624, pp. 485–488, 1997.

[78] D. L. Sparkes, S. J. Holme, and O. Gaju, "Does light quality initiate tiller death in wheat?" *European Journal of Agronomy*, vol. 24, no. 3, pp. 212–217, 2006.

[79] J. J. Casal and V. O. Sadras, "Effects of end-of-day red/far-red ratio on growth and orientation of sunflower leaves," *Botanical Gazette*, vol. 148, pp. 463–467, 1987.

[80] J. J. Casal, R. A. Sanchez, and D. Gibson, "The significance of changes in the red/far-red ratio, associated with either neighbour plants or twilight, for tillering in *Lolium multiflorum* Lam.," *The New Phytologist*, vol. 116, no. 4, pp. 565–572, 1990.

[81] J. J. Casal and H. Smith, "The function, action and adaptive significance of phytochrome in light-grown plants," *Plant, Cell and Environment*, vol. 12, pp. 855–862, 1989.

[82] J. J. Casal, R. A. Sánchez, and V. A. Deregibus, "The effect of light quality on shoot extension growth in three species of grasses," *Annals of Botany*, vol. 59, no. 1, pp. 1–7, 1987.

[83] J. J. Casal and M. A. Alvarez, "Blue light effects on the growth of *Lolium multiflorum* Lam. leaves under natural radiation," *The New Phytologist*, vol. 109, pp. 41–45, 1988.

[84] M. C. Rousseaux, A. J. Hall, and R. A. Sánchez, "Far-red enrichment and photosynthetically active radiation level influence leaf senescence in field-grown sunflower," *Physiologia Plantarum*, vol. 96, no. 2, pp. 217–224, 1996.

[85] M. C. Rousseaux, A. J. Hall, and R. A. Sánchez, "Basal leaf senescence in a sunflower (*Helianthus annuus*) canopy: responses to increased R/FR ratio," *Physiologia Plantarum*, vol. 110, no. 4, pp. 477–482, 2000.

[86] T. Kinoshita, M. Doi, N. Suetsugu, T. Kagawa, M. Wada, and K. I. Shimazaki, "phot1 and phot2 mediate blue light regulation of stomatal opening," *Nature*, vol. 414, no. 6864, pp. 656–660, 2001.

[87] H. E. Boccalandro, C. V. Giordano, E. L. Ploschuk et al., "Phototropins but not cryptochromes mediate the blue light-specific promotion of stomatal conductance, while both enhance photosynthesis and transpiration under full sunlight," *Plant Physiology*, vol. 158, pp. 1475–1484, 2012.

[88] H. E. Boccalandro, M. L. Rugnone, J. E. Moreno et al., "Phytochrome B enhances photosynthesis at the expense of water-use efficiency in arabidopsis 1[W][OA]," *Plant Physiology*, vol. 150, no. 2, pp. 1083–1092, 2009.

[89] S. A. Casson, K. A. Franklin, J. E. Gray, C. S. Grierson, G. C. Whitelam, and A. M. Hetherington, "Phytochrome B and *PIF4* regulate stomatal development in response to light quantity," *Current Biology*, vol. 19, no. 3, pp. 229–234, 2009.

[90] J. Liu, F. Zhang, J. Zhou et al., "Phytochrome B control of total leaf area and stomatal density affects drought tolerance in rice," *Plant Molecular Biology*, vol. 78, pp. 289–300.

[91] C. V. González, S. E. Ibarra, P. N. Piccoli et al., "Phytochrome B increases drought tolerance by enhancing ABA sensitivity in *Arabidopsis thaliana*," *Plant, Cell and Environment*, vol. 35, no. 11, pp. 1958–1968, 2012.

[92] J. A. Jarillo, H. Gabrys, J. Capel, J. M. Alonso, J. R. Ecker, and A. R. Cashmore, "Phototropin-related NPL1 controls chloroplast relocation induced by blue light," *Nature*, vol. 410, no. 6831, pp. 952–954, 2001.

[93] T. Kagawa, T. Sakai, N. Suetsugu et al., "Arabidopsis NPL1: a phototropin homolog controlling the chloroplast high-light avoidance response," *Science*, vol. 291, no. 5511, pp. 2138–2141, 2001.

[94] T. Sakai, T. Kagawa, M. Kasahara et al., "Arabidopsis nph1 and npl1: blue light receptors that mediate both phototropism and chloroplast relocation," *Proceedings of the National Academy of*

Sciences of the United States of America, vol. 98, no. 12, pp. 6969–6974, 2001.

[95] M. Kasahara, T. E. Swartz, M. A. Olney et al., "Photochemical properties of the flavin mononucleotide-binding domains of the phototropins from *Arabidopsis*, rice, and *Chlamydomonas reinhardtii*," *Plant Physiology*, vol. 129, no. 2, pp. 762–773, 2002.

[96] P. A. Davis, S. Caylor, C. W. Whippo et al., "Changes in leaf optical properties associated with light-dependent chloroplast movements," *Plant, Cell and Environment*, vol. 34, pp. 2047–2059, 2011.

[97] M. J. Kasperbauer and P. G. Hunt, "Shoot/root assimilate allocation and nodulation of *Vigna unguiculata* seedlings as influenced by shoot light environment," *Plant and Soil*, vol. 161, no. 1, pp. 97–101, 1994.

[98] F. J. Salisbury, A. Hall, C. S. Grierson, and K. J. Halliday, "Phytochrome coordinates Arabidopsis shoot and root development," *Plant Journal*, vol. 50, no. 3, pp. 429–438, 2007.

[99] A. Suzuki, L. Suriyagoda, T. Shigeyama et al., "*Lotus japonicus* nodulation is photomorphogenetically controlled by sensing the red/far red (R/FR) ratio through jasmonic acid (JA) signaling," *Proccedings of the National Academy of Sciencesof the United States of America*, vol. 108, pp. 16837–16842, 2011.

[100] G. F. Deitzer, R. Hayes, and M. Jabben, "Kinetics and time dependence of the effect of far-red light on the photoperiodic induction of flowering in wintex Barley," *Plant Physiology*, vol. 64, pp. 1015–1021, 1979.

[101] J. J. Casal, V. A. Deregibus, and R. A. Sánchez, "Variations in tiller dynamics and morphology in *Lolium multiflorum* lam. vegetative and reproductive plants as affected by differences in red/far-red irradiation," *Annals of Botany*, vol. 56, no. 4, pp. 553–559, 1985.

[102] E. R. Cober and H. D. Voldeng, "Low R:FR light quality delays flowering of E7E7 soybean lines," *Crop Science*, vol. 41, no. 6, pp. 1823–1826, 2001.

[103] P. D. Cerdán and J. Chory, "Regulation of flowering time by light quality," *Nature*, vol. 423, no. 6942, pp. 881–885, 2003.

[104] K. J. Halliday, M. G. Salter, E. Thingnaes, and G. C. Whitelam, "Phytochrome control of flowering is temperature sensitive and correlates with expression of the floral integrator FT," *Plant Journal*, vol. 33, no. 5, pp. 875–885, 2003.

[105] S. E. Sanchez, J. I. Cagnola, M. Crepy, M. J. Yanovsky, and J. J. Casal, "Balancing forces in the photoperiodic control of flowering," *Photochemical and Photobiological Sciences*, vol. 10, no. 4, pp. 451–460, 2011.

[106] J. C. Heindl and W. A. Brun, "Light and shade effects on abscission and ^{14}C-photoassimilate partitioning among reproductive structures in soybean," *Plant Physiology*, vol. 73, pp. 434–439, 1983.

[107] Y. Zhen-wen, D. A. V. Sanford, and D. B. Egli, "The effect of population density on floret initiation, development and abortion in winter wheat," *Annals of Botany*, vol. 62, no. 3, pp. 295–302, 1988.

[108] P. Bancal, "Positive contribution of stem growth to grain number per spike in wheat," *Field Crops Research*, vol. 105, no. 1-2, pp. 27–39, 2008.

[109] M. Takano, N. Inagaki, X. Xie et al., "Phytochromes are the sole photoreceptors for perceiving red/far-red light in rice," *Proceedings of the National Academy of Sciences of the United States of America*, vol. 106, no. 34, pp. 14705–14710, 2009.

[110] I. Cerrudo, M. M. Keller, M. D. Cargnel et al., "Low red/far-red ratios reduce arabidopsis resistance to *Botrytis cinerea*

and jasmonate responses via a COI1-JAZ10-dependent, salicylic acid-independent mechanism," *Plant Physiology*, vol. 158, pp. 2042–2052, 2012.

[111] M. M. Izaguirre, C. A. Mazza, M. Biondini, I. T. Baldwin, and C. L. Ballaré, "Remote sensing of future competitors: impacts on plants defenses," *Proceedings of the National Academy of Sciences of the United States of America*, vol. 103, no. 18, pp. 7170–7174, 2006.

[112] J. E. Moreno, Y. Tao, J. Chory, and C. L. Ballaré, "Ecological modulation of plant defense via phytochrome control of jasmonate sensitivity," *Proceedings of the National Academy of Sciences of the United States of America*, vol. 106, no. 12, pp. 4935–4940, 2009.

[113] R. McGuire and A. A. Agrawal, "Trade-offs between the shade-avoidance response and plant resistance to herbivores? Tests with mutant *Cucumis sativus*," *Functional Ecology*, vol. 19, no. 6, pp. 1025–1031, 2005.

[114] C. L. Ballaré, A. L. Scopel, A. E. Stapleton, and M. J. Yanovsky, "Solar ultraviolet-B radiation affects seedling emergence, DNA integrity, plant morphology, growth rate, and attractiveness to herbivore insects in *Datura ferox*," *Plant Physiology*, vol. 112, no. 1, pp. 161–170, 1996.

[115] C. A. Mazza, J. Zavala, A. L. Scopel, and C. L. Ballaré, "Perception of solar UVB radiation by phytophagous insects: behavioral responses and ecosystem implications," *Proceedings of the National Academy of Sciences of the United States of America*, vol. 96, no. 3, pp. 980–985, 1999.

[116] J. A. Zavala, A. L. Scopel, and C. L. Ballaré, "Effects of ambient UV-B radiation on soybean crops: impact on leaf herbivory by *Anticarsia gemmatalis*," *Plant Ecology*, vol. 156, no. 2, pp. 121–130, 2001.

[117] M. M. Izaguirre, A. L. Scopel, I. T. Baldwin, and C. L. Ballaré, "Convergent responses to stress. Solar ultraviolet-B radiation and *Manduca sexta* herbivory elicit overlapping transcriptional responses in field-grown plants of *Nicotiana longiflora*," *Plant Physiology*, vol. 132, no. 4, pp. 1755–1767, 2003.

[118] X. Z. Xie, Y. J. Xue, J. J. Zhou et al., "Phytochromes regulate SA and JA signaling pathways in rice and are required for developmentally controlled resistance to *Magnaporthe grisea*," *Molecular Plant*, vol. 4, pp. 688–696, 2011.

[119] T. Genoud, A. J. Buchala, N. H. Chua, and J. P. Métraux, "Phytochrome signalling modulates the SA-perceptive pathway in Arabidopsis," *Plant Journal*, vol. 31, no. 1, pp. 87–95, 2002.

[120] A. Faigón-Soverna, F. G. Harmon, L. Storani et al., "A constitutive shade-avoidance mutant implicates TIR-NBS-LRR proteins in Arabidopsis photomorphogenic development," *Plant Cell*, vol. 18, no. 11, pp. 2919–2928, 2006.

[121] M. Y. Markham and D. E. Stoltenberg, "Corn morphology, mass, and grain yield as affected by early-season red: far-red light environments," *Crop Science*, vol. 50, no. 1, pp. 273–280, 2010.

[122] I. Rajcan and C. J. Swanton, "Understanding maize-weed competition: resource competition, light quality and the whole plant," *Field Crops Research*, vol. 71, no. 2, pp. 139–150, 2001.

[123] E. R. Page, W. Liu, D. Cerrudo, E. A. Lee, and C. J. Swanton, "Shade avoidance influences stress tolerance in maize," *Weed Science*, vol. 59, no. 3, pp. 326–334, 2010.

[124] E. R. Page, M. Tollenaar, E. A. Lee, L. Lukens, and C. J. Swanton, "Does the shade avoidance response contribute to the critical period for weed control in maize (*Zea mays*)?" *Weed Research*, vol. 49, no. 6, pp. 563–571, 2009.

[125] J. G. Liu, K. J. Mahoney, P. H. Sikkema, and C. J. Swanton, "The importance of light quality in crop-weed competition," *Weed Research*, vol. 49, no. 2, pp. 217–224, 2009.

[126] E. Green-Tracewicz, E. R. Page, and C. J. Swanton, "Light quality and the critical period for weed control in soybean," *Weed Science*, vol. 60, pp. 86–91, 2012.

[127] P. R. H. Robson, A. C. McCormac, A. S. Irvine, and H. Smith, "Genetic engineering of harvest index in tobacco through overexpression of a phytochrome gene," *Nature Biotechnology*, vol. 14, no. 8, pp. 995–998, 1996.

[128] C. L. Ballaré, A. L. Scopel, E. T. Jordan, and R. D. Vierstra, "Signaling among neighboring plants and the development of size inequalities in plant populations," *Proceedings of the National Academy of Sciences of the United States of America*, vol. 91, no. 21, pp. 10094–10098, 1994.

[129] P. J. Aphalo, C. L. Ballaré, and A. L. Scopel, "Plant-plant signalling, the shade-avoidance response and competition," *Journal of Experimental Botany*, vol. 50, no. 340, pp. 1629–1634, 1999.

[130] H. E. Boccalandro, E. L. Ploschuk, M. J. Yanovsky, R. A. Sánchez, C. Gatz, and J. J. Casal, "Increased phytochrome B alleviates density effects on tuber yield of field potato crops," *Plant Physiology*, vol. 133, no. 4, pp. 1539–1546, 2003.

[131] S. Schittenhelm, U. Menge-Hartmann, and E. Oldenburg, "Photosynthesis, carbohydrate metabolism, and yield of phytochrome-b-overexpressing potatoes under different light regimes," *Crop Science*, vol. 44, no. 1, pp. 131–143, 2004.

[132] S. G. Kong, D. S. Lee, S. N. Kwak, J. K. Kim, J. K. Sohn, and I. S. Kim, "Characterization of sunlight-grown transgenic rice plants expressing Arabidopsis phytochrome A," *Molecular Breeding*, vol. 14, no. 1, pp. 35–45, 2004.

[133] A. K. Garg, R. J. H. Sawers, H. Wang et al., "Light-regulated overexpression of an Arabidopsis phytochrome A gene in rice alters plant architecture and increases grain yield," *Planta*, vol. 223, no. 4, pp. 627–636, 2006.

[134] F. Andrés and G. Coupland, "The genetic basis of flowering responses to seasonal cues," *Nature Reviews Genetics*, vol. 13, pp. 627–639, 2012.

[135] Z. Xia, H. Zhai, B. Liu et al., "Molecular identification of genes controlling flowering time, maturity, and photoperiod response in soybean," *Plant Systematics and Evolution*, vol. 298, pp. 1217–1227, 2012.

14

Resistance to Phomopsis Seed Decay in Soybean

Shuxian Li[1] and Pengyin Chen[2]

[1] *United States Department of Agriculture-Agricultural Research Service, Crop Genetics Research Unit, Stoneville, MS 38776, USA*
[2] *University of Arkansas, Fayetteville, AR 72701, USA*

Correspondence should be addressed to Shuxian Li; shuxian.li@ars.usda.gov

Academic Editors: J. A. Casaretto, E. Perez-Artes, and J. Ransom

Phomopsis seed decay (PSD) of soybean is caused primarily by the fungal pathogen *Phomopsis longicolla* Hobbs along with other *Phomopsis* and *Diaporthe* spp. This disease causes poor seed quality and suppresses yield in most soybean-growing countries. Infected soybean seeds can be symptomless, but are typically shriveled, elongated, cracked, and have a chalky white appearance. Development of PSD is sensitive to environmental conditions. Hot and humid environments favor pathogen growth and disease development. Several control strategies have been used to manage PSD and reduce its impact; however, the use of resistant cultivars is the most effective method for controlling PSD. Efforts have been made to identify sources of PSD resistance in the past decades. At least 28 soybean lines were reported to have certain levels of PSD resistance in certain locations. Inheritance of resistance to PSD has been studied in several soybean lines. In this paper, general information about the disease, the causal agent, an overview of research on evaluation and identification of sources of resistance to PSD, and inheritance of resistance to PSD are presented and discussed.

1. Introduction

Phomopsis seed decay (PSD) of soybean, *Glycine max* (L.) Merrill, is caused primarily by the fungal pathogen *Phomopsis longicolla* Hobbs et al. [1] along with other *Phomopsis* and *Diaporthe* spp. This disease causes poor seed quality and suppresses yield in most soybean-growing countries, especially in the mid-southern region of the United States [2, 3]. PSD severely affects soybean seed quality due to reduction in seed viability and oil content, alteration of seed composition, and increased frequencies of moldy and/or split beans [4–7]. Development of PSD is sensitive to environmental conditions. Hot and humid environmental conditions, especially during the period from the pod fill through harvest stages, favor pathogen growth and disease development [2, 8–11].

PSD has been reported to cause significant economic losses [4, 12]. Annual yield loss caused by PSD in the United States from 1996 to 2007 ranged from 0.38 to 0.43 million metric tons (MMT) [13]. PSD has been exacerbated in recent years because of the use of the early soybean production system (ESPS), which generally involves planting of early-maturing cultivars in April to avoid the late-season droughts typical of the region in the mid-southern USA [14].

Unfortunately, seed infection by *P. longicolla* was high in some cultivars that matured in July or August, when high temperatures and high humidity were conducive to PSD development [15]. In 2009, due to the prevalence of hot and humid environments from pod fill to harvest in the southern United States, PSD caused over 0.33 MMT loss in 16 states [16]. The identification and utilization of sources of resistance to PSD for breeding programs are important, since planting resistant cultivars would be one of the most economical and effective ways to manage the disease [2, 10, 17–20].

The research on PSD including pathogen characterization, germplasm screening, and genetic resistance in the USA were summarized and presented at the World Soybean Research Conference VII in 2009 [21].

2. Disease Symptoms and the Causal Pathogens

Soybean seeds infected by *P. longicolla* can be symptomless [22] but are typically shriveled, elongated, or cracked, and often appear chalky white (Figure 1). Infected seeds either fail to germinate or germinate more slowly than healthy seeds.

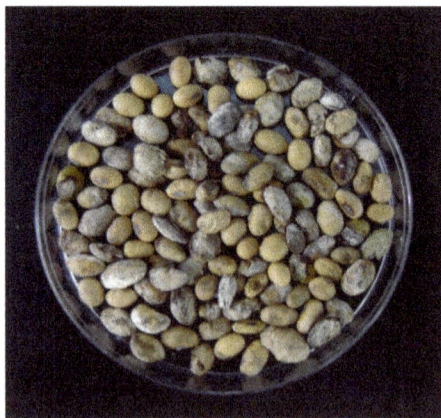

Figure 1: Phomopsis seed decay caused primarily by the fungus *Phomopsis longicolla*.

Seed infection causes pre- and postemergence damping-off, and under severe conditions, stands can be reduced to the point of lowering yield [2, 3, 10, 23]. Soybean pods can be infected at any time after they form.

The culture morphology of *P. longicolla* on potato dextrose agar is floccose, dense, and white with occasional greenish yellow areas. The undersides of cultures are colorless with large, black, spreading stromata (Figure 2). *P. longicolla* has been found to account for more than 85% of all *Phomopsis* seed infection [1]. This pathogen is primarily known as a seedborne pathogen, but it can be isolated from all parts of plant and is the predominant species isolated from diseased plants [24, 25], and from both discolored and nondiscolored mature soybean stems [26]. It has been found that *P. longicolla* is more prevalent in seeds from pods at the bottom of plant than at the top [23].

Although *P. longicolla* is the primary cause of PSD, other fungal members in the *Diaporthe-Phomopsis* complex may also be associated with PSD [23]. The *Diaporthe-Phomopsis* complex consists of *P. longicolla* and three varieties of *Diaporthe phaseolorum* (Cooke and Ellis) Sacc. (anamorph *P. phaseoli* (Desmaz.) Sacc.), in which *D. phaseolorum* var. *caulivora* K. L. Athow and R. M. Caldwell and *D. phaseolorum* var. *meridionalis* Fernández cause stem canker of soybean while *D. phaseolorum* var. *sojae* (S. G. Lehman) Wehmeyer causes pod and stem blight [2, 27, 28]. *P. longicolla* differs from others in the *Diaporthe-Phomopsis* complex in its morphology. It also does not have a known teleomorph [1].

Understanding the nature and the variability of the pathogen is essential for understanding its population diversity, and such information will also be important for selecting isolates to develop broad-based disease resistance in soybean lines. However, there are only a few publications on the variability of aggressiveness on soybean among *P. longicolla* isolates. The term "pathogen aggressiveness," as used in this paper, is based on colonization of and damage to soybean [29, 30].

In a greenhouse study, Li et al. [32] evaluated 48 isolates from the National Soybean Pathogen Collection Center at the University of Illinois at Urbana-Champaign using the cut-stem inoculation method. Isolates used in this study included 35 *P. longicolla* isolates from soybean in eight states in the U.S., along with the type culture of *P. longicolla* (Fau 600, ATCC 64802) from soybean in Ohio [1], two *P. longicolla* isolates from velvetleaf in Illinois [33], and 11 other *Phomopsis* spp. isolates from other hosts in four states in the USA, as well as from Canada and Costa Rica. Besides morphological identification, the identities of the isolates were also verified by sequence analysis of the ITS regions and the mitochondrial small subunit rRNA genes ([33–35] (http://nt.ars-grin.gov/fungaldatabases/specimens/specimens.cfm)).

There were significant ($P \leq 0.0001$) differences among isolates based on stem length and stem lesion length. The *P. longicolla* soybean isolate PL16, from Mississippi, caused the shortest stem length while the nonsoybean isolate P9, from Illinois, caused the greatest stem lesion length. The type isolate of *P. longicolla*, PL31 (Fau 600), was one of the three most aggressive isolates among the 48 isolates evaluated. The velvetleaf isolate P9 from Illinois was the most aggressive among 13 isolates from nonsoybean hosts. This study provided the first evaluation of aggressiveness of *P. longicolla* isolates from different geographic origins and the first demonstration that *Phomopsis* spp. isolated from cantaloupe, eggplant, and watermelon were able to infect soybean [32]. This study provided important information that is very useful for selecting isolates for screening in breeding broad-based resistance in soybean lines to PSD.

3. Evaluation and Identification of Sources for Resistance to PSD

Screening soybean lines for resistance to PSD is the first step toward developing PSD-resistant cultivars. Efforts have been made to identify sources of PSD resistance in the past decades. In a 4-year project carried out by the USDA-ARS scientists, seeds of 208 representative maturity group V soybean plant introductions (PIs) obtained from the USDA Soybean Germplasm Collection in 2006 were plated on cultural medium and assayed for the percentage of *Phomopsis* seed infection [31]. The general procedure for seed plating is shown in Figure 3. Briefly, the seeds were first surface-disinfected in 0.5% sodium hypochlorite for 3 min, rinsed in sterile distilled water, and then placed on potato dextrose agar that was acidified (pH 4.8) with 25% lactic acid (APDA). Five seeds per Petri dish and five Petri dishes per sample were used. After 4 days of incubation at 24°C, the incidence of *P. longicolla* growing on the APDA was recorded.

From the seed plating data in 2006, 122 PIs without seed infection were selected and field screened under natural infection in Stoneville, MS, USA, in 2007. On the basis of assays of naturally infected seeds from 2006 and 2007, 14 PIs were selected for further evaluation with *Phomopsis*-inoculated and non-inoculated treatments in 2008 and 2009 (Table 1). PI 424324B was identified as the most PSD-resistant line. It had no *Phomopsis* infection in the seed plating assays from 2006, 2007, and 2008. In 2009, frequent rainfall during seed maturation led to high levels of seed infection by

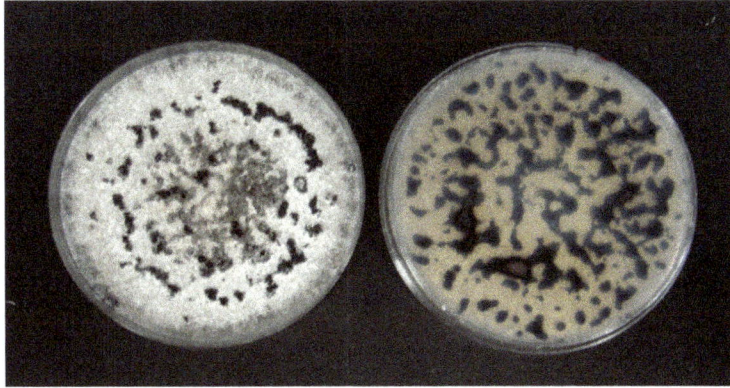

FIGURE 2: *Phomopsis longicolla* isolate MS10-6 collected from Mississippi Delta in 2010 and grown on acidified potato dextrose agar (pH 4.5) at 24°C for 50 days.

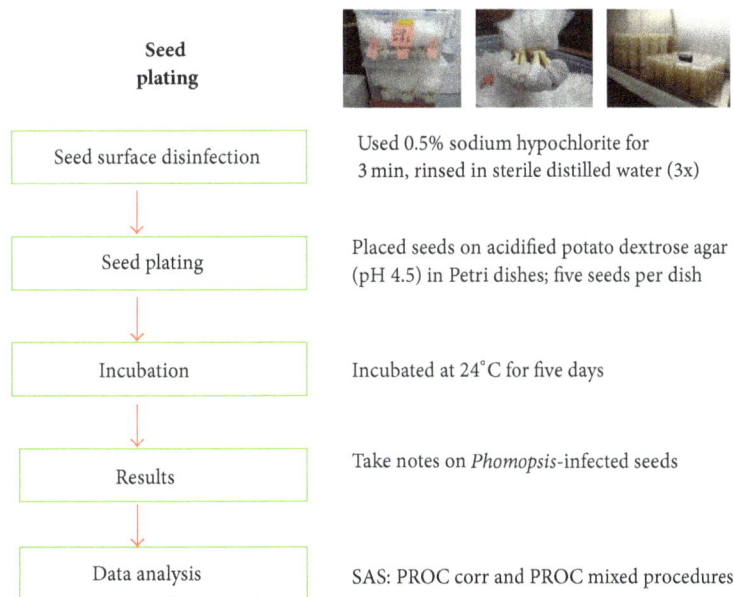

Seed plating

Seed surface disinfection	Used 0.5% sodium hypochlorite for 3 min, rinsed in sterile distilled water (3x)
Seed plating	Placed seeds on acidified potato dextrose agar (pH 4.5) in Petri dishes; five seeds per dish
Incubation	Incubated at 24°C for five days
Results	Take notes on *Phomopsis*-infected seeds
Data analysis	SAS: PROC corr and PROC mixed procedures

FIGURE 3: General procedure of seed plating.

Phomopsis (up to 80%) and other fungal pathogens on most of the soybean lines tested in Stoneville, MS, USA. However, only 1% and 2% of the seeds from PI 424324B were infected by *P. longicolla* in the non-inoculated and inoculated treatments, respectively [31]. In addition, PI 458130 was also resistant to PSD, with no seed infection from the naturally infected trials in 2006 and 2007, and less than 3% *Phomopsis* seed infection in the 2008 and 2009 inoculated trails [31]. Eight of 14 accessions selected for resistance (PI 424324B, PI 458130, PI 506647, PI 567270C, PI 567381B, PI 567521, PI 567635, and PI 594858A) had an average percent seed infection of less than 5% in 2008 and 2009, and this level was significantly less than that of 5601T, a well-adapted cultivar in Mississippi (Table 1). Some of these resistant accessions have been used to develop populations for genetics studies and breeding for resistance to PSD.

A 6-year project on "Screening germplasm and breeding for resistance to Phomopsis seed decay in soybean" has been funded by the United Soybean Board (USB), a group of farmer directors that administers the organization's soybean checkoff program, which seeks to strengthen soybean marketing, production technology, and research on new, value-added uses. In this project, hundreds of soybean germplasm accessions representing 28 regions or origins and from MG III to V, breeding lines, and commercial cultivars collected around the world were screened for resistance to PSD in three states (Arkansas, Mississippi, and Missouri) of the USA starting in 2009. The seeds, 33 seeds m^{-1} of row, were planted in 3.04 m long single-row plots with a 0.96 m row spacing. The experimental design was a randomized complete block with four replications. The seeds were harvested from each plot when the plants were mature. The seeds from each plot were

TABLE 1: Means of percent seed infected by *Phomopsis longicolla* of 16 soybean lines in replicated field tests with inoculated and non-inoculated treatments at Stoneville, Mississippi, in 2008 and 2009 [31].

Soybean line	2008		2009		2008 and 2009		
	Non	Inoc.	Non	Inoc.	Non	Inoc.	Mean
5601T	3.0	9.0	6.7	9.8	4.8	9.4	7.1
PI 304217	7.0	12.0	7.8	8.3	7.4	10.2	8.8
PI 398745	8.0	9.0	4.0	6.0	6.0	7.5	6.8
PI 399045	18.0	18.0	11.8	12.4	14.9	15.2	15.1
PI 408305	6.0	7.0	5.0	5.6	5.5	6.3	5.9
PI 424324B	0.0	0.0	1.0	2.0	0.5	1.0	0.8
PI 458130	0.0	3.0	2.0	2.7	1.0	2.8	1.9
PI 506647	2.0	4.0	3.0	3.0	2.5	3.5	3.0
PI 507290	0.0	1.0	5.8	6.2	2.9	3.6	3.3
PI 549020	2.0	7.0	18.2	18.3	10.1	12.6	11.4
PI 549021B	1.0	3.0	1.0	10.8	1.0	6.9	4.0
PI 567270C	3.0	7.0	1.0	2.0	2.0	4.5	3.3
PI 567381B	0.0	1.0	3.7	5.0	1.8	3.0	2.4
PI 567521	1.0	2.0	4.0	5.7	2.5	3.8	3.2
PI 567635	2.0	3.0	3.0	4.5	2.5	3.8	3.1
PI 594858A	3.0	3.0	2.0	4.3	2.5	3.7	3.1
Mean	3.5	5.6	5.0	6.7	4.2	6.1	5.2
LSD ($P \le 0.05$)	7.5	10.5	2.9	3.2	4.5	5.5	3.3

Non: non-inoculated control sprayed with distilled water, Inoc.: inoculated with spore suspension of *Phomopsis longicolla* (2×10^5) at the R5 stage twice with a 10-day interval between inoculations, Mean: overall mean of non-inoculated and inoculated treatments across both 2008 and 2009, LSD: Fisher's protected least significant difference at $P \le 0.05$.

tested for percent seed infected by *Phomopsis* spp., percent seed germination [42], and visual quality using a scale of 1 to 5 where 1 = excellent (no bad/infected seed); 2 = good (less than 10% bad/infected seed); 3 = fair (11–30% bad/infected seed); 4 = poor (31–50% bad/infected seed); and 5 = very poor (more than 50% bad/infected seed). Factors considered in estimating seed quality were seed wrinkling, molding, mottling, and discoloration. Frequent rainfall during seed maturation led to high levels of seed infection by a number of fungi. Significant differences in seed infection by *Phomopsis* spp. were observed among soybean lines with some lines having no infection, while others had infection levels as high as 90% [43–45]. These differences between lines were also reflected in visual seed quality and seed germination [43–45]. Soybean lines with low infection incidences, good visual quality, and high germination rates at all locations have potential to be resistant to PSD. Further tests of those lines are under progress.

Besides evaluating germplasm lines, 50 soybean cultivars were planted at Stoneville, Mississippi, to determine their reaction to PSD in 2007 [46]. Two lines, SS93-6012 and SS93-6181, previously reported to be PSD-resistant in Missouri were included [19], and the cultivars Hill and Williams 82 were used as susceptible checks. The seeds of soybean lines selected for planting were generally healthy. Of 50 lines tested, six lines had 100% germination, 30 lines had germination rates ranging from 80% to 97%, and 12 lines had rates ranging from 63% to 77%. Two lines had germination rates of 50% and 53%, respectively. In the seed plating assay of non-inoculated seed, 37 lines had no *P. longicolla*-infected seed and 10

and three lines had *P. longicolla* incidences of 3% and 7%, respectively. Incidence of *P. longicolla* in seed from inoculated field plots differed significantly ($P \le 0.05$) ranging from 6% to 50% among soybean lines. Several soybean cultivars with lower disease incidence than the PSD resistant lines SS93-6181 and SS93-6013 were identified [46].

To date, at least 28 soybean lines have been reported to have certain levels of PSD resistance (Table 2). However, some resistant lines identified in other regions were susceptible in Arkansas, Mississippi, or Missouri, USA (Li et al., unpublished). It is not known if there is the possibility of diversity between populations of the pathogen in different locations, but there is a need to identify new sources of resistance to PSD.

4. Inheritance of Resistance to Phomopsis Seed Decay

Inheritance of resistance to PSD has been studied in several soybean lines including PI 417479, MO/PSD-0259, PI 80837, and PI 360841 (Table 1). In all cases, PSD resistance was characterized as qualitative traits and conditioned by major dominant genes [17, 18, 47–49]. However, allelism of PSD resistance is still not clear; therefore, no gene symbol has been assigned for PSD resistance yet.

Soybean PI 417479 was identified as a source of resistance to PSD from screening approximately 3,000 soybean introductions in MGs III and IV from 1983 through 1985 at the

TABLE 2: A list of reported sources of resistance to Phomopsis seed decay.

	Soybean line	Reference
1	PI 548438 (Arksoy)	[17]
2	PI 80837	[36], [37]
3	PI 82264	[38]
4	PI 88264*	[17]
5	PI 181550	[39]
6	PI 200501	[40]
7	PI 200510	[17]
8	PI 204331	http://www.soydiseases.illinois.edu/
9	PI 205089	http://www.soydiseases.illinois.edu/
10	PI 205907	http://www.soydiseases.illinois.edu/
11	PI 205908	http://www.soydiseases.illinois.edu/
12	PI 205912	http://www.soydiseases.illinois.edu/
13	PI 209908	[17]
14	PI 219635*	http://www.soydiseases.illinois.edu/
15	PI 227687	[17]
16	PI 229358	[17]
17	PI 259539	http://www.soydiseases.illinois.edu/
18	PI 279088	http://www.soydiseases.illinois.edu/
19	PI 341249	http://www.soydiseases.illinois.edu/
20	PI 360835	http://www.soydiseases.illinois.edu/
21	PI 360841	[36]
22	PI 385942	http://www.soydiseases.illinois.edu/
23	PI 417419	[36]
24	PI 423903**	http://www.soydiseases.illinois.edu/
25	PI 562694	[41]
26	PI 424324B	[31]
27	PI 458130	[31]
28	PI 567381B	[31]

*Named/assigned by the authors (see reference), not the accession number of the USDA Soybean Germplasm Collection.
**Equal to PI 385942 (personal communication with Dr. Randy Nelson).

Agronomy Research Center of the University of Missouri, Columbia, MO, USA, and at the Isabela Substation of the University of Puerto Rico at Mayaguez [36]. To study the inheritance of PSD resistance in PI 417479, crosses were made between PI 417419 and two PSD-susceptible genotypes [49]. By analyzing PSD incidence on plants from five generations (F_1, F_2, F_3, BC_1, and BC_2, in which BC_1 represented a backcross between the F_1 and the resistant parent and BC_2 represented a backcross between the F_1 and the susceptible parents), it was concluded that the PSD resistance in PI 417479 was controlled by two complementary dominant nuclear genes. The two resistance genes can thus be transferred using a backcross procedure in a breeding program [49].

It was reported that the PSD-resistant line MO/PSD-0259 derived its resistance from PI 417479 [18]. PI 80837 had low levels of PSD infection in field trials [37] and also exhibited resistance to soybean mosaic virus and purple seed stain [50–52]. To characterize the inheritance of PSD resistance in PI 80837 and to determine if it differs from resistance in MO/PSD-0259, PI 80837 was crossed with two PSD-susceptible lines Agripro 350 (AP350) and PI 9113, and with MO/PSD-0259 [18] (Table 3). The derived genetic populations were screened in the field and *Phomopsis* infection was assayed by plating seed. Results showed that seed infection of reciprocal F_1 plants of AP350 × PI 80837 was not different from that of PI 80837. Data from F_2 populations of AP350 × PI 80837 and PI 91113 × PI 80837 and $F_{2:3}$ lines from AP350 × PI 80837 fit yhe models for a single dominant gene in PI 80837 that confers PSD resistance. Likewise, F_2 population data from AP350 × MO/PSD-0259 fit a model for single dominant gene resistance in MO/PSD-0259. However, data from an F_2 population and $F_{2:3}$ lines of PI 80837 × MO/PSD-0259 fit a model for two different dominant genes. Based on those results, Jackson et al. [18] concluded that PSD resistance in PI 80837 is conferred by a single dominant gene under nuclear control that is different from the gene in MO/PSD-0259.

Studies on the inheritance of PSD resistance in soybean PI 360841 were conducted to test populations developed from the crosses between PI 360841 and two PSD-susceptible (S) genotypes, Agripro 350 (AP350) and PI 91113 (Figure 4), to determine the number of genes for PSD resistance [48]. Other crosses were made between PI 360841 and the PSD-resistant (R) parents MO/PSD-0259 and PI 80837 to test the allelic relationships of the resistance genes [48]. Seeds from the parents and the F_2 population were assayed for *Phomopsis* infection. Chi-square analysis of F_2 data from the resistant × susceptible crosses indicated a good fit to a 9R:7S model for two complementary dominant genes conferring PSD resistance in PI 360841. F_2 data from MO/PSD-0259 × PI 360841 showed a good fit to a 57R:7S model for two complementary dominant genes from PI 360841 and a different dominant gene from MO/PSD-0259. Since there was no apparent segregation for resistance in the F_2 population derived from PI 360841 × PI 80837, except for one suspicious susceptible plant, it suggested that one of the genes in PI 360841 is allelic to a PSD resistance gene in PI 80837 [48]. This was the first report of a new gene in PI 360841 for PSD resistance. This gene, along with other different PSD-resistance genes in MO/PSD-0259 and PI 80837, is useful to breeders for developing lines with a high level of resistance to PSD. Further studies have identified simple sequence repeat (SSR) markers linked to genes for PSD resistance in PI 80837 and MO/PSD-0259. Jackson et al. (2009) reported that PSD resistance in MO/PSD-0259 and PI 80837 is controlled by two different single dominant genes [47]. The gene in PI 80837 is located in the vicinity of Sat_177 (4.3 cM) and Sat_342 (15.8 cM) on molecular linkage group (MLG) B2 (chromosome 14). The gene that conditions resistance in MO/PSD-0259 is linked to Sat_317 (5.9 cM) and Sat_120 (12.7 cM) on MLG F (chromosome 13). Identification of these loci and markers linked to them will facilitate marker-assisted selection in breeding programs.

TABLE 3: Genetics of resistance to Phomopsis seed decay in soybean.

Crosses	Type	Generation	Ratio	Reference
AP350 × PI 80837	S × R	F_2	3R : 1S	[18, 47]
PI 91113 × PI 80837	S × R	F_2	3R : 1S	[18]
AP 350 × MO/PSD-0259	S × R	F_2	3R : 1S	[18]
AP350 × PI 80837	S × R	$F_{2:3}$	1R : 2H : 1S	[18]
PI 80837 × MO/PSD-0259	R × R	F_2	15R : 1S	[18]
PI 80837 × MO/PSD-0259	R × R	$F_{2:3}$	7R : 8H : 1S	[18]
PI 360841 × AP350	R × S	F_2	9R : 7S	[48]
PI 91113 × PI 360841	S × R	F_2	9R : 7S	[48]
MO/PSD-0259 × PI 360841	R × R	F_2	57R : 7S	[48]
PI 360841 × PI 80837	R × R	F_2	63 : 1S	[48]
PI 417479 × AP350	R × S	F_2	3R : 1S	[49]
PI 417479 × AP350	R × S	F_3	5R : 3S	[49]
(PI 417479 × AP350) × AP 350	(R × S) × S	BC_1	1R : 1S	[49]
PI 417479 × PI 91113	R × S	F_2	3R : 1S	[49]
PI 417479 × PI 91113	R × S	F_3	5R : 3S	[49]
(PI 417479 × PI 91113) × PI 91113	(R × S) × S	BC_1	1R : 1S	[49]

R: resistant, S: susceptible, H: heterozygous and segregating for PSD reaction, BC: backcross.

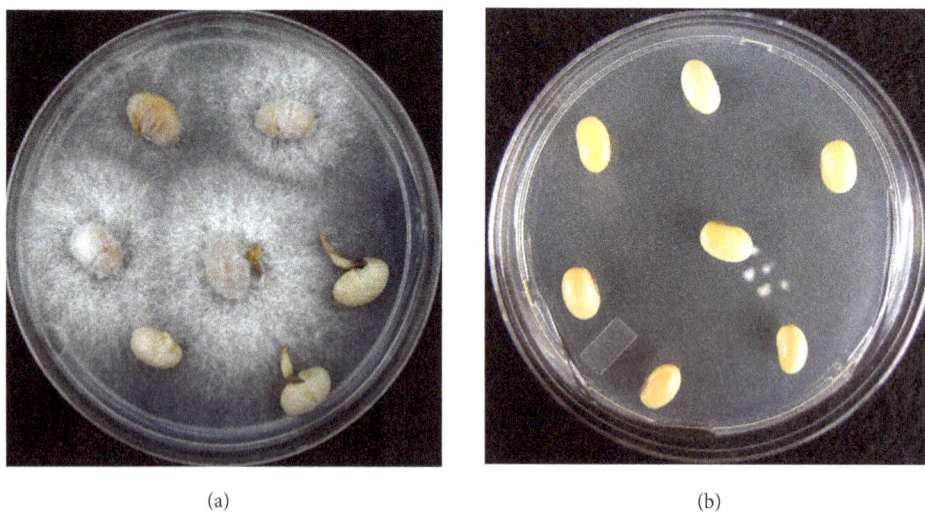

(a)   (b)

FIGURE 4: Susceptible parent PI 91113 and resistant parent PI 360841 with and without *Phomopsis* spp. infection, respectively.

5. Conclusions

Phomopsis seed decay (PSD) of soybean causes poor seed quality and suppresses yield in most soybean-growing countries. The disease is caused primarily by the fungal pathogen *Phomopsis longicolla* along with other *Phomopsis* and *Diaporthe* spp. Infected seeds range from symptomless to shriveled, elongated, cracked, and often appear chalky white. Seed quality is poor due to reduction in seed viability and oil content, alteration of seed composition, and increased frequencies of moldy and/or split beans. Hot and humid environments, especially during the period from pod fill through harvest, favor pathogen growth and disease development. The use of resistant cultivars is the most effective method for controlling PSD. Extensive screening for PSD resistance has resulted in the identification of resistant sources. MO/PSD-0259 carries a single dominant gene for PSD resistance derived from PI 417479, whereas resistance in PI 80837 is conferred by a different gene. PI 360841 carries two complementary dominant genes for PSD resistance; both are different from the gene in MO/PSD-0259 but one of them maps to the same region as the gene in PI 80837. Simple sequence repeat (SSR) markers linked to PSD resistance genes in PI 80837 and MO/PSD-0259 have been identified. These SSR markers should be useful in selection for resistant genotypes in breeding programs.

Acknowledgments

The authors appreciate the support of the United Soybean Board for the research Projects #9261 (2009–2012) and #2261 (2012–2015) on "Screening germplasm and breeding for resistance to Phomopsis seed decay in soybean," and the Mississippi Soybean Promotion Board for the research on "Identification of soybean varieties with resistance to Phomopsis seed decay to enhance soybean seed quality" as well as the support from Drs. John Rupe, Allen Wrather, and Gabe Sciumbato for their collaboration in the projects. The research was also supported by USDA-ARS Project 6402-21220-010-00D. Special thanks go to Drs. Richard Joost, Stephen Muench, Kelly Whiting, Larry G. Heatherly, and Lawrence Young for the valuable discussion about the research on Phomopsis seed decay. The mention of a trademark or proprietary product does not constitute a guarantee or warranty of the product by the US Department of Agriculture and does not imply approval or the exclusion of other products that may also be suitable.

References

[1] T. W. Hobbs, A. F. Schmitthenner, and G. A. Kuter, "A new Phomopsis species from soybean," Mycologia, vol. 77, no. 4, pp. 535–544, 1985.

[2] S. Li, "Phomopsis seed decay of soybean," in Soybean—Molecular Aspects of Breeding, A. Sudaric, Ed., pp. 277–292, InTech, Vienna, Austria, 2011.

[3] J. B. Sinclair, "Phomopsis seed decay of soybeans—a prototype for studying seed disease," Plant Disease, vol. 77, pp. 329–334, 1993.

[4] P. R. Hepperly and J. B. Sinclair, "Quality losses in Phomopsis—infected soybean seeds," Phytopathology, vol. 68, pp. 1684–1687, 1978.

[5] J. C. Rupe, "Effects of temperature on the rate of infection of soybean seedlings by Phomopsis longicolla," Canadian Journal of Plant Pathology, vol. 12, no. 1, pp. 43–47, 1990.

[6] J. C. Rupe and R. S. Ferriss, "Effects of pod moisture on soybean seed infection by Phomopsis sp.," Phytopathology, vol. 76, pp. 273–277, 1986.

[7] J. A. Wrather, J. G. Shannon, W. E. Stevens, D. A. Sleper, and A. P. Arelli, "Soybean cultivar and foliar fungicide effects on Phomopsis sp. seed infection," Plant Disease, vol. 88, no. 7, pp. 721–723, 2004.

[8] A. J. Balducchi and D. C. McGee, "Environmental factors influencing infection of soybean seeds by Phomopsis and Diaporthe species during seed maturation," Plant Disease, vol. 71, pp. 209–212, 1987.

[9] K. T. Kmetz, C. W. Ellett, and A. F. Schmitthenner, "Soybean seed decay: sources of inoculum and nature of infection," Phytopathology, vol. 69, pp. 798–801, 1979.

[10] G. L. Hartman, J. B. Sinclair, and J. C. Rupe, Compendium of Soybean Diseases, American Phytopathological Society, St. Paul, Minn, USA, 4th edition, 1999.

[11] J. A. Wrather, D. A. Sleper, W. E. Stevens, J. G. Shannon, and R. F. Wilson, "Planting date and cultivar effects on soybean yield, seed quality, and Phomopsis sp. seed infection," Plant Disease, vol. 87, no. 5, pp. 529–532, 2003.

[12] R. E. Baird, S. Abney, and B. G. Mullinix, "Fungi associated with pods and seeds during the R6 and R8 stages of four soybean cultivars in Southwestern Indiana," Phytoprotection, vol. 82, no. 1, pp. 1–11, 2001.

[13] J. A. Wrather and S. R. Koenning, "Effects of diseases on soybean yields in the United States 1996 to 2007," Plant Health Progress, 2009.

[14] L. G. Heatherly, "Early soybean production system (ESPS)," in Soybean Production in the Midsouth, L. G. Heatherly and H. F. Hodges, Eds., pp. 103–118, CRC Press, Boca Raton, Fla, USA, 1999.

[15] W. L. Mayhew and C. E. Caviness, "Seed quality and yield of early-planted, short-season soybean genotypes," Agronomy Journal, vol. 86, no. 1, pp. 16–19, 1994.

[16] S. R. Koenning, "Southern United States soybean disease loss estimate for 2009," in Proceedings of the Southern Soybean Disease Workers 37th Annual Meeting, p. 1, Pensacola Beach, Fla, USA, 2010.

[17] H. C. Minor, E. A. Brown, and M. S. Zimmerman, "Developing soybean varieties with genetic resistance to Phomopsis spp.," Journal of the American Oil Chemists' Society, vol. 72, no. 12, pp. 1431–1434, 1995.

[18] E. W. Jackson, P. Fenn, and P. Chen, "Inheritance of resistance to Phomopsis seed decay in soybean PI 80837 and MO/PSD-0259 (PI 562694)," Crop Science, vol. 45, no. 6, pp. 2400–2404, 2005.

[19] M. S. Pathan, K. M. Clark, J. A. Wrather et al., "Registration of soybean germplasm SS93-6012 and SS93-6181 resistant to Phomopsis seed decay," Journal of Plant Registrations, vol. 3, no. 1, pp. 91–93, 2009.

[20] K. W. Roy, B. C. Keith, and C. H. Andrews, "Resistance of hard seeded soybean lines to seed infection by Phomopsis, other fungi and soybean mosaic virus," Canadian Journal of Plant Pathology, vol. 16, no. 2, pp. 122–128, 1994.

[21] S. Li, P. Chen, G. L. Hartman, J. Smith, and R. Nelson, "Research highlights on soybean Phomopsis seed decay in the US: pathogen characterization, germplasm screening, and genetic resistance," in Proceedings of the 8th Abstracts of 2009 World Soybean Research Conference, pp. 91–92, Beijing, China, August 2009.

[22] R. K. Velicheti, C. Lamison, L. M. Brill, and J. B. Sinclair, "Immunodetection of Phomopsis species in asymptomatic soybean plants," Plant Disease, vol. 77, pp. 70–73, 1993.

[23] M. M. Kulik and J. B. Sinclair, "Phomopsis seed decay," in Compendium of Soybean Diseases, G. L. Hartman, J. B. Sinclair, and J. C. Rupe, Eds., pp. 31–32, American Phytopathological Society, St. Paul, Minn, USA, 1999.

[24] A. Mengistu, L. Castlebury, R. Smith, and J. Ray, "Seasonal progress of Phomopsis longicolla infection on soybean plant parts and its relationship to seed quality," Plant Disease, vol. 93, no. 10, pp. 1009–1018, 2009.

[25] A. G. Xue, M. J. Morrison, E. Cober et al., "Frequency of isolation of species of Diaporthe and Phomopsis from soybean plants in Ontario and benefits of seed treatments," Canadian Journal of Plant Pathology, vol. 29, no. 4, pp. 354–364, 2007.

[26] T. C. Harrington, J. Steimel, F. Workneh, and X. B. Yang, "Molecular identification of fungi associated with vascular discoloration of soybean in the North Central United States," Plant Disease, vol. 84, no. 1, pp. 83–89, 2000.

[27] S. G. Lehman, "Pod and stem blight of the soybean," Annals of the Missouri Botanical Garden, vol. 10, no. 2, pp. 111–178, 1923.

[28] J. B. Sinclair, "Diaporthe—Phomopsis," in Compendium of Soybean Diseases, G. L. Hartman, J. B. Sinclair, and J. C. Rupe, Eds., p. 31, American Phytopathological Society, St. Paul, Minn, USA, 1999.

[29] G. N. Agrios, *Plant Pathology*, Academic Press, San Diego, Calif, USA, 1999.

[30] M. C. Shurtleff and C. W. Averre III, *Glossary of Plant-Pathological Terms*, American Phytopathological Society, St. Paul, Minn, USA, 1997.

[31] S. Li, J. Smith, and R. Nelson, "Resistance to Phomopsis seed decay identified in maturity group V soybean plant introductions," *Crop Science*, vol. 51, no. 6, pp. 2681–2688, 2011.

[32] S. Li, G. L. Hartman, and D. L. Boykin, "Aggressiveness of *Phomopsis longicolla* and other *Phomopsis* spp. on soybean," *Plant Disease*, vol. 94, no. 8, pp. 1035–1040, 2010.

[33] S. Li, C. A. Bradley, G. L. Hartman, and W. L. Pedersen, "First report of *Phomopsis longicolla* from velvetleaf causing stem lesions on inoculated soybean and velvetleaf plants," *Plant Disease*, vol. 85, no. 9, p. 1031, 2001.

[34] A. W. Zhang, G. L. Hartman, L. Riccioni, W. D. Chen, R. Z. Ma, and W. L. Pedersen, "Using PCR to distinguish *Diaporthe phaseolorum* and *Phomopsis longicolla* from other soybean fungal pathogens and to detect them in soybean tissues," *Plant Disease*, vol. 81, no. 10, pp. 1143–1149, 1997.

[35] A. W. Zhang, L. Riccioni, W. L. Pedersen, K. P. Kollipara, and G. L. Hartman, "Molecular identification and phylogenetic grouping of *Diaporthe phaseolorum* and *Phomopsis longicolla* isolates from soybean," *Phytopathology*, vol. 88, no. 12, pp. 1306–1314, 1998.

[36] E. A. Brown, H. C. Minor, and O. H. Calvert, "A soybean genotypes resistant to Phomopsis seed decay," *Crop Science*, vol. 27, no. 5, pp. 895–898, 1987.

[37] L. D. Ploper, T. S. Abney, and K. W. Roy, "Influence of soybean genotype on rate of seed maturation and its impact on seedborne fungi," *Plant Disease*, vol. 76, pp. 287–292, 1992.

[38] H. J. Walters and C. E. Caviness, "Breeding for improved soybean seed quality," *Arkansas Farm Research*, vol. 22, article 5, 1973.

[39] K. L. Athow, "Fungal diseases," in *Soybeans: Improvement, Production, and Uses*, J. R. Wlcox, Ed., vol. 16 of *Agronomy Monograph*, pp. 687–727, ASA, CSSA, and SSSA, Madison, Wis, USA, 2nd edition, 1987.

[40] J. P. Ross, "Registration of eight soybean germplasm lines resistant to seed infection by *Phomopsis* spp.," *Crop Science*, vol. 26, article 210, 1986.

[41] H. C. Minor, E. A. Brown, B. L. Doupnik Jr., R. W. Elmore, and M. S. Zimmerman, "Registration of Phomopsis seed decay-resistant soybean germplasm MO/PSD-0259," *Crop Science*, vol. 33, no. 5, p. 1105, 1993.

[42] L. O. Copeland, "Rules for testing seeds," in *Association of Official Seed Analysis*, Stone Printing, Lansing, Mich, USA, 1981.

[43] S. Li, P. Chen, J. Rupe, and A. Wrather, "Reaction of maturity group V soybean plant introductions to Phomopsis seed decay in Arkansas, Mississippi, and Missouri, 2009," *Plant Disease Management Reports*, vol. 4, Article ID ST034, 2010.

[44] S. Li, J. Rupe, P. Chen, and A. Wrather, "Reaction of maturity group IV soybean plant introductions to Phomopsis seed decay in Arkansas, Mississippi, and Missouri, 2009," *Plant Disease Management Reports*, vol. 4, Article ID ST035, 2010.

[45] S. Li, A. Wrather, J. Rupe, and P. Chen, "Reaction of maturity group III soybean plant introductions to Phomopsis seed decay in Arkansas, Mississippi, and Missouri, 2009," *Plant Disease Management Reports*, vol. 4, Article ID ST036, 2010.

[46] S. Li, D. Boykin, G. Sciumbato, A. Wrather, G. Shannon, and D. Sleper, "Reaction of soybean cultivars to Phomopsis seed decay in the Mississippi Delta, 2007," *Plant Disease Management Reports*, 2009.

[47] E. W. Jackson, C. Feng, P. Fenn, and P. Chen, "Genetic mapping of resistance to Phomopsis seed decay in the soybean breeding line MO/PSD-0259 (PI562694) and plant introduction 80837," *Journal of Heredity*, vol. 100, no. 6, pp. 777–783, 2009.

[48] S. Smith, P. Fenn, P. Chen, and E. Jackson, "Inheritance of resistance to Phomopsis seed decay in PI 360841 soybean," *Journal of Heredity*, vol. 99, no. 6, pp. 588–592, 2008.

[49] M. S. Zimmerman and H. C. Minor, "Inheritance of Phomopsis seed decay resistance in soybean PI 417479," *Crop Science*, vol. 32, pp. 96–100, 1993.

[50] G. R. Buss, T. J. Smith, and H. M. Camper, "Registration of Ware soybean cultivar," *Crop Science*, vol. 19, article 564, 1979.

[51] K. W. Roy and T. S. Abney, "Colonization of pods and infection of seeds by *Phomopsis longicolla* in susceptible and resistant soybean lines inoculated in the greenhouse," *Canadian Journal of Plant Pathology*, vol. 10, no. 4, pp. 317–320, 1988.

[52] J. R. Wilcox, F. A. Laviolette, and R. J. Martin, "Heritability of purple seed stain resistance in soybean," *Crop Science*, vol. 15, no. 4, pp. 525–526, 1975.

15

Spectral Indices: In-Season Dry Mass and Yield Relationship of Flue-Cured Tobacco under Different Planting Dates and Fertiliser Levels

Ezekia Svotwa,[1] J. Anxious Masuka,[2] Barbara Maasdorp,[3] and Amon Murwira[3]

[1] *Department of Crop Science, University of Zimbabwe, Harare, Zimbabwe*
[2] *Department of Geography and Environmental Studies, University of Zimbabwe, Harare, Zimbabwe*
[3] *Tobacco Research Board/Kutsaga Research Station, Harare, Zimbabwe*

Correspondence should be addressed to Ezekia Svotwa; esvotwa2@gmail.com

Academic Editors: O. Ferrarese-Filho and J. Hatfield

This experiment investigated the relationship between tobacco canopy spectral characteristics and tobacco biomass. A completely randomized design, with plantings on the 15th of September, October, November, and December, each with 9 variety × fertiliser management treatments, was used. Starting from 6 weeks after planting, reflectance measurements were taken from one row, using a multispectral radiometer. Individual plants from the other 3 rows were also measured, and the above ground whole plants were harvested and dried for reflectance/dry mass regression analysis. The central row was harvested, cured, and weighed. Both the maximum NDVI and mass at untying declined with later planting and so was the mass-NDVI coefficient of determination. The best fitting curves for the yield-NDVI correlations were quadratic. September reflectance values from the October crop reflectance were statistically similar ($P > 0.05$), while those for the November and the December crops were significantly different ($P < 0.05$) from the former two. Mass at untying and NDVI showed a quadratic relationship in all the three tested varieties. The optimum stage for collecting spectral data for tobacco yield estimation was the 8–12 weeks after planting. The results could be useful in accurate monitoring of crop development patterns for yield forecasting purposes.

1. Background

Crop yield estimation in many countries is based on conventional techniques of data collection and ground-based field reports [1]. A variety of mathematical models relating to crop yield have also been proposed in recent years for many crops [2, 3]. In Zimbabwe crop surveys are mostly used in estimating crop yield [4]. The method is costly, time consuming, and is prone to large errors due to incomplete ground observations, leading to poor crop yield assessment and crop area estimations [1].

Remote sensing data has the potential and the capacity to provide spatial information in a global scale of features and phenomena on earth on an almost real-time basis [1]. Use of remote sensing techniques has the potential to provide quantitative and timely information on agricultural crops over large areas, and many different methods have been developed to estimate crop yields [5–7]. In general, the use of remote sensing was aimed at reducing the number of samples of ground surveys, making it less expensive [8]. With the application of remote sensing in agriculture, there is potential not only in identifying crop classes but also in estimating crop yield [1].

Spectral measurements from crops can be used in estimating crop parameters such as leaf area index [9], plant population, and even canopy total nitrogen status during the growth cycle of the crop [10]. Vegetation indices are algorithms aimed at simplifying data from multiple reflectance bands to a single value correlating with physical vegetation parameters, such as biomass, productivity, leaf area index, or percent vegetation ground cover [11]. Single band reflectance is combined into a vegetation index in order to minimize the effect of such factors as optical properties of the soil background and illumination and view geometric as well as meteorological factors on the canopy radiometric properties [12].

Vegetation indices, as summarized by Gross [13], are based on the characteristic reflection of plant leaves in the visible and near-infrared portions of light. By applying a "vegetation index" to the satellite imagery, concentration of green leaf vegetation can be quantified [14]. As explained by Liew [15] healthy vegetation has low reflection of visible light (from 0.4 to 0.7 μm), since the visible light is strongly absorbed by chlorophyll for photosynthesis and, at the same time, there is high reflection of near-infrared light (from 0.7 to 1.1 μm). The portion of reflected near-infrared light depends on the cell structure of the leaf [16]. In fading or unhealthy leaves, photosynthesis decreases and cell structure collapses resulting in an increase of reflected visible light and a decrease of reflected near-infrared light [13].

The normalized difference vegetation index (NDVI) has been considered to be a useful way for crop yield assessment models, using various approaches such as simple integration, to reflect vegetation greenness [17]. The index responds to changes in the amount of green biomass, chlorophyll content, and canopy water stress and, hence, is most commonly used in assessing crop vigor, vegetation cover, and biomass production from multispectral satellite data [18–20]. The NDVI is calculated from the near-infrared (NIR) and red (R) bands of either handheld or satellite sensors using the formula NDVI = (NIR − Red)/(NIR + Red).

The validity of crop yield models with NDVI is determined by the strength of association between the two variables included in the model [21]. It is also essential to have an understanding of the correlation existing between yield and NDVI at different phonological stages of crop for selecting appropriate date of satellite pass to include in the model [22].

Tobacco crop plays an important role in the economy of Zimbabwe, and in the 2012/2013 marketing season, 144 million kg of tobacco was sold, earning the country $525 million [23]. Crop area and yield forecasts play an important role in stabilizing tobacco prices at the auction floors. Crop forecasting is the art of predicting crop yields and production before the harvest actually takes place, typically a couple of months in advance. Zimbabwe mostly relies on crop statistical forecasting/estimation, crop reports/field visits from extension officers, and statistical crop forecasts for crop yield forecasts [4]. However, data from crop estimates, which are obtained through surveys conducted after harvests, are in most countries available quite lately for early warning purposes.

An overestimation of the crop would jeopardize the grower's profit in that it causes fall in prices when supply exceeds the estimated volume. Underestimation, on the other hand, causes unnecessary panic and competition among buyers of the crop, causing a rise in the price of the crop. The timely evaluation of potential crop yields in general becomes important because of the huge economic impact crops have on the world markets [5] and in particular on the economy of Zimbabwe.

Remotely sensed measurements can be used in monitoring the effects of agronomic practices, which are considered in developing yield prediction models [24]. A more direct

TABLE 1: Variety-fertilizer treatments.

Treatment	Description
(1)	K RK 26—50% recommended fertiliser
(2)	K RK 26—recommended fertiliser
(3)	K RK 26—150% recommended fertiliser
(4)	T 66—50% recommended fertiliser
(5)	T 66—recommended fertiliser
(6)	T 66—150% recommended fertiliser
(7)	K E1—50% recommended fertiliser
(8)	K E1—recommended fertiliser
(9)	K E1—150% recommended fertiliser

remote sensing data yield, described in simple formulae, without deeper physiological background, is simpler to use and easier to understand [25] and would be applicable in tobacco, where the target, the leaf, is the harvestable part. This experiment investigated the relationship between canopy spectral characteristics of three tobacco varieties established on three planting dates and, under three fertilizer regimes, in-season dry matter and final yield. It was assumed in this study that the most suitable stage to predict yield is that where the canopy NDVI was most positively correlated with in-season dry mass, and a model relating the NDVI for this stage to cured leaf mass would be established. It was also hypothesized that the strength of the relationship between in-season dry mass and yields expressed as mass at untying with NDVI is not affected by tobacco variety, planting date, and fertiliser application rate. The results for the project will be used to select the most appropriate stage of collecting remote sensing data for field level and national tobacco crop area and yield forecasting. This information could be very useful in relating the reflectance measured from the tobacco cropped lands to in-season crop condition and final yield and quality predictions using remote sensing.

2. Method

Study Area. The experiment was conducted at Kutsaga Research Station in Zimbabwe in the 2010–2012 cropping seasons. Kutsaga is located between longitude 31° 08′E, latitude 17° 55′S, and at an altitude of 1000 m to 1500 m [10]. The long-term annual average rainfall is 850 mm.

The experimental plots were located on well-drained granitic sands. During February of 2009 and 2010 the plots were disked after a three-year Katambora grass fallow period to incorporate grass. Agricultural lime was applied using recommendations given from soil test results to raise the soil pH from 5.3 to 6.3 optimum for tobacco production. For the three years preceding the 2008 experiments, the sites were under Katambora grass to control nematodes. Recommended cultural and management practices were followed [10], except regard N:P:K levels and planting times, which were treatments in the experiment.

2.1. Fertilizer Treatments. In order to establish the relationship between spectral data and yield, there was a need to create variable growth conditions [26], and three varieties,

Spectral Indices: In-Season Dry Mass and Yield Relationship of Flue-Cured Tobacco under Different Planting
Dates and Fertiliser Levels

129

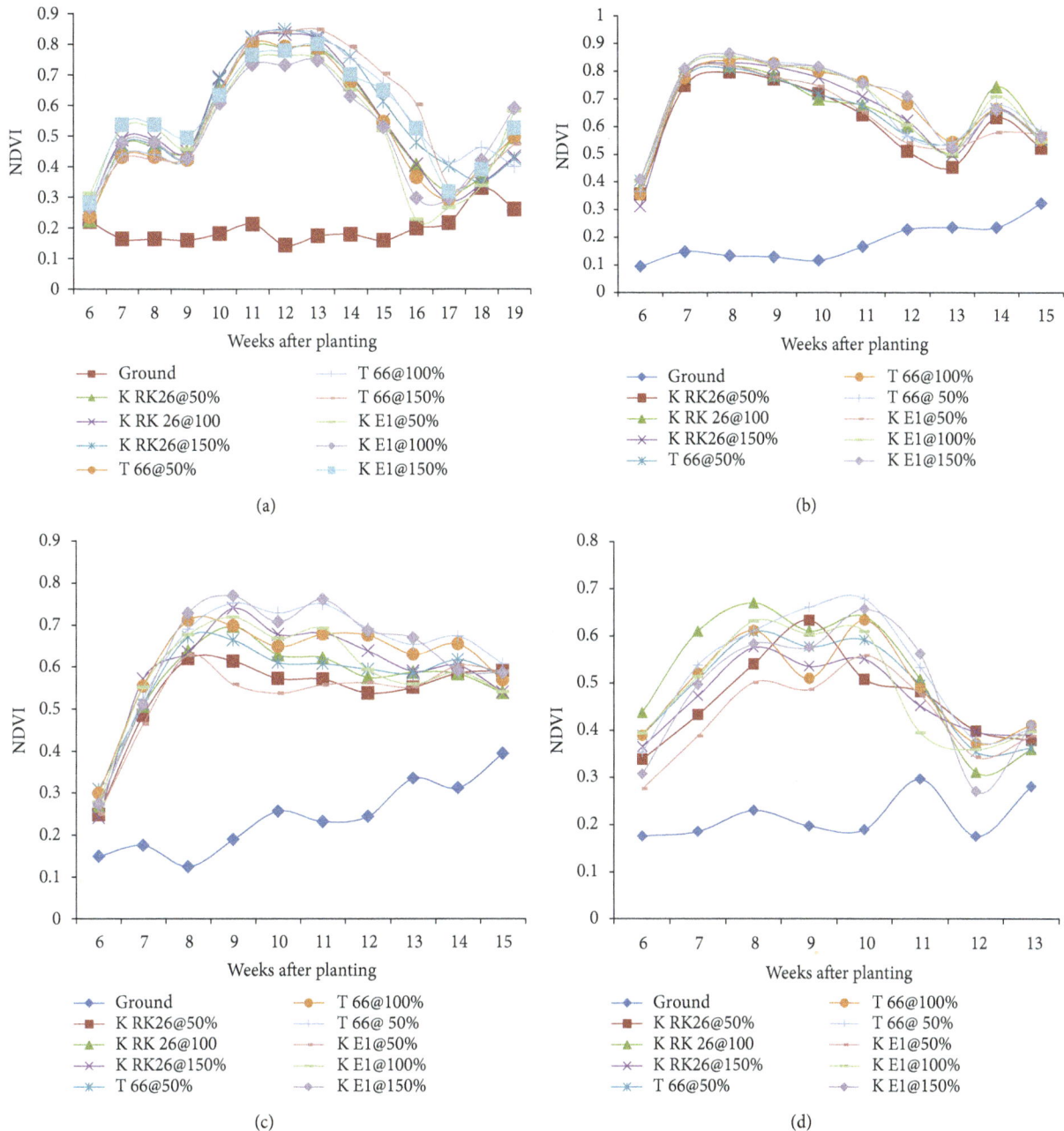

FIGURE 1: The NDVI temporal profiles for the (a) September, (b) October, (c) November, and (d) December planted crops.

four planting dates, and three fertiliser levels were tested. The variety-fertilizer treatments were applied by hands as shown in Table 1. The N:P:K treatment was hand-applied in bands about 10 cm deep and 30 cm to each side of a row at planting, while N treatments were applied at about 4 weeks after transplanting and after topping (at 6 weeks after planting).

The experiment was laid out in a completely randomized design with plantings on September 15, October 15, November 15, and December 15 each with 9 variety × fertiliser management treatments (Table 1). Three tobacco varieties K RK26, T 66, and K E1, developed by Kutsaga Research Station, were used, while three fertiliser management levels

(50%, 100%,and 150% recommended) were applied by hand (Table 1). The recommended compound fertiliser rate from soil test results was 700 kg/ha, while that for ammonium nitrate (34.5% N) was 96 kg/ha at 4 weeks after planting and 75 kg/ha after topping.

2.2. Procedure. Radiometric measurements were made weekly from the age of 6 weeks after planting on 5 m × 5 m square sampling plots, using a handheld multispectral radiometer (Cropscan MSR-5, 450–1750 nm), with the FOV centering over rows. All treatment applications had been completed at this stage of development.

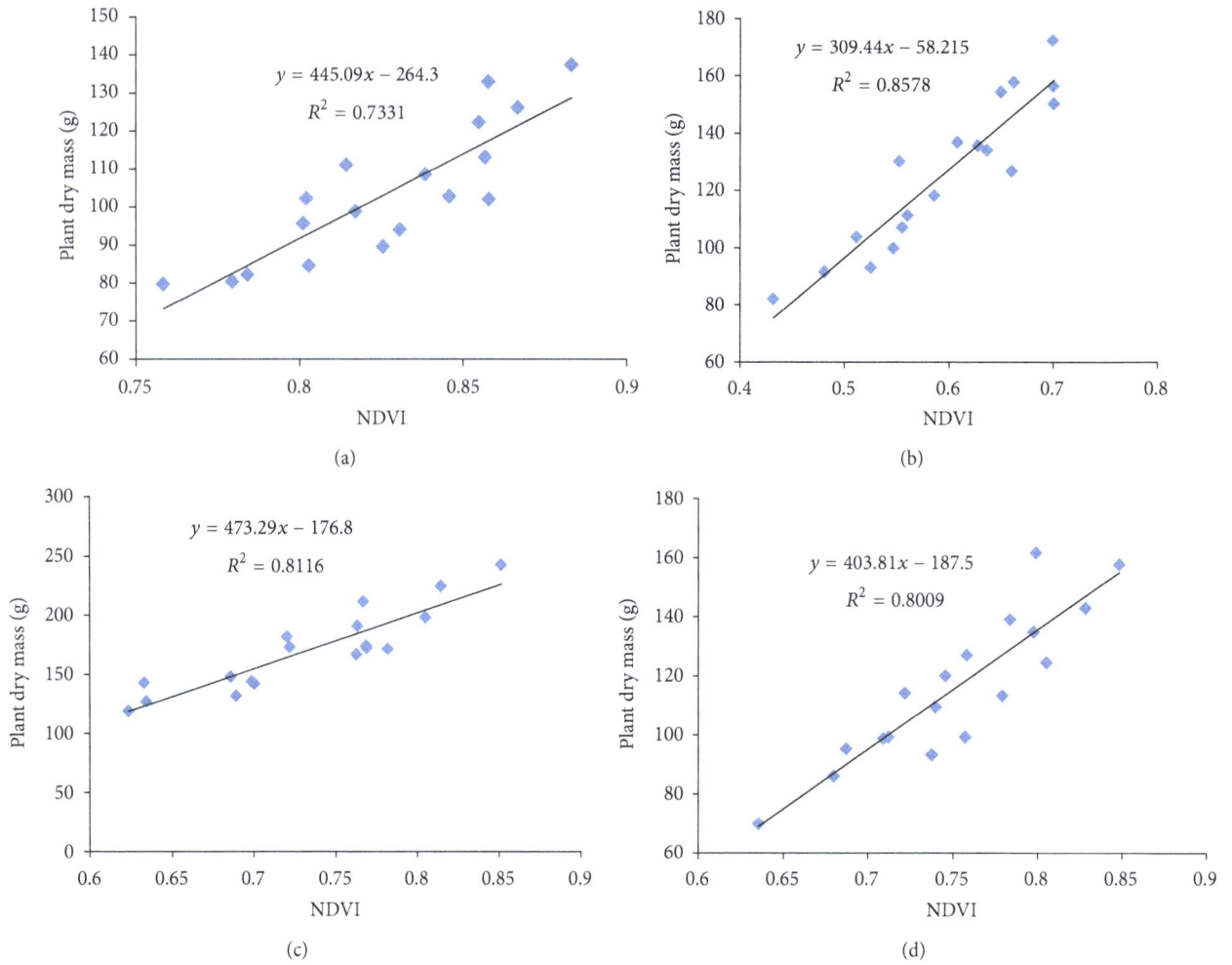

FIGURE 2: Dry-mass-reflectance correlations for the flue-cured tobacco samples collected from the (a) September, (b) October, (c) November, and (d) December planted crops.

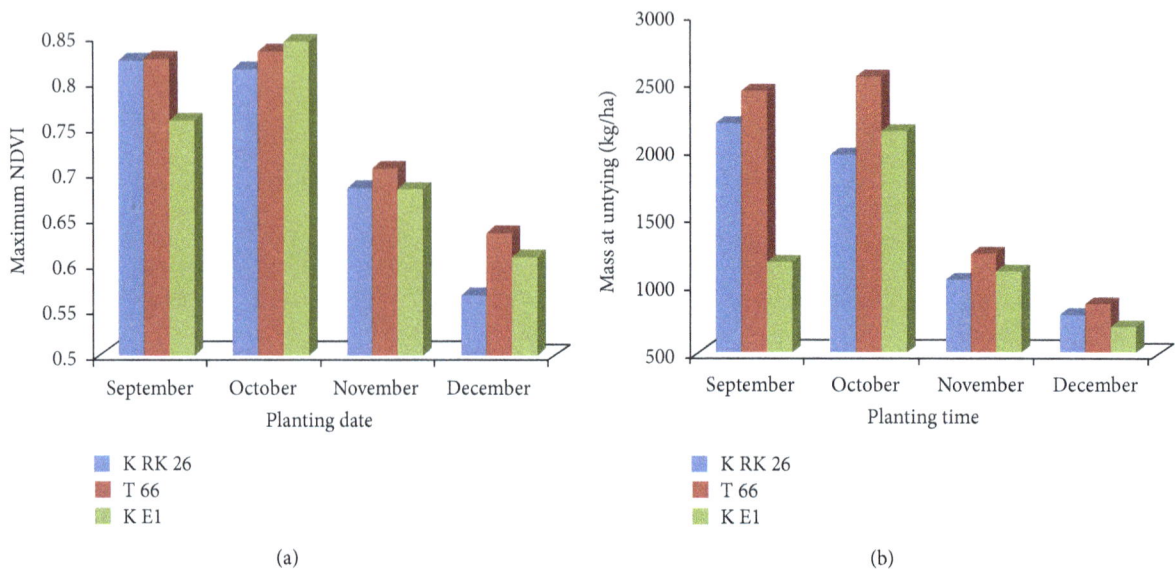

FIGURE 3: (a) Tobacco varieties' maximum NDVI and (b) mass at untying response to planting date.

Spectral Indices: In-Season Dry Mass and Yield Relationship of Flue-Cured Tobacco under Different Planting
Dates and Fertiliser Levels

131

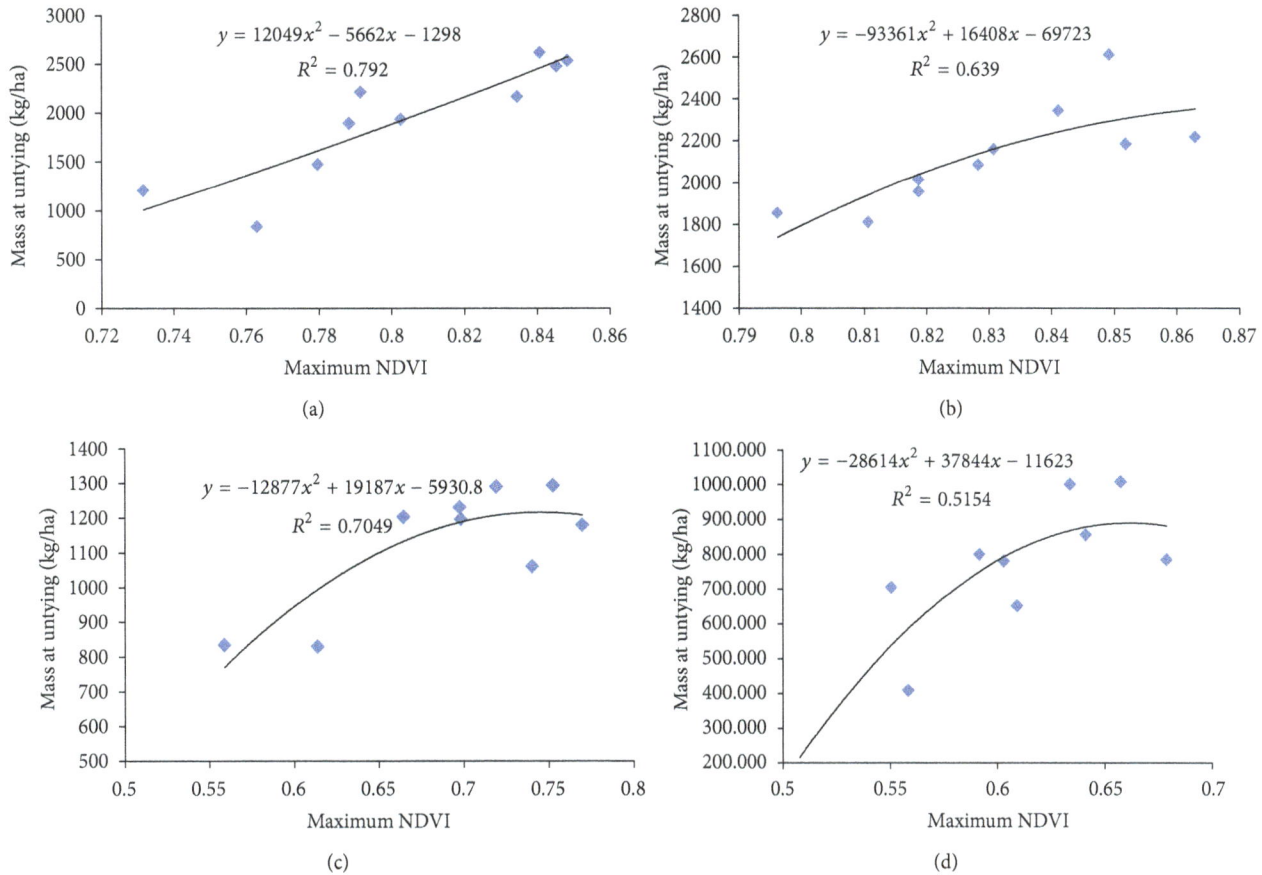

FIGURE 4: The relationship between the (a) September, (b) October, (c) November, and (d) December planted tobacco mass at untying and maximum NDVI.

Spectral data from the 4 planting dates were used to construct temporal NDVI profiles and one, with uninterrupted growth was selected for the in-season dry-mass-NDVI regression analysis. Above ground samples were collected after a corresponding canopy reflectance measurement had been obtained. Some 10 plants were sampled from each variety × fertiliser × planting time treatment after spectral data collection at 8, 10, 12, and 14 weeks after planting. The data collection timing, during midday and under cloudless conditions across the growing season, could include bare soil, early crop growth stage, peak crop greenness, and crop maturity imagery [27].

Each sampling plot measured consisted 5 rows of each with 32 plants spaced at 56 cm. The interrow distance was 1.2 m. normalized difference vegetation indices (NDVI) was calculated from the spectral bands obtained in Channels 3 and 4 of the MSR 5 which correspond to the visible (VIS) and near-infra red (NIR), respectively, using the following formula:

$$\text{NDVI} = \frac{\text{nir} - \text{red}}{\text{nir} + \text{red}}. \tag{1}$$

The multispectral radiometer (MRS 5) was positioned facing vertically downward at 1 m above crop canopies, and measurements were taken around solar noon to minimize the effect of diurnal changes in solar zenith angle. In total, 10 measurements were taken per sampling area and reflectance measurements were then averaged for each sampling plot to estimate a single reflectance value. Mature leaves were harvested from one row, cured, and yield determined before handle losses during crop grading as applied by Zhang et al. [28].

Reflectance measurements were also taken on individual plants from the other 3 rows. After taking reflectance readings, the above ground whole plants were harvested and packed in khaki bags and dried in microbarns. Dry matter measurements were later taken for reflectance/DM regression analysis. Three rows were also harvested and cured and mass was determined just after curing, before handling losses were incurred. The NDVI for the growth stage where there is the highest in-season dry-mass-NDVI correlation was selected for determining the mass-at-untying NDVI correlations.

Three-dimensional positions, latitude, longitude, and altitude, for the whole experimental area and for each treatment plot will be taken using a Garmin personal navigator (GPS V) to enable repeated sampling at the same location. Yield data were collected at harvest.

2.3. Data Analysis. NDVI data was analysed by analysis of variance and statistically significant treatment effects were

(a)

(b)

(c)

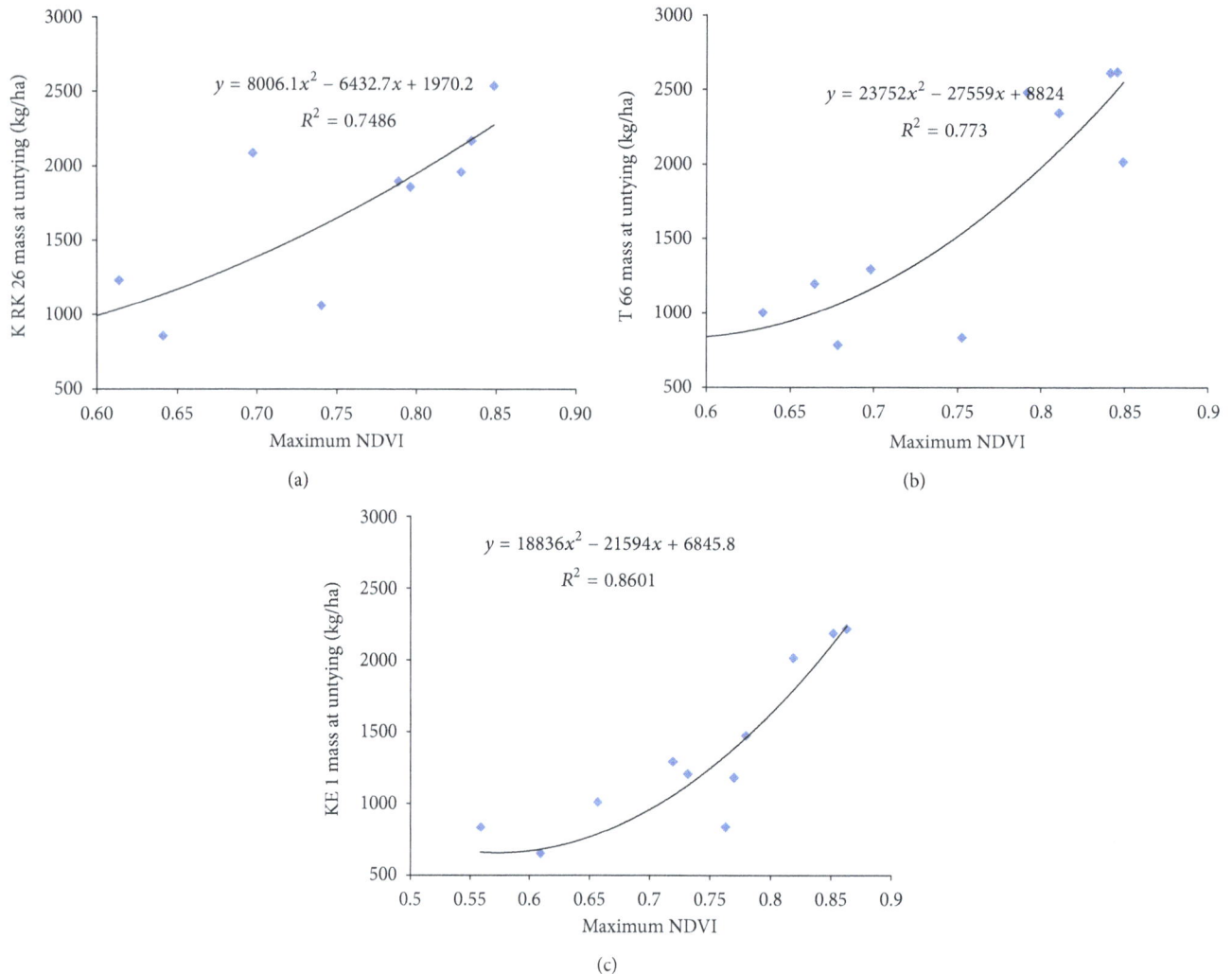

FIGURE 5: The relationship between tobacco varieties (a) K RK26, (b) T 66, and (c) K E1 mass at untying and maximum NDVI.

separated using least significant differences (LSDs). The data was analysed using the Genstat 9.2 statistical package at 5% level of significance. Student's t-test calculations were done to compare the planting date effect, and graphs were plotted using Excel 2007.

3. Results

The NDVI for all four planting date treatments crop rose from week 6 after planting to peak from 9 to 12 weeks after planting (Figure 1). At eight weeks after planting all the variety × fertiliser treatments in all, except for the December crop, started showing significant ($P < 0.05$) treatment differences. Beyond the peak, the NDVI also fell gently to reach the minimum at 13 weeks of age. The October crop (Figure 1(b)) was sampled for NDVI versus in-season dry mass analysis. This crop was selected because it was not subjected to long dry conditions after planting and had a good establishment. In addition temporal NDVI profile for the planting date had the highest NDVI value, which would enable a wide variation of the DM-NDVI relationships (Figure 1(b)).

The correlation between NDVI and in-season dry mass became stronger from the first sampling date (8 weeks after planting), reaching the highest at 10 weeks and later declined (Figure 2). Plants were not sampled after week 14 because reaping had become intense and some plants had already been stripped.

The NDVI response to variety × fertiliser treatment was similar to that for mass at untying (Figure 3). There was a general decline in both maximum NDVI and mass at untying with later planting, with the least values attained in December planting.

The mass at untying-NDVI coeficient of determination also decreased with later planting, with least being that of December (Figure 4). In all the four planting date treatments, the best fitting curves for mass at untying and NDVI correlations were quadratic. The September coefficient of determination ($r^2 = 0.79$) was the highest as compared to October ($r^2 = 0.594$), November ($r^2 = 0.695$), and December ($r^2 = 0.515$).

The September and the October crop reflectance values were statistically similar ($P > 0.05$), while the November and

Spectral Indices: In-Season Dry Mass and Yield Relationship of Flue-Cured Tobacco under Different Planting
Dates and Fertiliser Levels

133

(a)

(b)

(c)

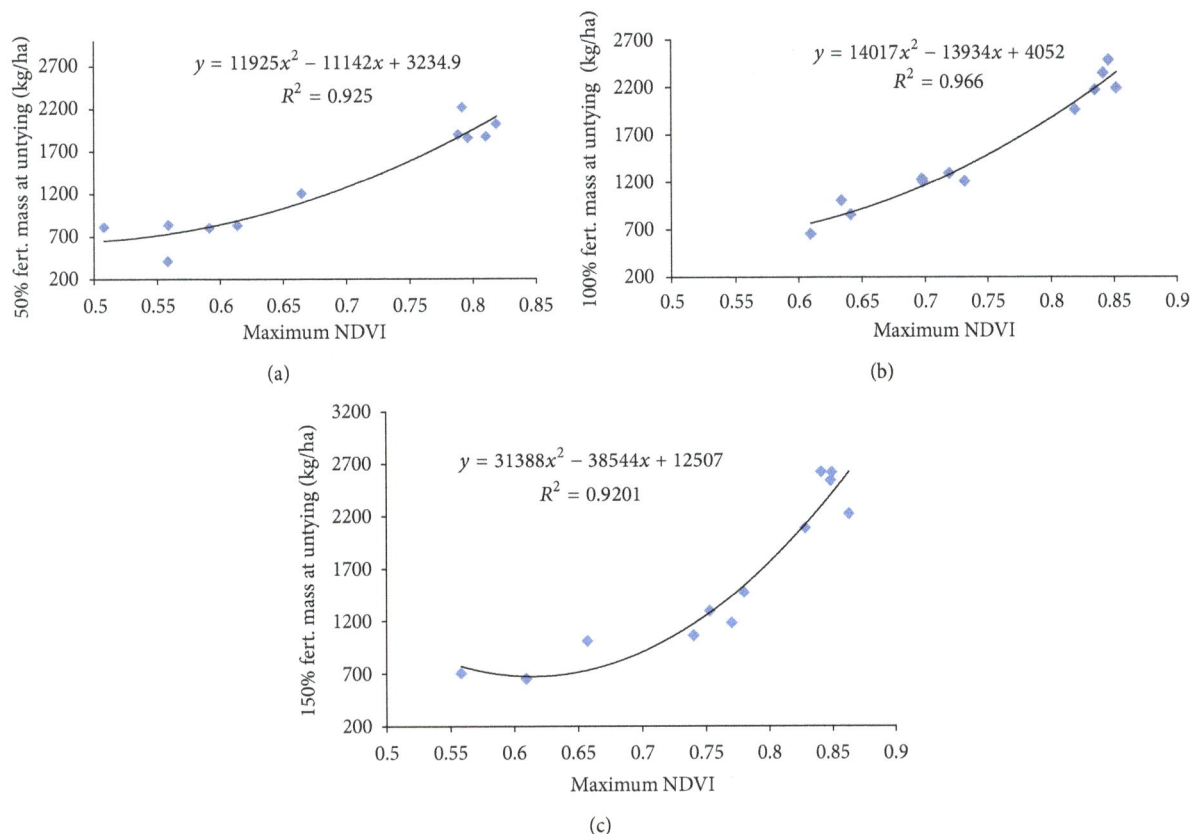

FIGURE 6: The relationship between tobacco mass at untying for the (a) 50%, (b) 100%, and (c) 150% fertiliser rate treatments and maximum NDVI.

TABLE 2: t-test for the comparison of mean maximum NDVI values for the different planting date treatments.

		P values	
	October	November	December
September	0.058204	0.000494993	$4.883E - 06$
October		$5.66702E - 06$	$1.08556E - 08$
November			0.000152201

TABLE 3: Variety yield-NDVI gradients at NDVI = 0.65, 0.7, and 0.75.

NDVI	0.65	0.7	0.75
		Yield-NDVI gradient	
K RK 26	4456.16	4776.4	5256.76
T66	4505.72	5448.8	6863.42
KE 1	4022.96	4776.4	5906.56

the December reflectance values were, for each, significantly different ($P < 0.05$) from all the rest (Table 2).

All the three varieties, K RK 26, T 66, and K E1, showed a quadratic relatioship between mass at untying and NDVI (Figure 5). K E1 had the highest ($r^2 = 0.86$) mass at untying NDVI coefficient of determination. The mass at untying NDVI correlations in K E1 ($r^2 = 0.748$) and T 66 ($r^2 = 0.773$) were comparable. In all the three varieties, the best fitting curves for mass at untying versus NDVI correlations were also quadratic, with comparable gradients at NDVI values between 0.65 and 0.75 (Table 3).

All the three fertiliser levels, 50% ($r^2 = 0.925$), 100% ($r^2 = 0.966$), 150% ($r^2 = 0.92$), displayed equally strong mass at untying NDVI relationship (Figure 6) with comparable gradients, again at NDVI value between 0.65 and 0.7 (Table 4).

TABLE 4: Fertiliser level yield-NDVI gradients at NDVI = 0.65, 0.7, and 0.75.

NDVI	0.65	0.7	0.75
		Yield-NDVI gradient	
50% fertiliser level	4360.5	5553.0	6745.5
100% fertiliser level	4288.1	5689.8	7091.5
150% fertiliser level	2260.4	5399.2	8538.0

The yield, expressed as mass at untying (kg/ha), for the September-October, November, and December planting can, therefore be estimated separately by the models in Figure 7:

(1) $y = -26708x^2 + 54365x - 24516,$

$$R^2 = 0.741, \tag{2}$$

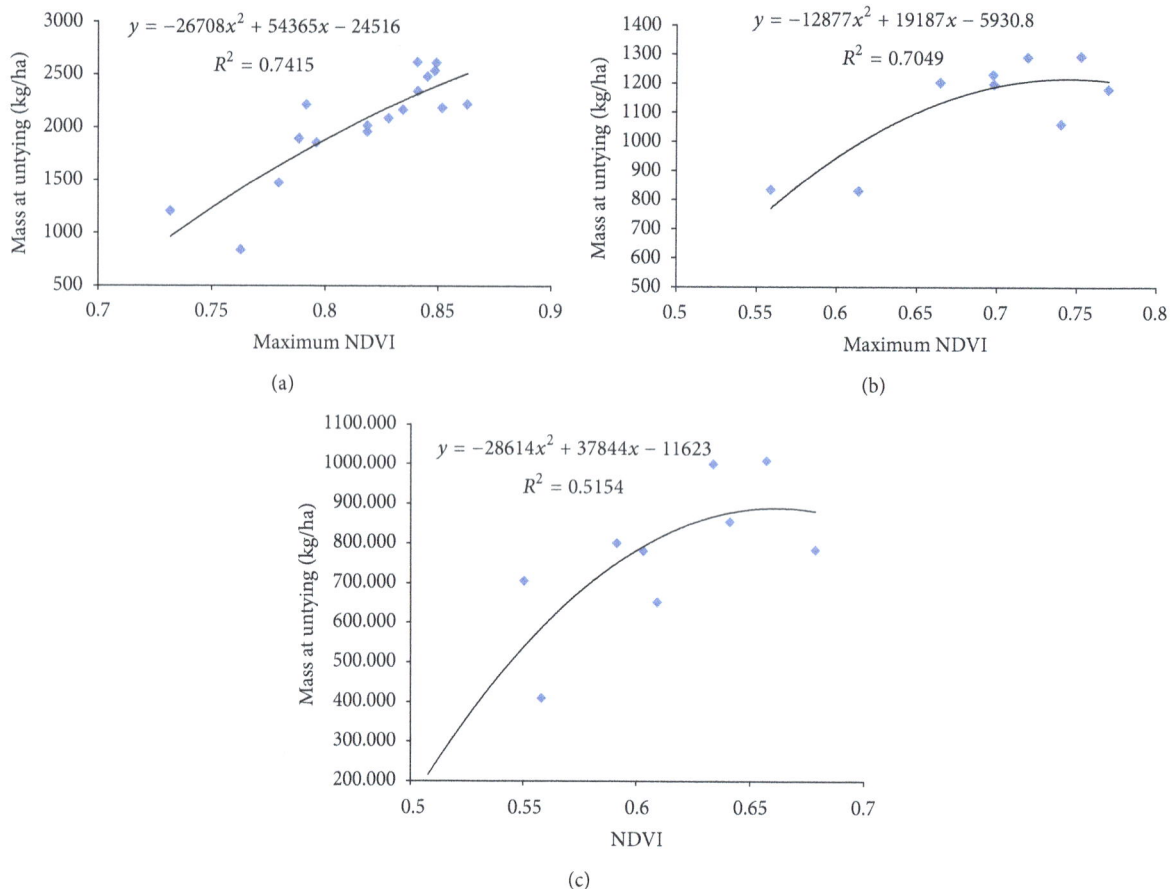

FIGURE 7: The relationship between tobacco mass at untying for the (a) September-October, (b) November, and (c) December planted crops and maximum NDVI.

(2) $y = -12877x^2 + 19187x - 5930,$

$R^2 = 0.704,$ (3)

(3) $y = -28614x^2 + 37844x - 11623,$

$R^2 = 0.515,$ (4)

where y is the mass at untying and x is the maximum normalised difference vegetative index (NDVI).

4. Discussions

The October crop was selected for in-season dry mass-NDVI analysis because of the clear cloudless conditions during the times of data collection. Thin cloud coverage, according to Nuarsa et al. [29], can lead to inconsistencies in the reflectance values, which will affect the NDVI, and, therefore, the selection of cloud-free images is one of the most important steps in the data analysis.

The NDVI-yield relationship increases with age up to 10–12-week after planting. As Nageswara-Rao et al. [30] explained it, the changes in spectral response of a crop are a function of phenological stages of the crop. The 10–12 weeks period, with the highest correlation, could be an indication

of the most suitable phenological stage to collect satellite data for yield forecasting.

Chlorophyll degradation related leaf ripening occurring during the ripening stage causes an increase in the red spectral reflectance which is normally absorbed by chlorophyll [31]. On the contrary, the NIR spectral reflectance is decreased due to a change in leaf internal structure [31] resulting in the fall of NDVI [32]. The fall in the in-season dry-mass-NDVI relationship after 14 weeks of age is related to the decrease in canopy reflectance spectra decrease at crop maturity stage that is brought about by reaping [33], while the final yield, in the data analysis, remains unchanged.

The decrease in tobacco mass at untying with later planting at all variety × fertiliser treatments was long since established [10]. Apparently the maximum NDVI in this experiment also followed the same trend, indicating a positive relationship between the two.

The similarity in the coefficients of determination between mass at untying and NDVI for the September and October planted crops could be an indication of the need to combine the three, when assessing area and yield using remote sensing, while the November and the December crops could each be assessed separately. The high coefficients of determination for all the three varieties and fertilizer levels could also be an indication of the possibility to disregard the

Spectral Indices: In-Season Dry Mass and Yield Relationship of Flue-Cured Tobacco under Different Planting
Dates and Fertiliser Levels

135

variety and fertiliser differences in the processes of developing yield forecasting models.

The coefficients of determination between mass at untying and NDVI for the September-October (0.741) and the November (0.704) planted crops were higher than the 0.65 reported by Povkh et al. [34] but lower than the r^2 = 0.90–0.98 that Jiang et al. [35] found between wheat grain yield and NDVI. The established coefficients in this experiment were, however, high enough for tobacco yield to be estimated using Cropscan calculated NDVI. The yield models derived were quadratic, similar to the findings of Jiang et al. [35] in wheat. The high value of R^2 indicated that the relationship between tobacco yield and the NDVI was consistent [29]. The December crop coefficient of determination was, however, low (0.515) meaning that yield estimation for this crop would not be accurately made using the model.

As the channels of the sensor used in the experiment is LANDSAT Thematic Mapper compatible [36], the models derived can be applicable in tobacco yield estimation using operation remote sensing data from the satellite.

However, more work is needed to establish the relationship between the Cropscan reflectance and those for selected Satellite platforms like Modis, Landsat 5 and Landsat TM which have been used for the same purpose in other crops [37].

5. Conclusions

The NDVI is positively related to in-season dry mass and, can be used to assess crop health, tobacco response to fertilizer, and accurate monitoring of crop development patterns for yield forecasting purposes. For yield forecasting purposes, the September and the October crops could be estimated together, while the November and the December crops each could also be estimated, separately. There was a strong positive correlation between NDVI and flue-cured tobacco yield at all fertiliser levels and for all the tested varieties, and, hence, for yield forecasting purposes, one may not separate these factors. There is, however, a need to establish the relationship between Cropscan multispectral radiometer 5 data and various satellite platforms before this information can be applied in satellite remote sensing.

Acknowledgments

The authors are grateful to the Tobacco Research Board/ Kutsaga Research Station for funding this series of experiments on developing-flue cured tobacco crop area and yield forecasting models using remote sensing and agronomic techniques.

References

[1] C. A. Reynolds, M. Yitayew, D. C. Slack, C. F. Hutchinson, A. Huetes, and M. S. Petersen, "Estimating crop yields and production by integrating the FAO Crop Specific Water Balance model with real-time satellite data and ground-based ancillary data," *International Journal of Remote Sensing*, vol. 21, no. 18, pp. 3487–3508, 2000.

[2] S. Landau, R. A. C. Mitchell, V. Barnett, J. J. Colls, J. Craigon, and R. W. Payne, "A parsimonious, multiple-regression model of wheat yield response to environment," *Agricultural and Forest Meteorology*, vol. 101, no. 2-3, pp. 151–166, 2000.

[3] T. R. Wheeler, P. Q. Craufurd, R. H. Ellis, J. R. Porter, and P. V. Vara Prasad, "Temperature variability and the yield of annual crops," *Agriculture, Ecosystems and Environment*, vol. 82, no. 1–3, pp. 159–167, 2000.

[4] SADC, *Selected Technical Papers: Methodology of Food Crop Forecasting in SAD*, SADC Secretariat, Gaborone, Botswana, 2009.

[5] P. C. Doraiswamy, B. Akhmedovb, L. Beardc, A. Sterna, and R. Mueller, "Operational prediction of crop yields using MODIS data and products," in *Proceedings of the International Achives of Photogrametry, Remote Sensing and Spatial Information Sciences*, vol. 38, pp. 45–50, 2011.

[6] F. Tao, M. Yokozawa, Z. Zhang, Y. Xu, and Y. Hayashi, "Remote sensing of crop production in China by production efficiency models: models comparisons, estimates and uncertainties," *Ecological Modelling*, vol. 183, no. 4, pp. 385–396, 2005.

[7] P. C. Doraiswamy, J. L. Hatfield, T. J. Jackson, B. Akhmedov, J. Prueger, and A. Stern, "Crop condition and yield simulations using Landsat and MODIS," *Remote Sensing of Environment*, vol. 92, no. 4, pp. 548–559, 2004.

[8] F. Rembold, C. Atzberger, I. Savin, and O. Rojas, "Using low resolution satellite imagery for Yield prediction. and yield anomaly detection," *Remote Sensing*, vol. 5, no. 4, pp. 1704–1733, 2013.

[9] A. D. Baez-Gonzalez, J. R. Kiniry, S. J. Maas et al., "Large-area maize yield forecasting using leaf area index based yield model," *Agronomy Journal*, vol. 97, no. 2, pp. 418–425, 2005.

[10] Tobacco Research Board (TRB), *Flue Cured Tobacco Recommendations*, TRB, Harare, Zimbabwe, 2012.

[11] C. J. Tucker, "Red and photographic infrared linear combinations for monitoring vegetation," *Remote Sensing of Environment*, vol. 8, no. 2, pp. 127–150, 1979.

[12] J. Verrelst, B. Koetz, M. Kneubühler, and M. Schaepman, "Directional sensitivity analysis of vegetation indices from multiangular CHRIS/PROBA data," in *Proceedings of the ISPRS Commission VII Mid-Term Symposium*, N. Kerle and A. Skidmore, Eds., Enschede, The Netherlands.

[13] D. Gross, *Monitoring Agricultural Biomass Using NDVI Time Series*, Food and Agriculture Organization of the United Nations (FAO), Rome, Italy, 2005.

[14] A. Viña, A. A. Gitelson, A. L. Nguy-Robertson, and Y. Peng, "Comparison of different vegetation indices for the remote assessment of green leaf area index of crops," *Remote Sensing of Environment*, vol. 115, no. 12, pp. 3468–3478, 2011.

[15] S. C. Liew, "Principles of Remote Sensing," Centre for Remote Imaging, Sensing and Processing, National University of Singapore, Singapore, 2001, http://www.crisp.nus.edu.sg/~research/tutorial/rsmain.htm.

[16] J. T. Woolley, "Reflectance and Transmittance of Light by Leaves," *Plant Physiology*, vol. 47, pp. 656–662, 1970.

[17] A. K. Prasad, L. Chai, R. P. Singh, and M. Kafatos, "Crop yield estimation model for Iowa using remote sensing and surface parameters," *International Journal of Applied Earth Observation and Geoinformation*, vol. 8, no. 1, pp. 26–33, 2006.

[18] J. U. H. Eitel, R. F. Keefe, D. S. Long, A. S. Davis, and L. A. Vierling, "Active ground optical remote sensing for improved monitoring of seedling stress in nurseries," *Sensors*, vol. 10, no. 4, pp. 2843–2850, 2010.

[19] X. Yin, A. McClure, and D. Tyler, "Relationships of plant height and canopy NDVI with nitrogen nutrition and yields of corn," in *Proceedings of the 19th World Congress of Soil Science, Soil Solutions for a Changing World*, Brisbane, Australia, August 2010.

[20] R. D. Jackson, P. N. Slater, and P. J. Pinter Jr., "Discrimination of growth and water stress in wheat by various vegetation indices through clear and turbid atmospheres," *Remote Sensing of Environment*, vol. 13, no. 3, pp. 187–208, 1983.

[21] C. S. Muthy, S. Jonna, P. V. Raju, S. Thurivengadachari, and K. A. Hakeem, "Crop Yield Prediction in Command Area using Satellite Data," GISdevelopment.net, AARS, ACRS 1994, Poster Session.

[22] T. Engel, G. Hoogenboom, J. W. Jones, and P. W. Wilkens, "AEGIS/WIN: a computer program for the application of crop simulation models across geographic areas," *Agronomy Journal*, vol. 89, no. 6, pp. 919–928, 1997.

[23] Tobacco Industries Marketing Board (T.I.M.B), *Annual Report and Accounts For the Year Ended 30 June, 2011*, T.I.M.B, Harare, Zimbabwe, 2012.

[24] C. Atzberger, "Advances in remote sensing of agriculture: context description, existing operational monitoring systems and major information needs," *Remote Sensing*, vol. 5, pp. 949–981, 2012.

[25] C. Ferencz, P. Bognár, J. Lichtenberger et al., "Crop yield estimation by satellite remote sensing," *International Journal of Remote Sensing*, vol. 25, no. 20, pp. 4113–4149, 2004.

[26] S. A. Mohammad, "Spectral indices and agronomic variables relationship of cotton (Gossypium sps.) under sowing dates and nitrogen levels," Asian Conference for Remote Sensing, 2008, http://www.gisdevelopment.net/application/agriculture/yield/mi08_305.htm.

[27] K. Dalsted and L. Queen, "Interpreting remote sensing data," The Site-Specific Management Guideline SSMG-26.

[28] H. Zhang, H. Chen, and G. Zhou, "The model of wheat yield forecast based on modis-ndvi: a case study of xinxiang," in *Proceedings of the ISPRS Annals of the Photogrammetry, Remote Sensing and Spatial Information Sciences Congress*, Melbourne, Australia, August 2012.

[29] I. W. Nuarsa, F. Nishio, and H. Chiharu, "Rice yield estimation using landsat ETM+ data and field observation," *Journal of Agricultural Science*, vol. 4, no. 3, pp. 45–56, 2012.

[30] R. C. Nageswara-Rao, J. H. Williams, M. V. K. Sivakumar, and K. D. R. Wadia, "Effect of water deficit at different growth phases of groundnut. II. Response to drought during pre-flowering phas," *Agronomy Journal*, vol. 80, pp. 431–438, 1988.

[31] W. Gunnula, M. Kosittrakun, T. L. Righetti, P. Weerathaworn, and M. Prabpan, "Normalized difference vegetation index relationships with rainfall patterns and yield in small plantings of rain-fed sugarcane," *Australian Journal of Crop Science*, vol. 5, no. 13, pp. 1852–1857, 2011.

[32] D. M. Gates, H. J. Keegan, J. C. Schleter, and V. R. Weidner, "Spectral properties of plants," *Applied Optics*, vol. 4, no. 1, pp. 11–20, 1965.

[33] H. Qiao, W. Mei, Y. Yang, W. Yong, J. Zhang, and Y. Hua, "Study on relationship between tobacco canopy spectra and LAI," *IFIP Advances in Information and Communication Technology*, vol. 345, no. 2, pp. 650–657, 2011.

[34] V. Povkh, G. L. Shljakhova, Garbuzov, and E. Vorobeychik, "Operational monitoring of the agricultural production based on the observational MODIS data as a support for improving regional planning," in *Proceedings Of The International Symposium On Remote Sensing Of Environment*, South Regional Information and Analytical Center (SRIA-Center), Rostov-on-Don, RUSSIA, 2005.

[35] D. Jiang, N.-B. Wang, X.-H. Yang, and J.-H. Wang, "Study on the interaction between NDVI profile and the growing status of crops," *Chinese Geographical Science*, vol. 13, no. 1, pp. 62–65, 2003.

[36] I. C. T. International, *Multispectral Radiometers*, ICT International, 2003.

[37] C. Yang, J. H. Everitt, and J. M. Bradford, "Using high resolution QuickBird satellite imagery for cotton yield estimation," in *Proceedings of the ASAE Annual International Meeting*, pp. 893–904, August 2004.

16

Effect of Carboxymethyl Cellulose and Alginate Coating Combined with Brewer Yeast on Postharvest Grape Preservation

Ren Yinzhe[1] and Zhang Shaoying[2]

[1] College of Chemistry and Materials, Shanxi Normal University, Linfen 041004, China
[2] College of Engineering, Shanxi Normal University, Linfen 041004, China

Correspondence should be addressed to Zhang Shaoying; zsynew@163.com

Academic Editors: O. Ferrarese-Filho and H. P. Singh

The effect of carboxymethyl cellulose and alginate coating combined with brewer yeast on postharvest grape preservation was investigated. The postharvest grapes were coated with 2% of alginate and 3% of carboxymethyl cellulose combined with 1.5×10^9 CFU/mL of brewer yeast. The combined treatment samples showed good sensory character on day 13 compared with control samples or only coated samples. The increase of weight loss and decrease of total soluble solids of combined treatment grapes were restrained. Furthermore, the protective enzymes including superoxide dismutase, peroxidase, and catalase of combined treatment sample showed higher activities. Accordingly, the increase of malonaldehyde content was also restrained and more vitamin C was preserved in combined treatment samples. At day 13, the weight loss rate and the total soluble solids of grape treated with coating + yeast were 23.6% lower and 20.6% higher than those of control samples, respectively. Coating grapes with 2% of alginate and 3% of carboxymethyl cellulose combined with brewer yeast of 1.5×10^9 CFU/mL was a well-proven method to preserve postharvest grapes.

1. Introduction

In recent years, conventional production systems of fruits and vegetables have been characterized by an excessive use of chemical compounds during pre- and postharvest treatments. The grape is a highly perishable nonclimacteric fruit with reduced shelf-life due to decay, weight loss, and nutrient degradation during the storage time. It is traditionally treated with different chemical products such as SO_2 to control the main postharvest pathogen [1]. Nevertheless, new consumer trends and subsequent legislative changes demand healthier, environmentally friendly food production systems.

Edible films can be used to protect perishable food products from deterioration by retarding dehydration, providing a selective barrier to moisture, oxygen, and carbon dioxide, suppressing respiration, improving textural quality, helping to retain volatile flavor compounds, and reducing microbial growth [2]. Carboxymethyl cellulose (CMC) is the most important water-soluble cellulose derivative, with many applications in the food industry and in cosmetics, pharmaceutcals, detergents, and so forth [3]. Alginate, a polysaccharide derived from marine brown algae, has been preponderant in making edible films due to its unique colloidal properties and its ability to form strong gels or insoluble polymers upon reaction with multivalent metal cations such as calcium [4]. At present, CMC and alginate are used in fruit preservation, such as fresh garlic, processed apples [5, 6].

Postharvest biocontrol is especially feasible for inhibiting postharvest fruit pathogens by the inoculation of antagonists. *Cryptococcus laurentii* has been investigated for the postharvest biological control of gray mold rot of apples, gray mold and blue mold rot of pears [7]. Brewer yeast is cultured from a one-celled fungus called *Saccharomyces cerevisiae* and is applied in beer industry. It also can be grown to make nutritional supplements. Brewer yeast is a rich source of minerals, particularly chromium, an essential trace mineral, which helps the body maintain normal blood sugar levels, selenium, protein, and the B-complex vitamins [8].

The objective of this work was to investigate the effect of carboxymethyl cellulose and alginate coating combined with brewer yeast on postharvest grape preservation during storage under ambient temperature. We tried to explore an integrated strategy to preserve fresh grape and provide reference for other vegetable and fruit preservation.

2. Materials and Methods

2.1. Materials. Grapes (*Vitis labrusca* L. kyoho) were purchased from an orchard in the vicinity of the Shanxi Normal University and picked at a preclimacteric but physiologically mature stage in the noon. Grapes with uniform shape, size, colour, and no defects were selected and quickly transported in open cartons to the laboratory. The grape particles were cut from fruiting pedicel and prepared for the following experiment.

Brewer yeast was provided by the microbiological laboratory of Northwest Sci-Tech University of Agriculture and Forestry. Carboxymethyl cellulose (food grade) was purchased from Beifang Chemical Limited Company (Renqiu, China). Alginate (food grade) was purchased from Datang Bioengineering Co., Ltd. (Hebei, China). Thiobarbituric acid, methionine, and nitroblue tetrazolium (biochemical reagent) were purchased from Sinopharm Chemical Reagent Co., Ltd. (Shanghai, China). Alfa Aesar Company (Tianjin, China) supplied other reagents, which were all of analytical grades.

PDA culture medium was prepared as the following method. 10 g of agar, 20 g of glucose, 200 g of peeled potato, and 1000 mL of deionized water were boiled for 30 min. The residual was removed through filtration. Thus, PDA medium was acquired. It was sterilized for 20 min at 121°C with high-pressure steam sterilization. After cooling, the medium in tube is placed into slope. If no bacteria were observed in two days, the medium may be used in the following experiment.

2.2. Fruit Treatment. Brewer yeast was taken out from refrigerator and was placed under ambient temperature for 3 hours. Afterward, it was cultivated with continuous transfers using PDA medium. Then the brewer yeast was diluted with several concentrations through 10 times dilution method with deionized water.

20 g of alginate and 30 g of carboxymethyl cellulose were diluted to 1 liter using deionized water with or without brewer yeast. The grapes were washed clean with tap water and then were dipped into different solutions for 2 min. The only washed grapes were served as control samples. In each treatment, about 350 fruits were coated and each treatment was repeated three times. A fan generating low-speed air was used to hasten the drying. The samples were then placed in plastic bags and stored under ambient temperature with 90% of relative humidity. The related parameters of the grapes were determined periodically.

2.3. Determination of Weight Loss and Total Soluble Solids. The weight loss was determined according to the method of Yu et al. [9]. In each treatment, 30 fruits were selected at random. The weight loss rate was calculated as follows: weight loss (%) = $[(m_0 - m_1)/m_0] \times 100$, where m_0 is the initial weight and m_1 is the weight measured during storage.

Total soluble solids were assayed according to the method of Qiuping and Wenshui with modifications [10]. Tissues (50 g) from six fruits were homogenized and then centrifuged at 8000 ×g for 20 min using an Eppendorf 5417R centrifuge (Germany). The supernatant was collected to measure total solids (Brix) using a refractometer (WYT-II, Qingyang Optical Instrument Co., Ltd., Chengdu, China).

2.4. Determination of Enzyme Activities. SOD (superoxide dismutase) activity was determined using a modified method [11]. About 2 g of fruit tissue from ten fruits was homogenized with 15 mL of 50 mmol/L sodium phosphate buffer (pH 7.8) and centrifuged at 8000 g for 15 min at 4°C with an Eppendorf 5417R centrifuge (Germany). The supernatant was collected as a crude enzyme of SOD. The reaction mixture (3 mL) containing 0.1 mL of enzyme extracts, 50 mmol/L sodium phosphate buffer (pH 7.8), 13 mmol/L methionine, 75 μmol/L nitroblue tetrazolium (NBT), 10 ηM EDTA, and 20 ηM riboflavin was illuminated using a fluorescent lamp (60 mol m^{-2} s^{-1}) for 20 min. The absorbance at 560 nm was recorded using a UV spectrophotometer (UV-1100, Shanghai Meipuda Instrument Co., Ltd., Shanghai, China). An aliquot of an identical solution was kept in the dark and served as the blank control. One unit of SOD activity was defined as the amount of enzyme that catalyzed a 50% decrease in the SOD-inhibitable NBT reduction.

POD (peroxidase) activity was analyzed using a modified method [12]. The crude enzyme of POD was prepared as the crude SOD enzyme was extracted. The assay mixture contained 1.5 mL of enzyme extract, 2 mL of 50 mmol/L sodium phosphate buffer (pH 7.8), 0.6 mL of 0.04 M guaiacol, and 0.1 mL of 15% H_2O_2. POD activity was measured by an increase in absorbance at 470 nm. One unit of POD activity was defined as a 0.01 increase in absorbance at 470 nm per gram in one minute.

CAT (catalase) activity was assayed according to the method described by Tejera García et al. [13]. Tissue (2 g) was homogenized with 15 mL of sodium phosphate buffer (pH 7.0) containing 1% polyvinyl polypyrrolidone (PVPP) and centrifuged at 8000 g for 15 min at 4°C. The supernatant was collected as the crude enzyme of CAT. CAT activity was measured by adding 0.6 mL of enzyme extract to 2 mL of sodium phosphate buffer (pH 7.0) containing 1 mL of 0.03% H_2O_2 as substrate. H_2O_2 decomposition was measured by the reduction in absorbance at 240 nm. One unit was defined as the change 0.1 absorbance per gram in one minute.

2.5. Determination of Malonaldehyde (MDA) Content. MDA was measured as previously described by Xing et al. [12]. Flesh tissue (2.0 g) from 10 fruits was homogenized with 10 mL of 10% trichloroacetic acid containing 0.5% (w/v) thiobarbituric acid. The mixture was then heated at 100°C for 10 min. After the rapid cooling of the sample to room temperature and centrifugation at 4000 g for 15 min at 25°C, the absorbance of the supernatant was measured at both 532 and 600 nm.

TABLE 1: Sensory character of grape on day 13.

Treatment	Character description of grape particles
Control	No brilliance, dull purple, serious decay, strong bad smell
Coating	No brilliance, dull purple, slight decay, obvious bad smell
Coating + 1.5×10^7 brewer yeast CFU/mL	Weak brilliance, dull purple, slight decay, a little bad smell
Coating + 1.5×10^8 brewer yeast CFU/mL	Weak brilliance, bright purple, slight soft, a little bad smell
Coating + 1.5×10^9 brewer yeast CFU/mL	Brilliance, bright purple, plump, little bad smell
Coating + 1.5×10^{10} brewer yeast CFU/mL	Slight dark, bright purple, slight soft, a little bad smell

Note: coating was 2% of alginate and 3% of carboxymethyl cellulose.

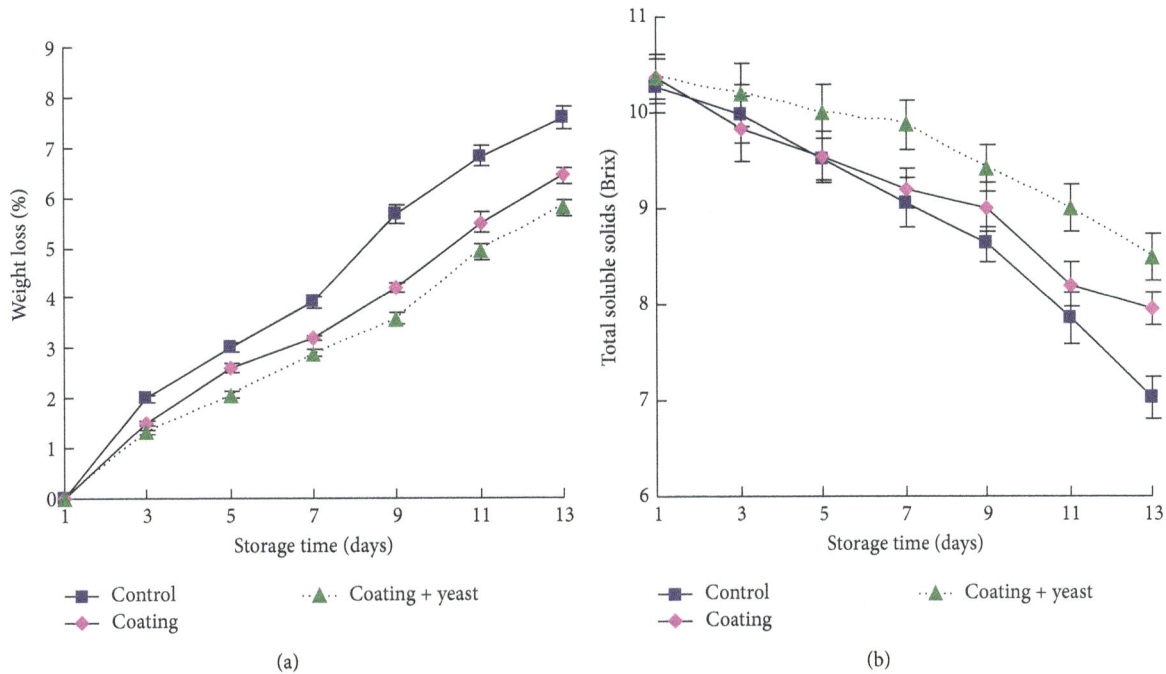

FIGURE 1: Effects of different treatment on the weight loss (a) and total soluble solids (b) of grape. Each point represents the mean value ± SD.

MDA concentration (μmoL g^{-1} fresh weight) was calculated by an extinction coefficient of 155 Mm^{-1} cm^{-1} through the formula $(OD_{532} - OD_{600}) \times 40/(0.155 \times$ formula weight).

2.6. Determination of Vitamin C. The vitamin C content was measured by 2,6-dichloriondophenol titration [14]. Briefly, tissue (2 g) from 10 fruits was immediately homogenized in 10 mL of 2% oxalic acid solution and then centrifuged at 8000 g for 15 min at 4°C. Afterwards, 2 mL of supernatant was titrated to a permanent pink colour using 0.1% of 2,6-dichlorophenolindophenol titration. The vitamin C concentration was calculated according to the titration volume of 2,6-dichloriondophenol.

2.7. Statistical Analysis. Experimental data were analyzed through ANOVA using the DPS7.05 statistical software (Refine Information Tech. Co., Ltd., Hangzhou, China). Experimental data were the means ± SD of three replicates of determinations for each sample. Mean separations were performed via Tukey's test; $P < 0.05$ was considered to indicate statistical significance.

3. Results and Analysis

3.1. Sensory Character. As shown in Table 1, the coating of 2% alginate and 3% carboxymethyl cellulose could increase the quality of postharvest grape during the storage. On day 13, the control sample was seriously rotted, had no brilliance, and gave off strong bad smell, while the coated sample was only slightly rotted, and the bad smell was lower compared with control samples. And brewer yeast was beneficial to grape preservation. With the concentration from 1.5×10^7 CFU/mL to 1.5×10^9 CFU/mL in coating, the quality of grape increased accordingly. On day 13, the sample of coating +1.5×10^9 CFU/mL of brewer yeast was brilliant, not rotten, the color was bright purple, and little bad smell was smelt. However, at the concentration of 1.5×10^{10}, sensory character deteriorated. So in the following experiment, the control, coating treatment (2% alginate and 3% carboxymethyl cellulose), and coating + yeast treatment (1.5×10^9) were further investigated.

3.2. Weight Loss and Total Soluble Solids. As shown in Figure 1(a), the weight loss of grape increased during the storage time. Compared with control samples, coating could

(a)

(b)

(c)

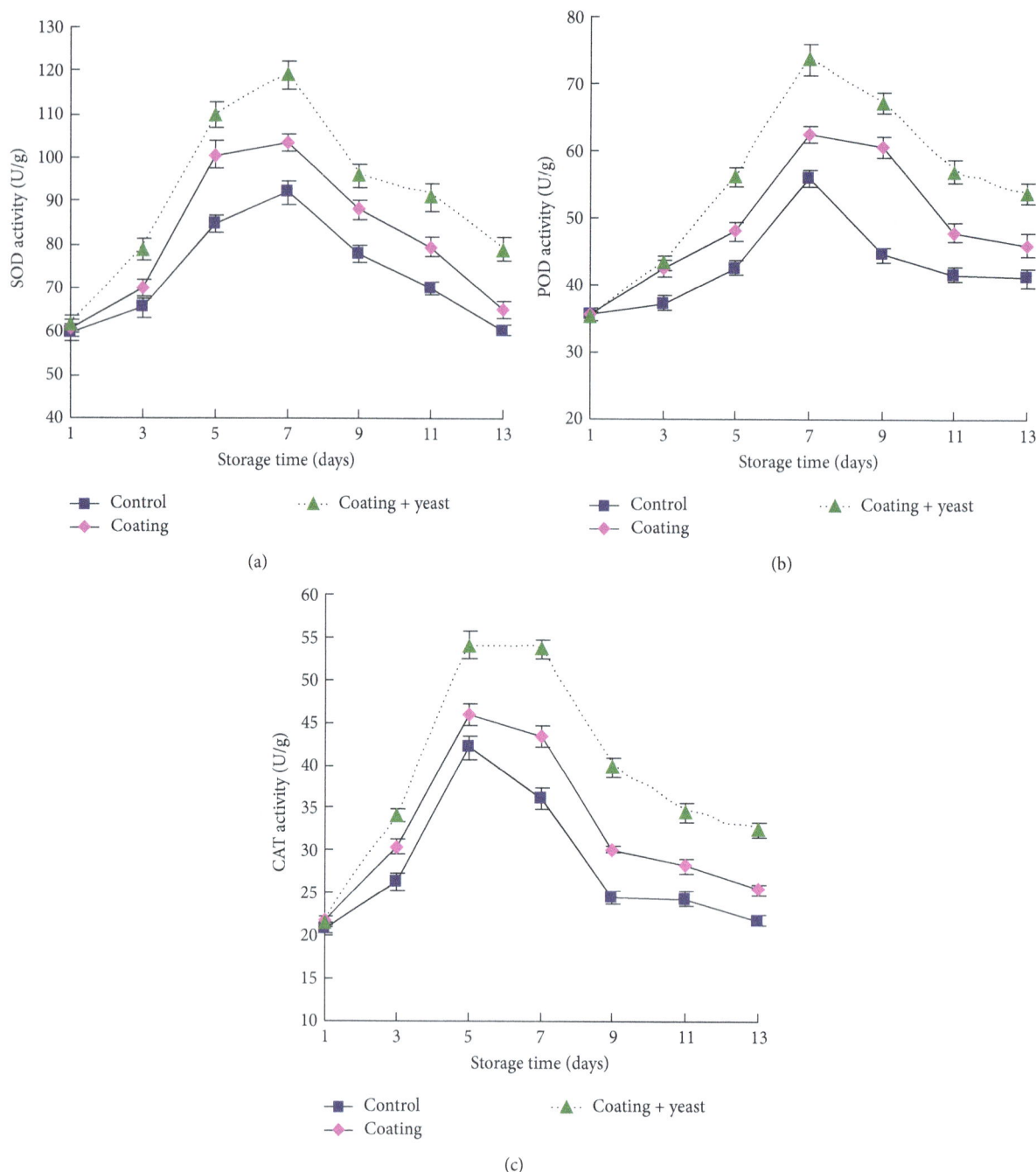

FIGURE 2: Effects of different treatments on the SOD (a), POD (b), and CAT (c) activities of grape. Each point represents the mean value ± SD.

restrain the increase of weight loss. Particularly the grape treated with coating + yeast demonstrated the lowest increase. At day 13, the weight loss rate of grape treated with coating + yeast was 5.79%, which was 23.6% lower than that of control samples. And there was significant difference between them (P < 0.05). The total soluble solids decreased in all samples (Figure 1(b)). The grape treated with coating + yeast showed the lowest decrease, the coated grape showed the lower decrease, and the control sample decreased the fast. On day 13, the total soluble solids of grape treated with

coating + yeast was 20.6% higher than that of control samples (P < 0.05).

3.3. SOD, POD, and CAT Activities. The SOD activities of all sample grapes increased before 7 days and then decreased from 7 to 13 days (Figure 2(a)). Compared with control samples, grape treated with coating + yeast showed the highest SOD activity, and the coated sample showed higher SOD activity. The peak value of control samples was 7.3% and

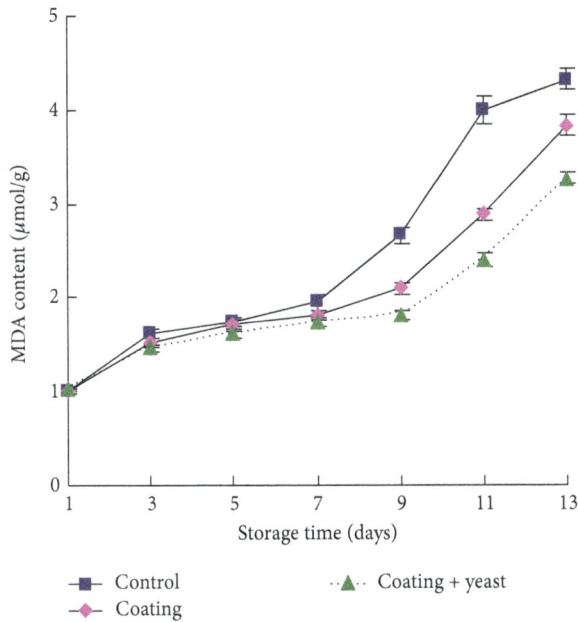

FIGURE 3: Effects of different treatments on the MDA content of grape. Each point represents the mean value ± SD.

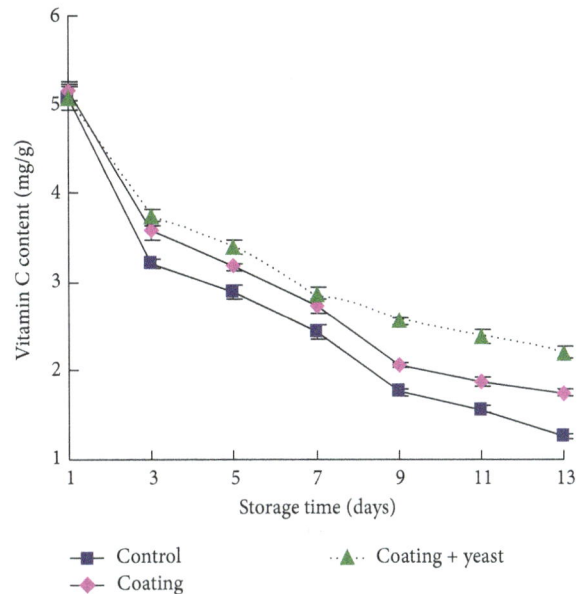

FIGURE 4: Effects of different treatments on the vitamin C content of grape. Each point represents the mean value ± SD.

23.6% lower than that of coated sample or grape treated with coating + yeast, respectively.

As shown in Figure 2(b), the POD activities of all sample grapes firstly increased and then decreased during the storage time. The coating or coating + yeast could maintain a higher level of the POD activity. And during the whole storage time, the grape treated with coating + yeast demonstrated the highest POD activity.

Similar to SOD activity or POD activity, the CAT activities of grapes also firstly increased and then decreased during the storage time (Figure 2(c)). The grape treated with coating + yeast showed the highest CAT activity, and the control sample showed the lowest activity during the whole storage. On day 13, the POD activity of grape treated with coating + yeast was 32.5 U/g, which was 48.4% and 27.5% higher than that of control sample or coated sample, respectively ($P < 0.05$).

3.4. MDA Content. The MDA content of all grapes increased during the storage time (Figure 3). Before 7 days, the MDA content of all samples slowly increases, and there were no differences among the treatments ($P > 0.05$). However, from 7 to 13 days, the MDA content of control sample fast increased. Though the MDA content of coated sample or grape treated with coating + yeast also increased from 7 to 13 days, they were lower than that of control sample. The grape treated with coating + yeast showed the lowest MDA content among all samples. At day 13, the MDA content of grape treated with coating + yeast was only 3.27 μmol/g, which was 24.3% lower than that of control samples ($P < 0.05$).

3.5. Vitamin C Content. As shown in Figure 4, the vitamin C content of grapes decreased during the storage time.

The control sample demonstrated the fastest decrease in vitamin C content, the coated sample showed the much faster decrease, and the grape treated with coating + yeast showed the slowest decrease. On day 13, the vitamin C content of control sample, coated sample, and grape treated with coating + yeast was 1.26, 1.75, and 2.21 mg/g, respectively. There were significant differences among them ($P < 0.05$).

4. Discussion

Postharvest grape is an active organism, undertaking metabolism ceaselessly. During the process, the inner substances of fruit were gradually exhausted. Thus, the weight loss and the total soluble solids decreased [7]. After treatment with coating + yeast, the increase of weight loss and the decrease of total soluble solids were partly restrained. All this suggested that coating+ yeast was beneficial to postharvest grape preservation.

Harvested grape generates free radicals, such as $O_2^{\bullet-}$ and H_2O_2, because of biochemical reactions. Free radicals can oxidize and destroy the cytoplasmic membrane, thereby accelerating senescence. The harm induced by free radicals is resisted by defence enzyme systems [15]. SOD can change $O_2^{\bullet-}$ into H_2O_2, and POD or CAT can eliminate H_2O_2. The united action of these three enzymes can reduce the harm to the cytoplasmic membrane [16, 17]. The SOD, POD, and CAT activities of the grape with coating + yeast demonstrated higher activities (Figure 2), which could efficiently eliminate $O_2^{\bullet-}$ and H_2O_2.

MDA originated from cytoplasmic membrane oxidation, and it may indicate the degree of cell consenescence [18]. The MDA content of the grape treated with coating + yeast was the lowest (Figure 3), and the reason was probably that higher activities of SOD, POD, and CAT could quickly eliminate the free radical. Thus, the harm to the cytoplasmic membrane by the free radical was minimized to the least degree.

Ascorbic acid of grape is an important nutrient as well as antioxidant to eliminate the active oxygen of fruit tissue. It has certain function to postpone the senescence of harvested fruit. The ascorbic acid content of grapes gradually decreased during storage (Figure 4). The reactive oxygen species of fruit tissue oxidizes ascorbic acid into MDHA (monodehydroascorbic acid) or DHAA (dehydroascorbic acid). Defense enzymes including SOD, CAT, and POD can protect ascorbic acid from degradation. Coating + yeast could enhance activities of SOD, CAT, and POD, which was advantageous to eliminate reactive oxygen rapidly. Thus, the ascorbic acid was retained due to reactive oxygen elimination [19].

The characteristics of the grapes treated with coating + yeast were superior to those with coating alone. Such characteristics include sensory character, weight loss, the activities of defense enzymes, and vitamin C content. Similar results were observed; TU Kang [20], who sprayed strawberry with antagonistic yeast before harvest, and found that the weight loss increase and vitamin C decrease of strawberry were significantly reduced.

5. Conclusion

The postharvest grape coated with 2% of alginate and 3% of carboxymethyl cellulose combined with brewer yeast of 1.5×10^9 CFU/mL showed good sensory character on day 13 compared with control sample or only coated sample. The increase of weight loss and decrease of total soluble solids of grape treated with coating + yeast were restrained. Furthermore, the defence enzymes including SOD, POD, and CAT of grape treated with coating + yeast showed higher activities. Accordingly, the increase of MDA content was also restrained and more vitamin C was preserved in grape treated with coating + yeast. At day 13, the weight loss rate and the total soluble solids of grape treated with coating + yeast were 23.6% lower and 20.6% higher than those of control samples, respectively. Coating grapes with 2% of alginate and 3% of carboxymethyl cellulose combined with brewer yeast of 1.5×10^9 CFU/mL was a well-proven method to preserve postharvest grapes.

Acknowledgments

This work was supported by the project of the National Natural Science Foundation of China under Grant no. 31101359 and was also supported by Program for the Innovative Talents of Higher Learning Institutions of Shanxi (2012) as well as by project of Natural Science Foundation of Shanxi under Grant no. 2012021025-3.

References

[1] L. Sanchez-Gonzalez, C. Pastor, M. Vargas, A. Chiralt, C. Gonzalez-Martinez, and M. Chafer, "Effect of hydroxypropylmethylcellulose and chitosan coatings with and without bergamot essential oil on quality and safety of cold-stored grapes," *Postharvest Biology and Technology*, vol. 60, no. 1, pp. 57–63, 2011.

[2] J. Y. Lee, H. J. Park, C. Y. Lee, and W. Y. Choi, "Extending shelf-life of minimally processed apples with edible coatings and antibrowning agents," *LWT—Food Science and Technology*, vol. 36, no. 3, pp. 323–329, 2003.

[3] H. Toğrul and N. Arslan, "Production of carboxymethyl cellulose from sugar beet pulp cellulose and rheological behaviour of carboxymethyl cellulose," *Carbohydrate Polymers*, vol. 54, no. 1, pp. 73–82, 2003.

[4] J. W. Rhim, "Physical and mechanical properties of water resistant sodium alginate films," *LWT—Food Science and Technology*, vol. 37, no. 3, pp. 323–330, 2004.

[5] Y. Li, A. J. Wu, and L. T. Song, "Coating technology of fresh garlic by CMC/isolated soybean protein," *Journal of Jilin Agricultural Sciences*, vol. 34, no. 6, pp. 60–62 (Chinese).

[6] G. I. Olivas, D. S. Mattinson, and G. V. Barbosa-Cánovas, "Alginate coatings for preservation of minimally processed "Gala" apples," *Postharvest Biology and Technology*, vol. 45, no. 1, pp. 89–96, 2007.

[7] Y. Fan, Y. Xu, D. Wang et al., "Effect of alginate coating combined with yeast antagonist on strawberry (*Fragaria × ananassa*) preservation quality," *Postharvest Biology and Technology*, vol. 53, no. 1-2, pp. 84–90, 2009.

[8] "Brewer's yeast," Oils and plants, http://www.oilsandplants.com/brewersyeast.htm.

[9] Y. W. Yu, S. Y. Zhang, Y. Z. Ren, H. Li, X. N. Zhang, and J. H. Di, "Jujube preservation using chitosan film with nano-silicon dioxide," *Journal of Food Engineering*, vol. 113, pp. 408–414, 2012.

[10] Z. Qiuping and X. Wenshui, "Effect of 1-methylcyclopropene and/or chitosan coating treatments on storage life and quality maintenance of Indian jujube fruit," *LWT—Food Science and Technology*, vol. 40, no. 3, pp. 404–411, 2007.

[11] Y. Zhao, K. Tu, J. Su et al., "Heat treatment in combination with antagonistic yeast reduces diseases and elicits the active defense responses in harvested cherry tomato fruit," *Journal of Agricultural and Food Chemistry*, vol. 57, no. 16, pp. 7565–7570, 2009.

[12] Y. Xing, X. Li, Q. Xu, Y. Jiang, J. Yun, and W. Li, "Effects of chitosan-based coating and modified atmosphere packaging (MAP) on browning and shelf life of fresh-cut lotus root (Nelumbo nucifera Gaerth)," *Innovative Food Science and Emerging Technologies*, vol. 11, no. 4, pp. 684–689, 2010.

[13] N. A. Tejera García, C. Iribarne, F. Palma, and C. Lluch, "Inhibition of the catalase activity from Phaseolus vulgaris and Medicago sativa by sodium chloride," *Plant Physiology and Biochemistry*, vol. 45, no. 8, pp. 535–541, 2007.

[14] O. A. Bessey and C. G. King, "The distribution of vitamin C in plant and animal tissues, and its determination," *Journal of Biological Chemistry*, vol. 103, pp. 687–698, 1933.

[15] K. F. Pirker, B. A. Goodman, E. C. Pascual, S. Kiefer, G. Soja, and T. G. Reichenauer, "Free radicals in the fruit of three strawberry cultivars exposed to drought stress in the field," *Plant Physiology and Biochemistry*, vol. 40, no. 6-8, pp. 709–717, 2002.

[16] G. K. Isamah, S. O. Asagba, and A. E. Thomas, "Lipid peroxidation, o-diphenolase, superoxide dismutase and catalase profile along the three physiological regions of Dioscorea rotundata Poir cv Omi," *Food Chemistry*, vol. 69, no. 1, pp. 1–4, 2000.

[17] S. Doğan, P. Turan, M. Doğan, O. Arslan, and M. Alkan, "Variations of peroxidase activity among Salvia species," *Journal of Food Engineering*, vol. 79, no. 2, pp. 375–382, 2007.

[18] J. Long, X. Wang, H. Gao et al., "Malonaldehyde acts as a mito-chondrial toxin: inhibitory effects on respiratory function and enzyme activities in isolated rat liver mitochondria," *Life Sciences*, vol. 79, no. 15, pp. 1466–1472, 2006.

[19] F. A. Loewus, "Biosynthesis and metabolism of ascorbic acid in plants and of analogs of ascorbic acid in fungi," *Phytochemistry*, vol. 52, no. 2, pp. 193–210, 1999.

[20] S. B. Mao, N. Zhu, Y. Y. Wei et al., "Effect of preharvest spraying antagonistic yeast Cryptococcus laurentii on the preservation of strawberry," *Science and Technology of Food Industry*, vol. 34, no. 4, pp. 344–348, 2013 (Chinese).

17

Analysis of Some Technological and Physical Characters of Mandarin (*Citrus reticulata*) Fruit in Iran

Abdollah Khadivi-Khub

Department of Horticultural Sciences, Faculty of Agriculture and Natural Resources, Arak University, Arak 38156-8-8349, Iran

Correspondence should be addressed to Abdollah Khadivi-Khub; akhadivi@ut.ac.ir, a-khadivi@araku.ac.ir

Academic Editors: S. Imhoff, Z. Yanqun, and M. Zhou

Knowledge of the physical properties of date fruit is necessary for the design of postharvesting equipment such as cleaning, sorting, grading, kernel removing, and packing. Also, the physical and mechanical properties are incorporated in the development of the grading machine as a case study. In this study, some physical and mechanical properties of three mandarin cultivars, as promising fruits, were analyzed to help the design of handling machines. According to results, the greatest dimensional characteristics were found for Page cultivar, whereas Onsho cultivar showed the lowest sphericity value (93%), and the highest sphericity was observed in Clementine cultivar (97%). The specific gravity of Page cultivar was 0/97, and this cultivar had the biggest fruit. Thus, it may be used for export. The volume measured was 1% higher than the calculated assumed shape of the spheroid ($V = 4.19a^2b$). The relationship between diameters and mass was linear, and the correlation was high for all studied cultivars and mixed cultivar (combined all data). There was a linear relation between mass and volume of the mixed cultivar of mandarin with a high coefficient of determination.

1. Introduction

Citrus are the major horticultural crops in Iran so that this country has a high annual production level of citrus fruit and was ranked the 15th producer in the world [1, 2]. The mandarin (*Citrus reticulata*) is a species of citrus fruit that is an evergreen tree (like other trees of this family) and slow growing. Mandarin begins to bear fruit when it is around three years old. Iranian mandarins are not exported because of variability in size and shape and lack of proper packaging [3]. Consumers prefer fruits with equal weight and uniform shape. Mass grading of fruit can reduce packaging and transportation costs and also may provide an optimum packaging configuration [4].

Fruit crops and food products have several unique characteristics which set them different from engineering materials. These properties determine the quality of the fruit, and identification of correlation among these properties makes quality control easier [5]. To design a machine for handling, cleaning, conveying, and storing, the physical, mechanical, and hydraulic properties of agricultural products must be known. Physical characteristics of fruit crops are

the most important parameters to determine the proper standards of design of grading, conveying, processing, and packaging systems [6]. Among these physical characteristics, mass, volume, and projected area are the most important ones in determining sizing systems [7]. Information regarding dimensional attributes is used in describing fruit shape which is often necessary in horticultural research for a range of differing purposes including cultivar descriptions in applications for plant cultivar rights or cultivar registers [8, 9]. Quality differences in mandarin fruit can often be detected by differences in density. When mandarin fruits are transported hydraulically, the design of fluid velocity is related to both density and shape. Volumes and projected area of fruits must be known for accurate modeling of heat and mass transfer during cooling and drying [10]. Hydrodynamic properties are very important characters in hydraulic transport and handling as well as hydraulic sorting of fruit crops. To provide basic data essential for development of equipment for sorting and sizing mandarin, it should be determine several properties of this fruit such as fruit density and terminal velocity of that [11, 12]. Many studies have reported on the chemical, physical and mechanical properties of fruits, such as wild plum [13], rose fruit [14] and sweet orange [7]. Also, chemical, and physical and mechanical properties of fruits in mandarin were reported in several studies [15, 16], but limited studies concerning hydrodynamic, physiomechanical, and technological properties of Iranian mandarins have been performed. There are two main objectives for this study. The first is to determine the hydrodynamic and physiomechanical properties of three mandarin cultivars in Iran (Clementine, Onsho, and Page). The second is to produce a convenient reference table with hydrodynamical, physical, and mechanical information suitable for fresh mandarin mechanization and progressing.

2. Materials and Methods

Three mandarin cultivars in Iran consisted of Clementine, Onsho, and Page were used in this study. A total of 165 fruits (55 from each cultivar) were tested in the biophysical and biological laboratories. The mandarins were picked up at random from their storage piles. Fruit mass (M) was determined with an electronic balance with 0.01 g sensitivity. To determine the average size of the fruits, three linear dimensions, namely, as length, width, and thickness, were measured by using a digital caliber with 0.1 mm sensitivity. Volume (V) was determined by the water displacement method [17]. For this purpose, a mandarin was submerged into a known volume of water, and the volume of water displaced was measured. Water temperature was kept at 25°C. Specific gravity of each mandarin was calculated by the mass of mandarins in air divided by the mass of displaced water. Three mutually perpendicular axes, a major, (the longest intercept), b intermediate (the longest intercept normal to a), and c minor, (the longest intercept normal to a, b) of mandarin were measured by Win Area-Ut-06 meter (Figure 1) developed by Mirasheh [18]. Geometric mean diameter, GM, was determined from the cubic roots of three diameters, $(abc)^{1/3}$, and percentage sphericity was equal to the geometric mean diameter divided by the longest diameter multiplied by 100 as suggested by Mohsenin [17]. The volume of mandarins was calculated assuming the shape of a prolate spheroid, an oblate spheroid, and an ellipsoid applying the following equations, respectively, $V = 0.52ab^2$, $V = 0.52a^2b$, and $V = 4.19$ (geometric mean diameter/2)^3. An average projected area as a criterion for the sizing machine was proposed. Three mutually perpendicular areas, Pa, Pb, and Pc were measured by a computer vision (diameter) Area-meter with high accuracy.

An average area projected (known as the criterion area, Ac, cm^2) was determined from

$$Ac = \frac{(Pa + Pb + Pc)}{3}. \tag{1}$$

Spreadsheet software, Microsoft Excel, 2010, was used to analyse data and determine regression models between the parameters. A typical linear multiple regression model is shown in

$$Y = a + b_1X_1 + b_2X_2 + \cdots + b_nX_n, \tag{2}$$

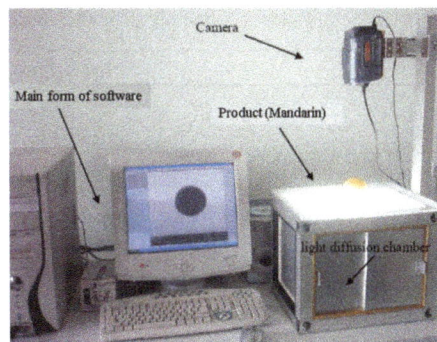

FIGURE 1: WinArea-UT-06 system.

where Y is a dependent variable, for example, mass, M, or a criterion area, Ac, is volume, $V, X_1, X_2, X_3, \ldots, X_n$-independent variable, for example, physical dimensions (mm), or volume, V, (cm^3), $b_1, b_2, \ldots, b_n$-regression coefficients, a-constant of regression. For example, mass is related to volume and can be estimated as a function of the volume measured as shown in

$$M = a + b_1V, \tag{3}$$

where V is the volume measured of mixed cultivars (combined all data) (cm^3).

3. Result and Discussion

A summary of the physical, mechanical, and hydrodynamic properties of Clementine, Onsho, and Page cultivars is shown in Table 1. According to these results, the greatest dimensional characteristics were found for Page cultivar with means of 65.33, 64.15, and 56.33 mm major, intermediate, and minor, respectively, whereas these values were 61.45, 60.32, and 57.50 mm and 60.90, 59.83, and 49.11 mm for Clementine, and Onsho cultivars, respectively. Erodğan et al. [19] reported that determining dimensional characteristics are essential to design a mechanism for mechanical harvesting.

An average of specific gravity of the Page cultivar was 0.97 and higher than others. The shape of studied cultivars is spheroid with a minimum probable error from the volume measured. Onsho cultivar showed the lowest sphericity value (93%) and the highest sphericity was observed in Clementine cultivar (97%). The specific gravity of Page cultivar was 0/97 and this cultivar had the biggest fruit. Thus, it may be used for export. Sphericity of mixed cultivar (combined all data) was 95%, and average of diameter of two diameters a and c was 1% less than the geometric mean diameter and with a similar coefficient of variation (8%). The knowledge related to geometric mean diameter would be valuable in designing the grading process [5]. The volume measured was 1% higher than the calculated assumed shape of the spheroid ($V = 4.19a^2b$). Volume and mass of each cultivar and also the mixed cultivar (combined all data) with three diameters were analyzed to determine the relationships between physical

TABLE 1: Some physical and engineering properties of mandarin fruit for studied cultivars.

Cultivar	Physical properties	Mean	Max.	Min.	SD	CV
Clementine	Major (mm)	61.45	73.70	54.20	4.07	6.62
	Intermediate (mm)	60.32	72.60	52.70	4.03	6.68
	Minor (mm)	57.50	67.30	79.90	4.37	7.54
	Mass (g)	82.13	124.70	60.60	13.22	16.09
	Volume (cc)	89.74	138.20	63.90	16.42	18.29
	Specific gravity (g/cm^3)	0.92	0.96	0.86	0.03	3.06
	Geometric mean (mm)	59.72	70.07	52.48	3.92	6.57
	Percent sphericity	97.00	98.00	94.00	0.02	1.74
Onsho	Major (mm)	60.90	72.60	53.40	4.03	6.62
	Intermediate (mm)	59.83	72.10	51.60	3.93	6.56
	Minor (mm)	49.11	58.30	43.90	3.32	6.76
	Mass (g)	68.52	97.40	48.80	11.81	17.23
	Volume (cc)	77.98	115.50	53.30	14.35	18.41
	Specific gravity (g/cm^3)	0.88	0.95	0.82	0.03	3.36
	Geometric mean (mm)	56.33	64.91	49.57	3.42	6.07
	Percent sphericity	93.00	96.00	88.00	0.02	1.95
Page	Major (mm)	65.33	74.30	53.90	4.31	6.59
	Intermediate (mm)	64.15	73.20	53.00	4.21	6.57
	Minor (mm)	56.33	64.20	47.10	4.25	7.54
	Mass (g)	103.82	147.3	61.70	18.37	17.70
	Volume (cc)	107.04	152.30	63.10	19.26	18.00
	Specific gravity (g/cm^3)	0.97	1.01	0.93	0.01	1.36
	Geometric mean (mm)	61.80	69.62	51.88	4.11	6.65
	Percent sphericity	95.00	99.00	89.00	0.02	1.68
Mixed cultivar (combined all data)	Major (mm)	62.56	74.30	53.40	4.56	7.30
	Intermediate (mm)	61.44	73.20	51.60	4.48	7.29
	Minor (mm)	54.31	67.30	43.90	5.44	10.02
	Mass (g)	84.82	147.3	48.80	20.67	24.37
	Volume (cc)	91.59	152.30	53.30	20.54	22.43
	Specific gravity (g/cm^3)	0.92	1.01	0.82	0.04	4.80
	Geometric mean (mm)	59.28	70.07	49.57	4.43	7.47
	Percent sphericity	95.00	99.00	88.00	0.03	2.69
	Ave. diameter $(a + c)/2$	58.44	69.00	48.80	4.55	7.78

TABLE 2: Relationship between volume and mass with three diameters in studied cultivars of mandarin.

	Volume $\ln V = k_1 \ln a + k_2 \ln b + K_3 \ln c + k_4$					Mass $M = K_1 a + k_2 b + k_3 c + k_4$				
Coefficient cultivar	k_1	k_2	k_3	k_4	R_2	k_1	k_2	k_3	k_4	R_2
Clementine	0.93	1.20	0.60	−6.70	0.99	0.65	2.07	0.58	−115.84	0.97
Onsho	0.27	1.98	0.60	−7.21	0.97	0.53	1.98	0.68	−115.66	0.95
Page	1.32	0.86	0.59	−6.80	0.99	1.74	1.55	1.12	−172.72	0.98
Mixed cultivar	0.97	1.32	0.70	−7.75	0.97	1.34	1.89	1.21	−180.44	0.89

properties (Table 2). Result showed high relationship between volume and high coefficient of determination, R_2, as shown in

$$\ln V = 0.97 \ln a + 1.32 \ln b + 0.7 \ln c - 7.75, \quad (4)$$

$$R^2 = 0.97.$$

Natural logarithm of volume with three diameters of all cultivars and mixed cultivar (combined all data) was high. The relationship between diameters and mass was linear, and the correlation was high for all studied cultivars and mixed cultivar (combined all data). Mass versus volume was plotted, and there was a linear relation between mass and volume of the mixed cultivar of mandarin with a high coefficient of determination, $R_2 = 0.96$ as shown in

$$M = 0.99V - 5.52. \tag{5}$$

Agamia et al. [20] reported that the average mass of fruit ranges from about 95 to 140 g, the fruit volume from 100 to 154 cm^3, and the diameter from 4 to 6.5 cm for Nareng, Clementine, Satsuma, Cleopatra, Mallawi, and Baladi Mandarins. Mousa (1998) [2] found that the mean values of diameter ranged from about 69 to 84 mm; height ranged from about 57 to 87 mm; mass ranged from about 160 to 208 g; volume ranged from 188 to 241 mm for Navel, Baladi, Acidless, and Valencia orange varieties.

Relation between the mean projected area and the volume of mandarin was determined from the plot and the coefficient of determination, between both was very high and close to unity. A nonlinear regression equation for the mixed cultivar of mandarin was determined as showed in

$$Ac = 1.48V^{0.65}, \qquad R^2 = 0.994. \tag{6}$$

Awady et al. [16] concluded that the physical properties of Minneola fruits which had oblong shape were as follows: diameter = 62–89 mm, height = 68–104 mm, mass = 201–345 g, volume = 120–342 cm^3, and projected area = 54–108 cm^2.

4. Conclusion

Some physical properties of Clementine, Onsho, and Page cultivars are presented in this study. From this study, it can be concluded that the highest and the lowest of length, geometric mean diameter, volume, mass, and specific gravity were obtained for Page cultivar, and it is the best for export. The lowest values for these traits were observed in Onsho cultivar. The mean percent sphericity of each mandarin cultivar resulted in different means, varying from 58.45 to 97.00%. Also, volume and diameter had a natural logarithmic relationship with three diameters. The physical and mechanical properties are incorporated in the design of the fruit hopper, revolving drums with holes (length and diameters of drums, diameter of holes, and number of holes), and exit chute of the designed grading machine.

References

[1] FAOSTAT, 2011, http://faostat.fao.org/site/567/DesktopDefauH.aspx?PageID=567.

[2] M. M. Mousa, *Engineering factors affecting the development of grading machine for citrus [Ph.D. thesis]*, Faculty of Agriculture Cairo University, 1998.

[3] A. Ganachari, K. Thangavel, S. M. Ali, U. Nidoni, and A. Ananthacharya, "Physical properties of Aonla fruit relevant to the design of processing equipments," *International Journal of Engineering, Science and Technology*, vol. 12, pp. 7562–7566, 2010.

[4] K. Peleg, *Produce Handling, Packaging and Distribution, Linear Form*, AVI Publishing, Westport, Conn, USA, 1985.

[5] A. Jannatizadeh, M. Naderi Boldaji, R. Fatahi, M. Ghasemi Varnamkhasti, and A. Tabatabaeefar, "Some postharvest physical properties of Iranian apricot (*Prunus armeniaca* L.) fruit," *International Agrophysics*, vol. 22, no. 2, pp. 125–131, 2008.

[6] A. Tabatabaeefar and A. Rajabipour, "Modeling the mass of apples by geometrical attributes," *Scientia Horticulturae*, vol. 105, no. 3, pp. 373–382, 2005.

[7] A. Topuz, M. Topakci, M. Canakci, I. Akinci, and F. Ozdemir, "Physical and nutritional properties of four orange varieties," *Journal of Food Engineering*, vol. 66, no. 4, pp. 519–523, 2005.

[8] H. Schmidt, J. V. Christensen, R. Watkins, and R. A. Smith, "Cherry descriptors," in *Plant Genetic Resources*, p. 23, ECSC, EEC, EAEC, Brussels, Luxembourg and International Board, Rome, Italy, 1995.

[9] M. Beyer, R. Hahn, S. Peschel, M. Harz, and M. Knoche, "Analysing fruit shape in sweet cherry (*Prunus avium* L.)," *Scientia Horticulturae*, vol. 96, no. 1–4, pp. 139–150, 2002.

[10] S. Peschel, R. Franke, L. Schreiber, and M. Knoche, "Composition of the cuticle of developing sweet cherry fruit," *Phytochemistry*, vol. 68, no. 7, pp. 1017–1025, 2007.

[11] R. W. Matthews, B. A. Stout, D. D. Dewey, and F. W. Bekker-Arkema, "Hydro handling of apple fruits," *Transactions of the American Society of Agricultural Engineers*, vol. 28, no. 3, pp. 65–130, 1965.

[12] D. H. Dewey, B. A. Stout, R. W. Matthews, and F. W. Bekker-Arkema, "Developing of hydrohandling system for sorting and sizing apples for storage in pallet boxes," USDA, Marketing Research Report 743, SDT, UDFS, 1966.

[13] S. Calisir, H. Haciseferogullari, M. Ozcan, and D. Arsalan, "Some nutritional and technological properties of wild plum (*Prunus* spp) fruit in turkey," *Journal of Food Engineering*, vol. 66, pp. 223–237, 2005.

[14] F. Demir and M. Özcan, "Chemical and technological properties of rose (*Rosa canina* L.) fruits grown wild in Turkey," *Journal of Food Engineering*, vol. 47, no. 4, pp. 333–336, 2001.

[15] A. H. Kinawy, *A comparative study on two mandarin cultivars [M.S. thesis]*, Horticulture Department, Faculty of Agriculture, Al-Azhar University, 1995.

[16] M. N. Awady, I. Yehia, M. A. Hassan, and A. M. El Lithy, "Some physical and mechanical properties of Minneola fruits," *Misr Journal of Agricultural Engineering*, vol. 21, no. 2, pp. 669–684, 2004.

[17] N. N. Mohsenin, *Physical Properties of Plant and Animal Materials*, Gordon and Breach, New York, NY, USA, 1986.

[18] R. Mirasheh, *Designing and making procedure for a machine determining olive image dimensions [M.S. thesis]*, Faculty of Biosystems Engineering. University of Tehran, Karaj, Iran, 2006.

[19] D. Erdoğan, M. Güner, E. Dursun, and I. Gezer, "Mechanical harvesting of apricots," *Biosystems Engineering*, vol. 85, no. 1, pp. 19–28, 2003.

[20] E. H. Agamia, M. M. Nageib, and M. M. Tamahy, "Evaluation of six varieties of mandarin," *Annals of Agricultural Science, Moshtohor*, vol. 18, pp. 225–233, 1982.

18

Remote Sensing Applications in Tobacco Yield Estimation and the Recommended Research in Zimbabwe

Ezekia Svotwa,[1] **Anxious J. Masuka,**[1] **Barbara Maasdorp,**[2] **Amon Murwira,**[3] **and Munyaradzi Shamudzarira**[1]

[1] *Tobacco Research Board, Kutsaga Research Station, Harare, Zimbabwe*
[2] *Department of Crop Science, University of Zimbabwe, Zimbabwe*
[3] *Department of Geography and Environmental Studies, University of Zimbabwe, Zimbabwe*

Correspondence should be addressed to Ezekia Svotwa; esvotwa2@gmail.com

Academic Editors: O. Ferrarese-Filho and C. Tsadilas

Tobacco crop area and yield forecasts are important in stabilizing tobacco prices at the auction floors. Tobacco yield estimation in Zimbabwe is currently based on statistical surveys and ground-based field reports. These methods are costly, time consuming, and are prone to large errors. Remote sensing can provide timely information on crop spectral characteristics which can be used to estimate crop yields. Remote sensing application on agriculture in Zimbabwe is still very limited. Research should focus on identifying suitable reflectance indices that are related to tobacco growth and yield. Varietal yield response to fertiliser and planting dates as well as suitable temporal windows for spectral data collection should be identified. The challenges of the different tobacco land sizes have to be overcome by identifying suitable satellite platform, with sufficient spectral resolution to separate the tobacco crop from the adjacent competing crops and noncrop vegetative surfaces. The identified suitable index should be strongly correlated with tobacco in season dry mass and yield. The suitable vegetative indices can be employed in establishing tobacco cropped area and then apply the long-term area yield relationship from government and nongovernmental statistical departments to estimate yield from remote sensing derived cropped area.

1. Background

Zimbabwe is the largest producer of tobacco in Africa and the world's fourth-largest producer of flue-cured tobacco (*Nicotiana tabacum*), after China, Brazil, and the United States of America. Tobacco production has been the leading driver behind the 34% growth in Zimbabwe's agriculture and one of the major sources of foreign currency [1]. Tobacco crop plays an important role in the economy of Zimbabwe and in the 2012/2013 marketing season, 144 million kg of tobacco was sold, earning the country $525 million [2].

Crop area and yield forecasts play an important role in stabilizing tobacco prices at the auction floors. Crop forecasting is the art of predicting crop yields and production before the harvest actually takes place, typically a couple of months in advance [2]. Zimbabwe mostly relies on crop statistical forecasting/estimation, crop reports/field visits from extension officers, and statistical crop forecasts for crop yield forecasts [3]. However, data from crop estimates, which are obtained through surveys conducted after harvests, are in most countries available quite late for early warning purposes.

Crop yield estimation in many countries is based on conventional techniques of data collection and ground-based field reports [4]. A variety of mathematical models relating to crop yield have also been proposed in recent years for many crops [4, 5]. In Zimbabwe crop surveys are mostly used in estimating crop yield [3]. The method is costly, time consuming, and prone to large errors due to incomplete ground observations, leading to poor crop yield assessment and crop area estimations [4].

2. Remote Sensing Applications in Crop Area Assessment

Remote sensing is defined as acquiring information about an object without physically getting into contact with it. Remote sensing has been used for some time to characterize properties of vegetation, to estimate yield, to estimate total biomass, and to monitor plant health and plant stress [6]. The interaction of the incident energy with the atomic structures of soil, rocks, plants, bodies of water, man-made objects, and so forth governs how much energy is absorbed and thus how much is reflected [7]. It is this reflected and absorbed energy that is picked up by the remote sensing devices, which is used to characterize the properties of a plant.

Visible (reflected light) and near-infrared (absorbed light) can be used to detect plant stress as a result of water shortages, nutrient deficiencies, and pests [8]. The contrast of light reflectance provides an assessment of the vegetation. Remote sensing can therefore provide a powerful tool for monitoring changes in the crop canopy over the growing season and can provide crop developmental information that is time-critical for site-specific crop management [9]. Remote sensing thus makes assessment objective faster, easier, and more reliable.

Remote sensing data has the potential and the capacity to provide spatial information at global scale of features and phenomena on earth on an almost real-time basis [4]. Use of remote sensing techniques has the potential to provide quantitative and timely information on agricultural crops over large areas, and many different methods have been developed to estimate crop yields [10, 11]. In general, the use of remote sensing is aimed at reducing the number of samples of ground surveys, making it less expensive [12]. With the application of remote sensing in agriculture, there is potential not only in identifying crop classes but also of estimating crop yield [4].

Remote sensing applications include monitoring deforestation, wildlife inventory, crop health status assessments, and yield forecasting [12]. Researchers have used remote sensing greatly to estimate fractional intercepted photosynthetically active radiation [13] and crop parameters like leaf chlorophyll, ground cover [12, 14], total dry-mass accumulation [14], plant greenness [15], yield [16], nitrogen status, and many other chemical properties of vegetation [17].

Spectral measurements from crops can be used in estimating crop parameters such as leaf area index [18], plant population, and even canopy total nitrogen status during the growth cycle of the crop [19]. Vegetation indices are algorithms which simplify data from multiple reflectance bands to a single value correlating to physical vegetation parameters, such as biomass, productivity, leaf area index, or percent vegetation ground cover [14]. Single refelectance bands are combined into a vegetation index in order to minimize the effect of such factors as optical properties of the soil background, illumination, and view geometric as well as meteorological factors on the canopy radiometric properties [20].

The current conventional tobacco yield forecasts rely on seed purchase records, land area, and visual assessment of the crop. Since farmers' records may not be exhaustive, the current forecast may not be accurate. Use of yield forecasting models can be employed to avoid these problems. Some existing models relate meteorological parameters to crop yield and production, while others are purely statistical in nature [3]. Although use of models has allowed fairly good forecasting capability to be conducted, the nature and relations between yield and some parameters may not be easily determined. Statistical models, for example, are location specific, and the use of averages in developing a yield model may not reflect conditions in extreme situations. The process of collecting data can be tedious and time consuming and thus preclude large-scale investigations [21].

Remote sensing can complement or even improve the current conventional tobacco yield prediction methods used. This is because remote sensing to provides useful information on real time crop condition as well as for yield forecasting. Every substance emits, absorbs, transmits, or reflects electromagnetic radiation in a manner characteristic of the substance [17] and depending on the chemical properties of the intercepting molecule, information from analysis of the energy of parts of the light spectrum absorbed or scattered by the atomic bonds, electrons, or atoms in the intercepting molecule can be used to predict yield [22].

The current conventional tobacco yield forecasts rely on seed purchase records, land area, and visual assessment of the crop. Since farmers' records may not be exhaustive the current forecast may not be accurate [21]. Use of yield forecasting models can be employed to avoid these problems. Some existing models relate meteorological parameters to crop yield and production, while others are purely statistical in nature [3]. Although use of models has allowed fairly good forecasting capability to be conducted, the nature and relations between yield and some parameters may not be easily determined.

3. Remote Sensing Science

By measuring the quantity of radiation in each of the wavelengths, the plant canopy characteristics can be defined [22]. The differences in leaf colours, textures, shapes or even how the leaves are attached to plants, determine the amount of reflected, absorbed, or transmitted energy, and such relationships are used to determine spectral signatures of individual plants, which are unique to plant species [23]. Spectral signatures make it possible to use remote sensing in studying changes in specific crop conditions in the field and relate these to final yield and quality [18].

The comparison of the reflectance values at different wavelengths is used to determine plant vigour [24]. The most common index that is used for this purpose is the normalized deviation vegetative index (NDVI) [8]. Vegetation indices, as summarized by Broge and Leblanc [25], are based on the characteristic reflection of plant leaves in the visible and near-infrared portions of light. By applying a "Vegetation Index" to the satellite imagery, concentration of green leaf vegetation can be quantified [26]. As explained by Broge and Leblanc [25], healthy vegetation has low reflection of visible light (from 0.4 to 0.7 μm), since it strongly absorbs chlorophyll for photosynthesis and, at the same time, there is high reflection of near-infrared light (from 0.7 to 1.1 μm). The portion of reflected near-infrared light depends on the cell structure of the leaf [27]. In fading or unhealthy leaves, photosynthesis decreases and cell structure collapses resulting in an increase of reflected visible light and a decrease of reflected near-infrared light [25].

The normalized difference vegetation index (NDVI) has been considered to be a useful way for crop yield assessment models, using various approaches such as simple integration, to reflect vegetation greenness [29]. The index responds to changes in the amount of green biomass, chlorophyll content, and canopy water stress and, hence, is the most commonly used in assessing crop vigor, vegetation cover, and biomass production from multispectral satellite data [30]. The NDVI is calculated from the near infrared (NIR) and red (R) bands of either handheld or satellite sensors using the formular: NDVI = (NIR − Red)/(NIR + Red). According to Kidwell [31] the NDVI value of each area on an image helps identify areas of varying levels of plant vigour within fields.

The validity of crop yield models with NDVI is determined by the strengths of association between the two variables included in the model [32]. It is also essential to have an understanding of the correlation existing between yield and NDVI at different phonological stages of crop for selecting appropriate date of satellite pass to include in the model [32].

Research has shown that the NDVI is directly related to the photosynthetic activity and hence energy absorption by plant canopy; typical examples include the leaf area index (LAI) and biomass chlorophyll concentration in leaves, plant productivity, and fractional vegetation cover [33]. These could also be considered when developing models for estimating seasonal biomass production for either individual species or communities. Remote sensing surveys have successfully been conducted elsewhere in forecasting yield in paddy rice [34], maize [18], and potatoes [35].

4. Need for a Remote Sensing Based Yield Estimation Model

Tobacco producers need to monitor crop growth and development and obtain early estimates of final yield [1]. In the current scenario, unavailability of a comprehensive method for estimating tobacco yield has often led to contradicting estimates, subjective national statistics, and general planning inefficiency by stakeholders. The current tobacco yield estimation is based on the Garvin model [21], seed tracking approach, and the statistical and crop condition assessment approaches. Such conventional methods of scouting are often labour intensive and are based on data collected from sampled area and, hence, their precision varied [21]. Variable crop conditions are only distinguishable to the very trained and experienced eye. A more objective and practical model for yield estimation could assist tobacco stakeholders with more precise data on tobacco growth characteristics, hectarage, and final yield that would be available for export [2].

Site-specific information on varieties, fertiliser management, and cultural practices may improve the accuracy of

yield crop forecasting and, offers the potential to provide quantitative and timely information on agricultural crops over large areas [36]. Multispectral imaging sensors are able to view more than one particular band of energy. These bands are selected in various regions of the electromagnetic spectrum, based on the optimum range of energy being reflected by the objects observed. In-season canopy images have also been found useful in predicting yields in maize, soybean, and cotton plant canopy [13, 16, 37].

Developing a model to forecast or estimate tobacco yield is very useful for decision making in the Zimbabwean tobacco industry [2]. A yield estimation model for tobacco could also assist stakeholders to accurately determine total energy requirements for curing [21]. For the government an accurate prediction of the crop size is a useful planning tool in view of foreign currency generated by tobacco, for determining import-export policies, government aid for farmers, and allocation of subsidies for agricultural programs [2].

By using satellite imagery instead of traditional sampling techniques, tobacco yield forecasts can be generated earlier than traditional estimates; and because they are based on images that can be constantly downloaded from the satellite, these forecasts can be updated frequently throughout the growing season, thus tracking growth response to different conditions as the season progresses [4]. The signatures from satellite imagery will then be fed into the model in order to come up with the volume estimate of the crop [16]. Remote sensing would enable observations over large areas at regular intervals, making it useful in large-scale crop modelling [22].

Use of satellite imagery would also enable the verification farmers' claims of seedbed area established, size of irrigated and dry land tobacco crop, varietal proportions in the field, and even monitor disease development adherence to legislation. Varietal distribution, nutrient, and cultural management effects can also be easily monitored and factored in the final yield forecast.

Tobacco yield estimates are essential for marketing of the crop as well as infrastructure development and policy making [2]. When estimates are overestimated, tobacco merchants supply larger volumes of money early in the season to accommodate an anticipated huge volume of leaf only for the prices to fall dramatically when the expectation is not met [2]. When crop yields are understated, merchants will typically avail less money for the leaf which will force market prices of the crop to drop in order to gain as much leaf as possible with the limited funds available. It goes without say as well that policy planning and infrastructure allocation become biased when based on tobacco estimates with a tendency to vary due to inaccuracy. Planted area estimates and field visits often present the challenge of not accurately representing the overall production picture because it is difficult to assess every farm every year due to accessibility challenges, financial constraints, and the temporal function of assessments.

There is need for a comprehensive and holistic approach to tobacco yield estimation for the nation. Remote sensing presents an interesting, cost effective, faster, comparatively cheaper, and more accurate means of determining planted area, crop vigour, and expected yield at a national level if suitably developed and applied. Satellite remote sensing specifically involves the use of space borne instruments to observe, analyze, and compare areas of interest for vegetative growth and development. Several platforms have become commercially available to aid the purpose of vegetative assessments. Satellite images collected over time specific crop periods can be easily accessed, compared, and used to provide key indications on crop vigour, biomass, and spatial distribution.

5. Recommended Research

Remote sensing application in Zimbabwe is still very limited, largely due to the perception that satellite data is expensive to obtain and complicated to process [38]. This could be as a result of research work that has focused on the high resolution spectral imagery from commercial satellites such as Quickbird which have been used extensively in more developed countries leading to wide adoption in the purposes of developing yield estimation procedures for various crops as was outlined by Wu et al. [35]. Indirect relationships between cereal yield and satellite derived vegetation indices have been developed and can accurately predict yields despite the economic organ not being directly assessed [39].

Satellite sensing presents the challenge of spectral confusion when imaging crops with planting dates spaced closely together or crops with near similar spectral signatures [40]. When the spectral resolution of a remote sensing instrument is comparatively low, it can be difficult to distinguish target crop species from other crops that may be in vegetative growth at the same time [41]. Spectral distinction of closely related species can be achieved by several methods such as using high spectral resolution sensors to identify specific wavelength regions that are unique to specific plant species [24]. Assuming the spectral resolution of the instrument in use is of adequate capacity such as high resolution Hyperion EO-1 platforms, discrimination can be based on the differences picked in specific wavelength regions affected by the growth and development of a species as demonstrated in the United States on trials to distinguish field peas, wheat, barley, and slashed wheat from the baseline soil reflectance [42].

Tobacco cultivation in Zimbabwe is guided by law [2] which states that tobacco can only be planted on or after the 1st of September up until the 31st of December of each growing season [19]. This narrows down the prospective window when land use change associated with tobacco area estimation can be done to practical time frames since any crop canopy reflectance's detected by satellite instruments before the 1st of September can easily be ruled out as those of flue cured tobacco. Most commercially grown crop species in Zimbabwe will not be in production by this time since they are dependent on rainfall distribution which does not normally begin until November [43]. The dominant reflection therefore detectable during September is that of the winter wheat that should be in senescence stage and ploughed lands in preparation of tobacco planting [38]. Senescencing wheat can be easily identified and separated from tobacco planted by the relatively higher reflection in the visible spectrum electromagnetic range than tobacco which would be in active growth and development. Bare soil displays a characteristic

spectral signature characteristic with an increasing linear profile, making it very easy to separate the September planted tobacco crop from adjacent bare fields [38].

When the October crop is planted, its growth profile appears similar to that of September planted crops with the distinct difference of temporal spacing [38]. Therefore the two crops appear to develop parallel to each other and this again makes separation and estimation relatively easier. The position of the red edge and corresponding reflective responses in specific wavelength sections can be used accurately to distinguish crop species that occur in the same temporal time frame but of different species [38].

Unlike cereal crops, there is a direct relationship between vegetative response and crop vigour, yield and biomass unlike in wheat, maize, and sorghum where biomass becomes a function of accumulated density but does not significantly contribute to canopy reflectance [41]. It becomes plausible to argue that due to this direct relationship, yield estimation and crop vigour assessments can be more accurately assessed by remote sensing instruments than cereal crops [38] and are less subjective to bias and anomalies in data interpretation [44]. In fact, the nature of yield-canopy reflectance relationships may be reversed in crop species whose economic yield function is inversely related to biomass, one such crop being cotton [28] even though the agronomic plant parameters such as stem height, leaf number, and vegetative overall plant biomass may be positively correlated to vegetative indices [45]. Interestingly enough, spatial resolution can significantly affect the ability of a satellite sensor to identify subtle differences in crops as was demonstrated by Toulios et al. [28]. Table 1 below shows the correlation strength variations that occur due to spatial variation differences applied to a cotton field to estimate yield.

Remote sensing skills developed for Zimbabwe's flue cured tobacco estimation should seek to address the challenges of yield conflicts that may arise from tobacco with different fertilizer management regimes [2]. Small scale farmers tend to apply lower rates of fertilizer than their large scale commercial counterparts; because of this, it becomes possible to overestimate crop yields of crops that lie under communal growers.

Research work should focus on whether the model should separate low fertilizer regime crops from the standard as well as those that might be subjected to over application of fertilizer [41]. Varietal differences should also be investigated to ascertain whether the influence of varietal differences grown in the country will affect the applicability of the method in predicting feasible tobacco yields [46]. The most applicable means of utilizing remote sensing instruments for yield estimation may lie in the temporal separation ability of instruments based on different planting dates [43].

According to Garvin, [21] yield estimation is possible with plant parameters such as plant height, leaf number, and drymass, it therefore becomes feasible to use NDVI derived from tobacco fields to estimate biomass and eventually derive final yield of flue cured tobacco. Garvin also argued that varietal differences do not significantly affect the prediction potential of tobacco varieties, thus making the NDVI-yield relationship independent of varietal expression.

TABLE 1: Correlation coefficients of spectral variables with cotton yield.

Spectral variables	Spectra 2 × 2 m versus Yield 2 × 2 m	Spectra 10 × 10 m versus Yield 10 × 10 m	Spectra 20 × 20 m versus Yield 20 × 20 m
12/07_NDVI	0.54	0.71	0.68
12/07_SAVI	0.54	0.71	0.68
12/07_IR	0.44	0.68	0.69
12/07_RED	0.15	0.57	0.67
12/07_Green	0.24	0.61	0.68
12/07_NDVI	0.55	0.72	0.7
12/07_SAVI	0.55	0.72	0.7
12/07_IR	0.46	0.68	0.69
12/07_RED	0.16	0.59	0.67
12/07_Green	0.27	0.63	0.68
12/07_NDVI	0.24	0.61	0.68
12/07_SAVI	0.24	0.61	0.68
12/07_IR	0.21	0.6	0.69
12/07_RED	0.24	0.62	0.69
12/07_Green	0.27	0.63	0.69

Source: Toulios et al. [28].

There are several vegetative indices that were developed for purposes of monitoring and quantifying crop growth and development. Among these are the NDVI [14], EVI, and SAVI [47], and correlation between these and the biophysical parameters of tobacco crop must be studied. The tobacco cropping season spans from September to April [19]. The planting periods for these are continuous from September to December [2]. Research should focus on identifying a suitable index that can separate the crops in the different planting regimes and then estimate yield separately. For each planting regime, there is also need for establishing the temporal window for collecting remote sensing data in order to achieve the best prediction ability [47].

The tobacco sector in Zimbabwe is divided into smallholder and commercial sector, with the former comprising 80% of total tobacco produced in the country [2]. The challenge of the different land sizes in the two sectors still has to be overcome by identifying suitable satellite platform, with sufficient spectral resolution to separate the tobacco crop from the adjacent competing crops and noncrop vegetative surfaces.

The identified suitable index should be strongly correlated with tobacco in season dry mass and yield for it to be suitable for use in crop yield forecasting [47]. However, another approach could be that of using the vegetative indices to establish tobacco cropped area [43] and then apply the long-term area yield relationship from government and nongovernmental statistical departments and develop models that can be used to estimate yield from remote sensing derived cropped area. For the experimentation purposes, use of hand held remote sensor like the multispectral radiometer can be useful in characterising the spectral response properties of the different phenological stage of tobacco, for the varieties and for different fertiliser levels. The information can then be applied in operational yield forecasting of tobacco. It

is recommended that further research be done to establish spectral ratios for different satellite platforms that can be utilised for the different land size situations where differently spectral resolutions could be required.

Acknowledgment

The authors are grateful to the Tobacco Research Board/Kutsaga Research Station for funding this series of experiments on "*Developing flue cured tobacco crop area and yield forecasting models using remote sensing and agronomic techniques.*" The authors declare that there is no conflict of interests. The Tobacco Research Board is a research institution and the first author of the paper is a registered DPhil student with the University of Zimbabwe, with the other three as supervisors. This paper is the third of the five papers developed from the five objectives of the DPhilAg studies. The TRB strives to publish all the research work that is carried out and not for financial gain.

References

[1] Tobacco Facts, "Zimbabwe Tobacco industry," 2009, http://www.tobacco-facts.net/tobacco-industry/zimbabwe-tobacco-industry.

[2] Tobacco Industries Marketing Board (T.I.M.B), "Annual report and Accounts for the Year Ended 30 June, 2004," T.I.M.B, Harare, Zimbabwe, 2005.

[3] FAO, *Agrometeorological Crop Forecasting*, Food and Agriculture Organization (FAO) of the United Nations, Rome, italy, 2008.

[4] C. A. Reynolds, M. Yitayew, D. C. Slack, C. F. Hutchinson, A. Huetes, and M. S. Petersen, "Estimating crop yields and production by integrating the FAO Crop Specific Water Balance model with real-time satellite data and ground-based ancillary data,"

International Journal of Remote Sensing, vol. 21, no. 18, pp. 3487–3508, 2000.

[5] T. R. Wheeler, P. Q. Craufurd, R. H. Ellis, J. R. Porter, and P. V. Vara Prasad, "Temperature variability and the yield of annual crops," *Agriculture, Ecosystems and Environment*, vol. 82, no. 1–3, pp. 159–167, 2000.

[6] J. Wu, D. Wang, and M. E. Bauer, "Assessing broadband vegetation indices and QuickBird data in estimating leaf area index of corn and potato canopies," *Field Crops Research*, vol. 102, no. 1, pp. 33–42, 2007.

[7] R. G. Keller, V. Harder, and J. Seeley, "Introduction to Basic Principles of Remote Sensing," Department of Geological Sciences. UTEP, 2008, http://www.geo.utep.edu/.

[8] R. D. Jackson, P. N. Slater, and P. J. Pinter Jr., "Discrimination of growth and water stress in wheat by various vegetation indices through clear and turbid atmospheres," *Remote Sensing of Environment*, vol. 13, no. 3, pp. 187–208, 1983.

[9] S. Landau, R. A. C. Mitchell, V. Barnett, J. J. Colls, J. Craigon, and R. W. Payne, "A parsimonious, multiple-regression model of wheat yield response to environment," *Agricultural and Forest Meteorology*, vol. 101, no. 2-3, pp. 151–166, 2000.

[10] J. A. Moran, A. K. Mitchell, G. Goodmanson, and K. A. Stockburger, "Differentiation among effects of nitrogen fertilization treatments on conifer seedlings by foliar reflectance: a comparison of methods," *Tree Physiology*, vol. 20, no. 16, pp. 1113–1120, 2000.

[11] M. E. Bausch, "The role of remote sensing in determining the distribution and yield of crop," *Advances in Agronomy*, vol. 27, pp. 271–304, 2000.

[12] P. C. Doraiswamy, T. R. Sinclair, S. Hollinger, B. Akhmedov, A. Stern, and J. Prueger, "Application of MODIS derived parameters for regional crop yield assessment," *Remote Sensing of Environment*, vol. 97, no. 2, pp. 192–202, 2005.

[13] L. Serrano, *Recent Advances in Quantitative Remote Sensing: Papers from the Second International Symposium*, Taylor & Francis, Bristol, Pa, USA, 2006.

[14] C. J. Tucker, "Red and photographic infrared linear combinations for monitoring vegetation," *Remote Sensing of Environment*, vol. 8, no. 2, pp. 127–150, 1979.

[15] G. Asrar, M. Fuchs, E. T. Kanemasu, and J. L. Hatfield, "Estimating absorbed photosynthetic radiation and leaf area index from spectral reflectance in wheat," *Agronomy Journal*, vol. 76, pp. 300–306, 1984.

[16] B. L. Ma, L. M. Dwyer, C. Costa, E. R. Cober, and M. J. Morrison, "Early prediction of soybean yield from canopy reflectance measurements," *Agronomy Journal*, vol. 93, no. 6, pp. 1227–1234, 2001.

[17] V. I. Adamchuk, R. L. Perk, and J. S. Schepers, "Applications of Remote Sensing in Site-Specific Management," Tech. Rep. EC 03-702, University of Nebraska Cooperative Extension Publication, 2003.

[18] A. D. Baez-Gonzalez, J. R. Kiniry, S. J. Maas et al., "Large-area maize yield forecasting using leaf area index based yield model," *Agronomy Journal*, vol. 97, no. 2, pp. 418–425, 2005.

[19] D. Haboudane, N. Tremblay, J. R. Miller, and P. Vigneault, "Remote estimation of crop chlorophyll content using spectral indices derived from hyperspectral data," *IEEE Transactions on Geoscience and Remote Sensing*, vol. 46, no. 2, pp. 423–437, 2008.

[20] M. Monteith, *Principles of Environmental Physics*, Edward Arnold, London, UK, 1990.

[21] R. T. Garvin, "Flue cured tobacco yield estimation in crop research," *Zimbabwe Science News*, vol. 19, no. 5-6, pp. 65–67, 1985.

[22] O. Mutanga, *Hyperspectral Remote Sensing of Tropical Grass Quality and Quantity [Ph.D. thesis]*, Wageningen University, ITC Dissertation Number 111, The Netherlands, 2004.

[23] J. Nowatzki, R. Andres, and K. Kyllo, "Agricultural Remote Sensing Basics," 2004, North Dakota State University Agriculture and University Extension, 2004, http://www.ag.ndsu.edu/.

[24] H. W. Gausman and D. E. Escobar, "Discrimination among plant nutrient deficiencies with reflectance measurements," *Annual Review of Crop Physiology*, vol. 244, 1973.

[25] N. H. Broge and E. Leblanc, "Comparing prediction power and stability of broadband and hyperspectral vegetation indices for estimation of green leaf area index and canopy chlorophyll density," *Remote Sensing of Environment*, vol. 76, no. 2, pp. 156–172, 2001.

[26] A. Viña, A. A. Gitelson, D. C. Rundquist, G. Keydan, B. Leavitt, and J. Schepers, "Monitoring maize (*Zea mays* L.) phenology with remote sensing," *Agronomy Journal*, vol. 96, no. 4, pp. 1139–1147, 2004.

[27] S. Moulin, A. Bondeau, and R. Delécolle, "Combining agricultural crop models and satellite observations: from field to regional scales," *International Journal of Remote Sensing*, vol. 19, no. 6, pp. 1021–1036, 1998.

[28] L. Toulios, D. Pateras, G. Zerva, T. A. Gemtos, and T. H. Markinos, "Combining satellite images and cotton yield maps to evaluate field variability inprecision farming," in *Proceedings of the World Cotton Research Conference*, F. M. Gillham, Ed., pp. 534–539, Athens, Greece, 1998.

[29] A. K. Prasad, L. Chai, R. P. Singh, and M. Kafatos, "Crop yield estimation model for Iowa using remote sensing and surface parameters," *International Journal of Applied Earth Observation and Geoinformation*, vol. 8, no. 1, pp. 26–33, 2006.

[30] S. Balaselvakumar and S. Saravanan, "Remote sensing techniques for agriculture survey," 2006, http://www.gisdevelopment.net/application/agriculture/overview/agrio014.htm.

[31] K. B. Kidwell, *Global Vegetation Index User's Guide*, U.S. Department of Commerce/National Oceanic and Atmospheric Administration/National Environmental Satellite Data and Information Service/National Climatic Data Center/Satellite Data Services Division, 1990.

[32] C. W. Jayroe, W. H. Baker, and A. B. Greenwalt, "Using multispectral aerial imagery to evaluate crop productivity," *Crop Management*, 2005.

[33] R. B. Myneni, F. G. Hall, P. J. Sellers, and A. L. Marshak, "Interpretation of spectral vegetation indexes," *IEEE Transactions on Geoscience and Remote Sensing*, vol. 33, no. 2, pp. 481–486, 1995.

[34] http://www.geospatialworld.net/paper/application/ArticleView.aspx?aid=255.

[35] J. Wu, D. Wang, and M. E. Bauer, "Assessing broadband vegetation indices and QuickBird data in estimating leaf area index of corn and potato canopies," *Field Crops Research*, vol. 102, no. 1, pp. 33–42, 2007.

[36] J. G. P. W. Clevers, "Application of a weighted infrared-red vegetation index for estimating leaf Area Index by Correcting for Soil Moisture," *Remote Sensing of Environment*, vol. 29, no. 1, pp. 25–37, 1989.

[37] V. Thomas, P. Treitz, J. H. Mccaughey, T. Noland, and L. Rich, "Canopy chlorophyll concentration estimation using hyperspectral and lidar data for a boreal mixedwood forest in

Northern Ontario, Canada," *International Journal of Remote Sensing*, vol. 29, no. 4, pp. 1029–1052, 2008.

[38] Tobacco Research Board (T.R.B), "Annual report and Accounts for the Year Ended 30 June, 2011," TRB, Harare, Zimbabwe, 2012.

[39] K. P. Gallo and T. K. Flesch, "Large-area crop monitoring with the NOAA AVHRR: estimating the silking stage of corn development," *Remote Sensing of Environment*, vol. 27, no. 1, pp. 73–80, 1989.

[40] E. B. Knipling, "Physical and physiological basis for the reflectance of visible and near-infrared radiation from vegetation," *Remote Sensing of Environment*, vol. 1, no. 3, pp. 155–159, 1970.

[41] C. P. Ferri, A. R. Formaggio, and M. A. Schiavinato, "Narrow band spectral indexes for chlorophyll determination in soybean canopies [*Glycine max* (L.) Merril]," *Brazilian Journal of Plant Physiology*, vol. 16, no. 3, pp. 131–136, 2004.

[42] J. L. Hatfield, A. A. Gitelson, J. S. Schepers, and C. L. Walthall, "Application of spectral remote sensing for agronomic decisions," *Agronomy Journal*, vol. 100, no. 3, supplement, pp. S117–S131, 2008.

[43] D. Manatsa, I. W. Nyakudya, G. Mukwada, and H. Matsikwa, "Maize yield forecasting for Zimbabwe farming sectors using satellite rainfall estimates," *Natural Hazards*, vol. 59, no. 1, pp. 447–463, 2011.

[44] C. Yang, J. H. Everitt, and J. M. Bradford, "Using high resolution QuickBird satellite imagery for cotton yield estimation," in *Proceedings of the ASAE Annual International Meeting*, Paper number 041119, pp. 893–904, August 2004.

[45] T. N. Carlson and D. A. Ripley, "On the relation between NDVI, fractional vegetation cover, and leaf area index," *Remote Sensing of Environment*, vol. 62, no. 3, pp. 241–252, 1997.

[46] F. M. Marumbwa, A. Murwira, E. K. Madamombe, S. Kusangaya, and F. Tererai, "Remotely sensing of irrigation water use in Mazowe Catchment, Zimbabwe," Zimbabwe National Water Authority (ZINWA), Department of Geography & Environmental Science, University of Zimbabwe, 2006.

[47] Z. Jiang, A. R. Huete, K. Didan, and T. Miura, "Development of a two-band enhanced vegetation index without a blue band," *Remote Sensing of Environment*, vol. 112, pp. 3833–3845, 2008.

19

Deterministic Imputation in Multienvironment Trials

Sergio Arciniegas-Alarcón,[1] Marisol García-Peña,[1] Wojtek Janusz Krzanowski,[2] and Carlos Tadeu dos Santos Dias[1]

[1] *Departamento de Ciências Exatas, Universidade de São Paulo/ESALQ, Cx.P.09, CEP. 13418-900, Piracicaba, SP, Brazil*
[2] *College of Engineering, Mathematics and Physical Sciences Harrison Building, University of Exeter, North Park Road, Exeter, EX4 4QF, UK*

Correspondence should be addressed to Sergio Arciniegas-Alarcón; sergio.arciniegas@gmail.com

Academic Editors: A. Escobar-Gutierrez and W. P. Williams

This paper proposes five new imputation methods for unbalanced experiments with genotype by-environment interaction ($G \times E$). The methods use cross-validation by eigenvector, based on an iterative scheme with the singular value decomposition (SVD) of a matrix. To test the methods, we performed a simulation study using three complete matrices of real data, obtained from $G \times E$ interaction trials of peas, cotton, and beans, and introducing lack of balance by randomly deleting in turn 10%, 20%, and 40% of the values in each matrix. The quality of the imputations was evaluated with the additive main effects and multiplicative interaction model (AMMI), using the root mean squared predictive difference (RMSPD) between the genotypes and environmental parameters of the original data set and the set completed by imputation. The proposed methodology does not make any distributional or structural assumptions and does not have any restrictions regarding the pattern or mechanism of missing values.

1. Introduction

In plant breeding, multienvironment trials are important for testing the general and specific adaptations of cultivars. A cultivar developed in different environments will show significant fluctuations of performance in production relative to other cultivars. These changes are influenced by different environmental conditions and are referred to as genotype-by-environment interactions, or $G \times E$. Often, $G \times E$ experiments are unbalanced because several genotypes are not tested in some environments. A common way of analyzing this type of study is by imputing the missing values and then applying established procedures on the completed data matrix (observed + imputed), for example, the additive main effects and multiplicative interaction model—AMMI—or factorial regression [1–5]. An alternative approximation is to work with the incomplete data using a mixed model with estimates based on maximum likelihood [6].

Several imputation methods have been suggested in the literature to solve the problem of missing values. One of the first was made by Freeman [7], who suggested imputing the missing values iteratively by minimizing the residual sum of squares and doing the $G \times E$ analysis on the completed table, reducing the degrees of freedom by the number of missing values. This work was developed by Gauch Jr. and Zobel [8], who made the imputations using the EM algorithm and the AMMI model or EM-AMMI. Some variants of this procedure using multivariate statistics (cluster analysis) were described in Godfrey et al. [9] and Godfrey [10]. Raju [11] proposed the EM-AMMI algorithm by treating the environments as random and suggested applying a robust statistic to the missing values in the stability analysis. Mandel [12] proposed the imputation to be made in incomplete two-way tables using linear functions of the rows (or columns). Other studies recommended by van Eeuwijk and Kroonenberg [13] as having good results in the case of missing values for $G \times E$ experiments were developed by Denis [14], Caliński et al. [15], and Denis and Baril [16]. They found that using imputations through alternating least squares with bilinear interaction models or AMMI estimates based on robust submodels could give results as good as those found with the EM algorithm. Additionally, Caliński et al. [17] introduced an algorithm that combines the singular value decomposition (SVD) of a matrix with the EM algorithm, obtaining results very useful for experiments in which the alternating least squares have some problems, for instance, convergence failures [18]. Recently, Bergamo et al. [19] proposed a distribution-free multiple imputation method that was assessed by Arciniegas-Alarcón [20] and compared by Arciniegas-Alarcón and Dias [21] with algorithms that use fixed effects models in a simulation study with real data. Meanwhile, a deterministic imputation method without structural or distributional assumptions for multienvironment experiments was proposed by Arciniegas-Alarcón et al. [22]. The method uses a mixture of regression and lower-rank approximation. Finally, other studies to analyze multienvironment experiments with missing values can be found in the literature. For example, methodologies for stability analysis have been studied by Raju and Bhatia [23] and Raju et al. [24, 25]. Recently, Pereira et al. [26], Rodrigues et al. [27], and Rodrigues [28] assessed the robustness of joint regression analysis and AMMI models without the use of data imputation.

Given the historical information about data imputation in experiments, and specifically in two-factor $G \times E$ experiments, the objective of the present paper is to propose a deterministic imputation algorithm without distributional or structural assumptions, using an extension of the cross-validation by eigenvector method presented by Bro et al. [29].

2. Materials and Methods

2.1. Data Imputation Using the Cross-Validation by Eigenvector Method. The cross-validation method was presented by Bro et al. [29] to find the optimum number of principal components in any data set that can be arranged in a matrix form. In this approximation, principal component analysis (PCA) models are calculated with one or several samples left out and the model is used to predict these samples. The method used cross-validation "leave-one-out" and the same study showed it to be more efficient than other well-known methodologies used in multivariate statistics, such as those presented by Wold [30] and Eastment and Krzanowski [31]. Because of this finding, Arciniegas-Alarcón et al. [32] used the method to determine the best AMMI models in $(G \times E)$ experiments. This methodology is now presented.

Step 1. Consider the $n \times p$ matrix $\mathbf{X}$ with elements x_{ij}, ($i = 1, \ldots, n$; $j = 1, \ldots, p$). The matrix is divided into disjoint groups, each group is deleted in turn (leave-one-out), and a PCA model $(\mathbf{T}, \mathbf{P})$ is obtained from the remainder by solving

$$\min \left\| \mathbf{X}^{(-i)} - \mathbf{TP}^T \right\|_m^2 \qquad (1)$$

with $m \leq \min(n-1, p-1)$. Here $\mathbf{X}^{(-i)}$ represents the matrix after deleting the ith group (leave-one-out), $\| \cdot \|^2$ defines the squared Frobenius norm, $\mathbf{P}^T\mathbf{P} = \mathbf{I}$, and $\mathbf{T}, \mathbf{P}$ are scores and loadings matrices with dimensions $(n-1) \times m$ and $p \times m$ respectively, where p is the number of columns and m is the number of components. Note that, in this method the deleted group corresponds to the ith row of $\mathbf{X}$ and according

to Smilde et al. [33] the model (1) can be rewritten in terms of the singular value decomposition (SVD)

$$\mathbf{X}^{(-i)} = \mathbf{UDV}^T = \sum_{k=1}^{m} \mathbf{u}_{(k)} d_k \mathbf{v}_{(k)}{}^T, \qquad (2)$$

where $\mathbf{U} = [\mathbf{u}_1, \mathbf{u}_2, \ldots, \mathbf{u}_m]$, $\mathbf{V} = [\mathbf{v}_1, \mathbf{v}_2, \ldots, \mathbf{v}_m]$, $\mathbf{D} = \text{diag}[d_1, d_2, \ldots, d_m]$, $\mathbf{T} = \mathbf{UD}$, and $\mathbf{P} = \mathbf{V}$.

Step 2. Estimate the score

$$\mathbf{t}^{(-j)T} = \mathbf{x}_i^{(-j)T} \mathbf{P}^{(-j)} \left(\mathbf{P}^{(-j)T} \mathbf{P}^{(-j)} \right)^{-1}, \qquad (3)$$

where $\mathbf{P}^{(-j)T}$ is the loading matrix found in Step 1 with the jth row excluded. $\mathbf{x}_i^{(-j)T}$ is a row vector containing the ith row of $\mathbf{X}$ except for the jth element.

Step 3. Estimate the element x_{ij} by

$$\widehat{x}_{ij}^{(m)} = \mathbf{t}^{(-j)T} \mathbf{p}_j^T, \qquad (4)$$

$\mathbf{p}_j$ is the jth row of $\mathbf{P}$.

Step 4. Find the prediction error of the (ij)th element, $e_{ij}^{(m)} = x_{ij} - \widehat{x}_{ij}^{(m)}$.

Step 5. Obtain the criterion value

$$\text{PRESS}(m) = \sum_{i=1}^{n} \sum_{j=1}^{p} \left(e_{ij}^{(m)} \right)^2. \qquad (5)$$

In order to make the imputation of missing values in the matrix from $(G \times E)$ experiments, a change is proposed in the method following the work of Krzanowski [34], Bergamo et al. [19], and Arciniegas-Alarcón et al. [22] using the singular value decomposition of a matrix [35].

Initially, suppose that $n \geq p$ and the matrix $\mathbf{X}$ has several missing values; in the case $n < p$, the matrix should first be transposed. The missing values are replaced by their respective column means $\overline{x}_j$, and after this has been done the matrix is standardized by columns, subtracting $\overline{x}_j$ and dividing by s_j (where $\overline{x}_j$ and s_j represent, resp., the mean and the standard deviation of the jth column). The eigenvector procedure using the SVD in expressions (2)–(4) is applied to the standardized matrix to find the imputation of the (i, j) element, denoted by $\widehat{x}_{ij}^{(m)}$. After the imputation, the matrix must be returned to its original scale, $x_{ij} = \overline{x}_j + s_j \widehat{x}_{ij}^{(m)}$.

At this point the matrix does not have any missing values, but the imputations are rather basic and need to be refined. In the works that previously mentioned an iterative scheme is advocated, iterations continuing until the imputations achieve convergence (i.e., there is stability in successive imputed values), but Caliński et al. [17] showed that this convergence is not always necessary when using a method that combines the EM algorithm with SVD. Therefore, taking this into account, we will also consider fixing in advance the number of iterations between 0 and 3, as well as permitting the process to run until convergence has been achieved. As regards to the computing effort, convergence can depend strongly on the size of matrix analyzed and also on the data structure (size of correlations, proportion of missing values, etc.), but in, for instance, the SVD method of Hastie et al. [36], convergence is achieved usually between 5 and 6 iterations, and in the Bergamo et al. [19] method it is achieved in between 20 and 50 iterations maximum.

On the other hand, the data imputation depends directly on (2) and (3). Equation (2) needs prior choice of the number of components (m) to extract from the SVD. Krzanowski [34] and Bergamo et al. [19] took $m = \min\{n-1, p-1\}$ with the objective of using the maximum amount of available information, but Hedderley and Wakeling [37] affirmed that if the estimation is based on the choice of a unique fixed number of dimensions, some of the lower dimensions may be essentially random. This can influence the imputation within an iterative scheme and can lead to the estimates becoming trapped in a cycle, hence preventing convergence. To solve this problem, they suggested including a test to check on the convergence rate, and in case a specific criterion is not being attained the number of dimensions should be reduced. Another option that has satisfactory results, suggested by Josse et al. [38] to choose an optimum m, is through cross-validation based uniquely on the observed data. However, the computational cost of this option is likely to be high.

Taking into account all the above mentioned in the present study, for imputation of each missing value of the matrix $\mathbf{X}$ the value of m in (2) is allowed to be different in each SVD calculated and is chosen according to the criterion used by Arciniegas-Alarcón et al. [22]. Thus, m is chosen such that $\left(\left(\sum_{k=1}^{m} d_k^2\right) / \left(\sum_{k=1}^{\min\{n-1,p-1\}} d_k^2\right)\right) \approx 0.75$. Moreover, in (3), the Moore-Penrose generalized inverse can be used instead of the classic inverse matrix as was studied in cross-validation by Dias and Krzanowski [39].

In this research, five imputation methods have been assessed. They are denoted Eigenvector0, Eigenvector1, Eigenvector2, Eigenvector3, and Eigenvector where the number indicates the number of iterations used while in the case of Eigenvector the process is iterated until convergence is achieved in the imputations.

These imputation methods are all deterministic imputations, and they have the advantage over other stochastic imputation methods (parametric multiple imputations) that the imputed values are uniquely determined and will always yield the same results when applied to a given data set. This is not necessarily true for the stochastic imputation methods [40].

2.2. The Data. To assess the imputation methods we used three data sets, published in Caliński et al. [41, page 227], Farias [42, page 115], and Flores et al. [43, page 274], respectively. In each case the data were obtained from a randomized complete block design with replication, and each reference offers an excellent description of the design if further details are required.

The first data set "Caliński" comprises an 18×9 matrix, for 18 pea varieties assessed in 9 different locations in Poland.

The experiment was conducted by the Research Center for Cultivar Testing, Slupia Wielka, and the studied variable was mean yield (dt/ha).

The second data set "Farias" was obtained from Upland cotton variety trials (Ensaio Estadual de Algodoeiro Herbáceo) in the agricultural year 2000/01, part of the cotton improvement program for the Cerrado conditions. The experiments assessed 15 cotton cultivars in 27 locations in the Brazilian states of Mato Grosso, Mato Grosso do Sul, Goiás, Minas Gerais, Rondônia, Maranhão, and Piauí. The studied variable was yield seed cotton (kg/ha).

The third data set "Flores" is in a 15×12 matrix, for 15 bean varieties assessed in 12 environments in Spain. The experiments were conducted by RAEA—Red Andaluza de Experimentación Agraria—where the studied variable was mean yield (kg/ha).

The three data matrices contained just the mean yield for each genotype in each environment, but the proposed methods work for any data set arranged in matrix form. For example, if information about the replications is available, an approach suggested by Bello [44] is to write the experiment in terms of a classic linear regression model in order to obtain the response vector and the design matrix, and then to join them into a single matrix and apply the proposed methods in this paper.

2.3. Simulation Study. Each original data matrix ("Caliński", "Farias", and "Flores") was submitted to random deletion of values at the three rates 10%, 20%, and 40%. The process was repeated in each data set 1000 times for each percentage of missing values, giving a total of 3000 different matrices with missing values. Altogether, therefore, 9000 incomplete data sets were available, and for each one the missing values were imputed with the 5 Eigenvector algorithms described above using computational code in R [45].

The random deletion process for a matrix $\mathbf{X}$ ($n \times p$) was conducted as follows. Random numbers between 0 and 1 were generated in R with the `runif` function. For a fixed r value ($0 < r < 1$), if the $(pi + j)$th random number was lower than r, then the element in the $(i+1, j)$ position of the matrix was deleted ($i = 0, 1, \ldots, n; j = 1, \ldots, p$). The expected proportion of missing values in the matrix will be r [34]. This technique was used with $r = 0.1, 0.2$ and 0.4 (i.e., 10%, 20%, and 40%).

2.4. Comparison Criteria. In general, the objective after imputation is to estimate model parameters from the complete table of information. One of the models frequently used in genotype-by-environmental trials is the AMMI model [46, 47], and for this reason the algorithms proposed in this paper will be compared through the genotypic and environmental parameters of the fitted AMMI models using the root mean squared predictive difference—RMSPD [39]. The AMMI model is first briefly presented.

The usual two-way ANOVA model to analyze data from genotype-by-environment trials is defined by

$$y_{ij} = \mu + a_i + b_j + (ab)_{ij} + e_{ij} \tag{6}$$

$(i = 1, \ldots, n; j = 1, \ldots, p)$ where μ, a_i, b_j, $(ab)_{ij}$, and e_{ij} are respectively, the overall mean, the genotypic and environmental main effects, the genotype-by-environment interaction, and an error term associated with the ith genotype and jth location. It is assumed that all effects except the error are fixed effects. The following reparametrization constraints are imposed: $\sum_i (ab)_{ij} = \sum_j (ab)_{ij} = \sum_i a_i = \sum_j b_j = 0$. The AMMI model implies that interactions can be expressed by the sum of multiplicative terms. The model is given by

$$y_{ij} = \mu + a_i + b_j + \theta_1 \alpha_{i1} \beta_{j1} + \theta_2 \alpha_{i2} \beta_{j2} + \cdots + e_{ij}, \quad (7)$$

where θ_l, α_{il}, and $\beta_{jl}(l = 1, 2, \ldots, \min(n - 1, p - 1))$ are estimated by the SVD of the matrix of residuals after fitting the additive part. θ_l is estimated by the lth singular value of the SVD, α_{il} and β_{jl} are estimated by the genotypic and environmental eigenvector values corresponding to θ_l.

Alternating regressions can be used in place of the SVD [48]; depending on the number of multiplicative terms, these models may be called AMMI0, AMMI1, and so forth.

An inherent requirement of the AMMI model is prior specification of the number of multiplicative components [49–51]. Rodrigues [28] made an exhaustive analysis of the related literature and concluded that usually two or three components can be used because, in general, one component is not enough to capture the entire pattern of response in the data, but with more than three components there are obvious visualization problems, and a huge quantity of noise is liable to be captured.

So, for the original matrices "Caliński", "Farias", and "Flores", we fitted the AMMI2 and AMMI3 models. The same models were then fitted for each one of the 9000 sets of data that had been completed by imputation, and each set of parameters was compared with its corresponding set from the original data by using the RMSPD in the following way:

$$\text{RMSPD}\,(gen) = \sqrt{\frac{\sum_{i=1}^{\text{NG}} \left(a_i - \widehat{a}_i\right)^2}{\text{NG}}};$$

$$\text{RMSPD}\,(env) = \sqrt{\frac{\sum_{j=1}^{\text{NE}} \left(b_j - \widehat{b}_j\right)^2}{\text{NE}}};$$

$$\text{RMSPD}_l\,(genmult) = \sqrt{\frac{\sum_{h=1}^{l} \sum_{i=1}^{\text{NG}} \left(\alpha_{ih} - \widehat{\alpha}_{ih}\right)^2}{(\text{NG})\,l}}; \quad (8)$$

$$\text{RMSPD}_l\,(envmult) = \sqrt{\frac{\sum_{h=1}^{l} \sum_{j=1}^{\text{NE}} \left(\beta_{jh} - \widehat{\beta}_{jh}\right)^2}{(\text{NE})\,l}}.$$

Here RMSPD(gen) represents the RMSPD among the estimated parameters for genotype main effects from the original data a_i and the corresponding parameters obtained from the completed data sets by imputation $\widehat{a}_i$. RMSPD(env) represents the RMSPD among the estimated parameters for environments main effects from the original data b_j and the corresponding parameters obtained from the completed data sets by imputation $\widehat{b}_j$. RMSPD$_l$($genmult$) represents the equivalent RMSPD for the pairs of estimated

parameters of genotype multiplicative components α_{ih}, $\widehat{\alpha}_{ih}$. RMSPD$_l$($envmult$) represents the equivalent RMSPD for the pairs of estimated parameters of environments multiplicative components β_{jh}, $\widehat{\beta}_{jh}$. In the statistics, NG represents the number of genotypes, NE the number of environments, and $l = 2$ or 3 depending on the considered model AMMI2 or AMMI3.

The best imputation method is the one with the lowest values of RMSPD in each case. Summarizing, in each simulated data set with missing values, we applied the methods Eigenvector, Eigenvector0, Eigenvector1, Eigenvector2, and Eigenvector3 and, then, in the completed data (observed + imputed) we fitted AMMI2, AMMI3 models for the calculation of the respectively RMSPD statistics. In order to visualize any differences more readily, the RMSPD values were standardized and the comparison was made directly. Note that because of the standardized scale, the values of the statistics can be either positive or negative.

3. Results

3.1. Polish Pea Data. Figure 1 shows the RMSPD(gen) distribution on the standardized scale for the "Caliński" data set, showing each imputation method and each percentage. It can be seen that the Eigenvector distribution is left asymmetric and this asymmetry increases as the missing values percentage increases. In general, the Eigenvector distribution has values above zero and when the number of missing values increases, it is concentrated above one. This means that this method had the biggest differences among the additive genotypic parameters of the real and completed (by imputation) data.

The best method according to RMSPD(gen) is Eigenvector1, the method with just one iteration. This method has the smallest median for the 10% and 20% percentages. In the 40% percentage the medians of Eigenvector0 and Eigenvector1 are practically the same in the figure, but Eigenvector1 continues be preferable because it has the smallest dispersion. So, Eigenvector1 gave the smallest differences between the additive genotypic parameters of the real and completed data.

Figure 2 shows the RMSPD(env) on the standardized scale for the "Caliński" data set. It shows very similar behaviour to that of RMSPD(gen). Again the Eigenvector method presents the biggest differences among the additive environment parameters of the real and completed data because of the algorithm that maximizes the RMSPD(env). In this case, the RMSPD(env) is minimized with Eigenvector0 and Eigenvector1, and in all the percentages of missing values the two have nearly equal medians. However, Eigenvector1 has the smallest dispersion and that makes this again the method of choice.

The box plot analysis was useful in determining the best imputation method for the RMSPD(gen) and RMSPD(env) distributions, but in the case of RMSPD$_2$($envmult$), RMSPD$_2$($genmult$), RMSPD$_3$($genmult$) and RMSPD$_3$($envmult$), a more formal analysis can be used to compare the distributions; for instance the Friedman nonparametric test and, if this is significant, then the Wilcoxon test [52].

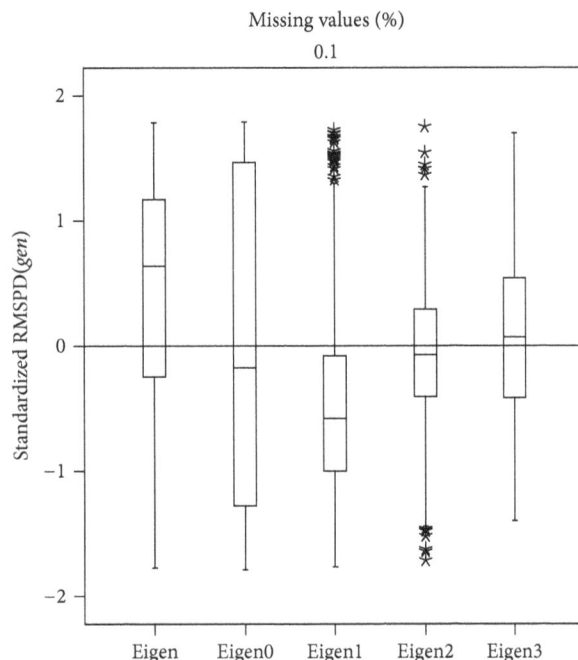

(a)

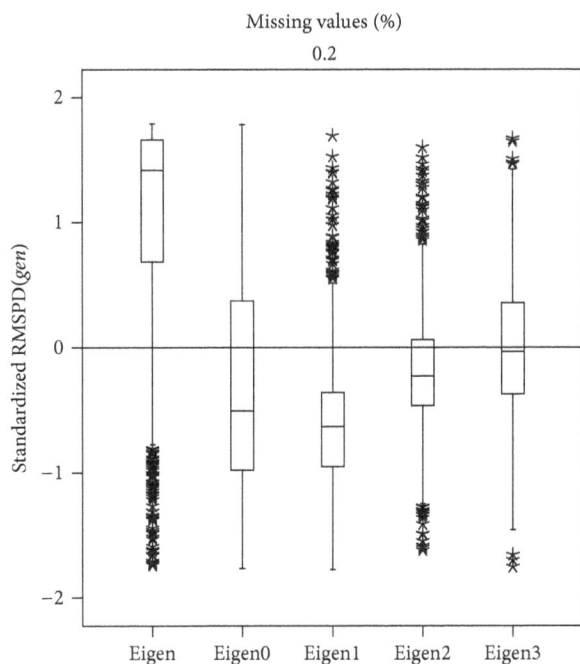

(b)

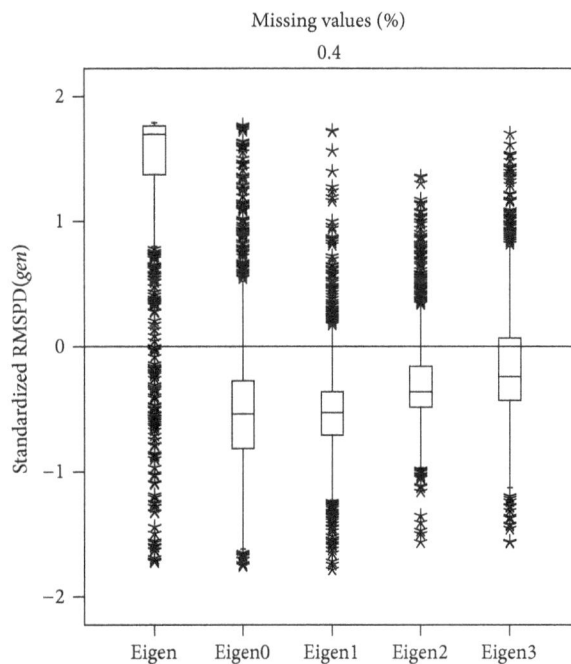

(c)

FIGURE 1: Box plot of the RMSPD(*gen*) distribution in Caliński data set.

Table 1 shows the Friedman test statistics. It can be seen that a significant difference exists among the imputation methods for the 10% and 20% percentage of missing values, but with 40% the five methods have equivalent results. After the general test, it is necessary to make multiple pairwise comparisons for the two lower percentages.

Table 2 shows the Wilcoxon test to find the methods that are different. When RMSPD$_2$(*genmult*) for 10% was used,

Eigenvector1 had significant differences with the other four methods. For 20%, Eigenvector1 was statistically different from Eigenvector, Eigenvector2, and Eigenvector3. For this percentage Eigenvector presents different results from Eigenvector0 and Eigenvector3. Joining the statistical differences found with the nonparametric test about RMSPD$_2$(*genmult*) and the correspond box plot in Figure 3, it can be said that for 10% and 20% the most efficient method is Eigenvector1,

(a)

(b)

(c)

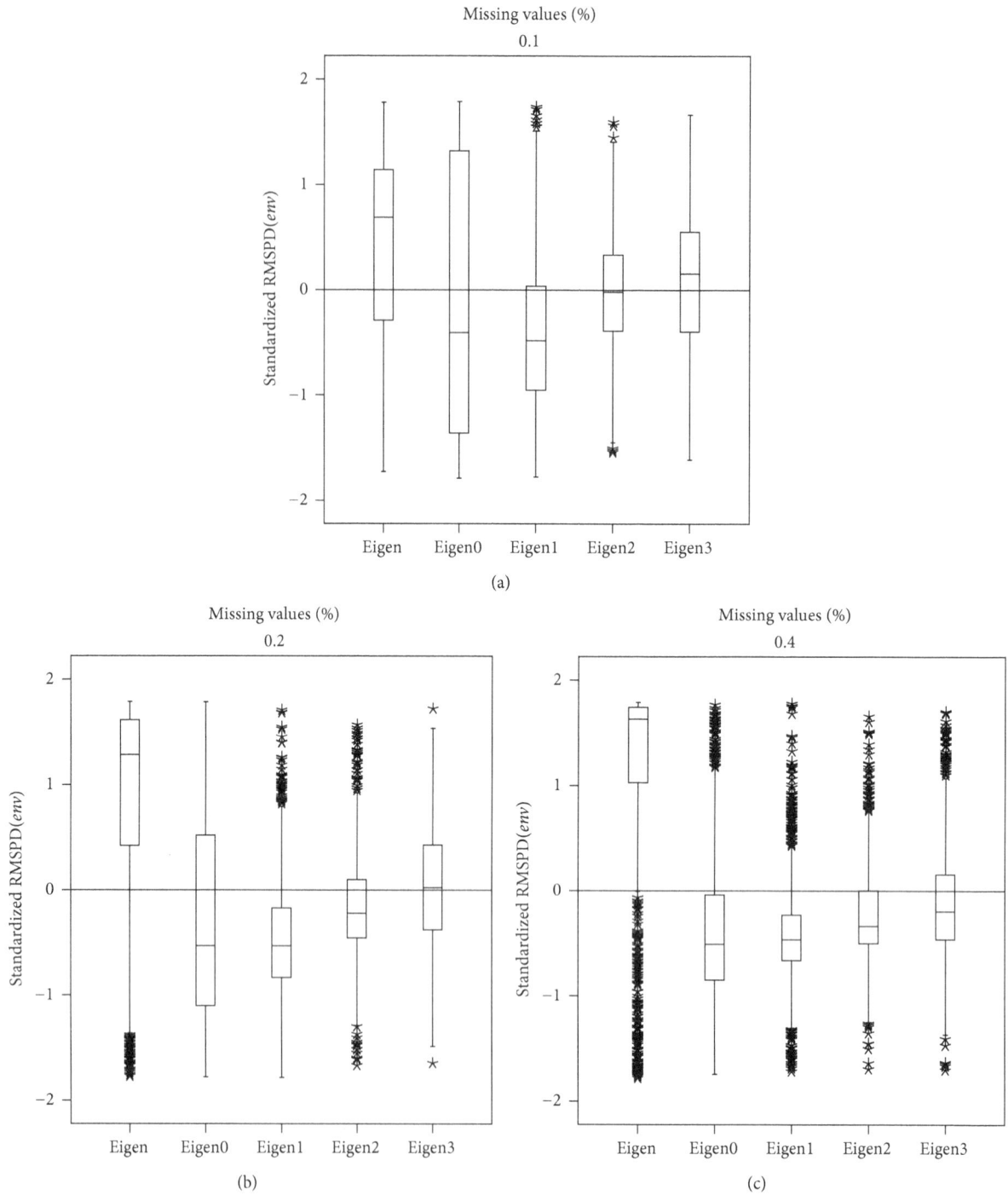

FIGURE 2: Box plot of the RMSPD(*env*) distribution in Caliński data set.

TABLE 1: Friedman test for the standardized RMSPD$_l$($\cdot$)—Caliński data set.

Perc.	Statistic							
	RMSPD$_2$(*genmult*)		RMSPD$_2$(*envmult*)		RMSPD$_3$(*genmult*)		RMSPD$_3$(*envmult*)	
	Friedman	P value	Friedman	P value	Friedman	P value	Friedman	P value
10%	15.6256	0.0036	34.4896	0.0000	34.9368	0.0000	30.4928	0.0000
20%	10.7848	0.0291	11.3688	0.0227	16.7144	0.0022	11.1104	0.0254
40%	2.8416	0.5847	2.5568	0.6345	4.9496	0.2925	5.9448	0.2033

TABLE 2: Wilcoxon test for the standardized $RMSPD_2(\cdot)$—Caliński data set.

| Percentage comparison | $RMSPD_2(genmult)$ | | $RMSPD_2(envmult)$ | |
	10% Wilcoxon	20% Wilcoxon	10% Wilcoxon	20% Wilcoxon
Eigen-Eigen0	−0.2913	−2.4166*	−0.8459	−1.7890
Eigen-Eigen1	−3.4322*	−2.6145*	−4.5972*	−1.5540
Eigen-Eigen2	−1.0087	−1.0783	−2.0225*	−0.1250
Eigen-Eigen3	−1.3178	−2.0335*	−2.4155*	−0.7970
Eigen1-Eigen0	−2.0468*	−0.1261	−2.8270*	−0.5490
Eigen2-Eigen0	−0.2997	−1.6703	−0.2598	−2.0420*
Eigen3-Eigen0	−0.3213	−1.3256	−0.0852	−1.6410
Eigen2-Eigen1	−3.3075*	−2.7537*	−4.3006*	−2.5030*
Eigen3-Eigen1	−3.5483*	−2.2389*	−5.0405*	−2.3590*
Eigen3-Eigen2	−0.4955	−0.0203	−0.9271	−0.6170

*Significant difference 5%.

TABLE 3: Wilcoxon test for the standardized $RMSPD_3(\cdot)$—Caliński data set.

| Percentage comparison | $RMSPD_3(genmult)$ | | $RMSPD_3(envmult)$ | |
	10% Wilcoxon	20% Wilcoxon	10% Wilcoxon	20% Wilcoxon
Eigen-Eigen0	−1.7875	−2.6026*	−2.1574*	−1.9962*
Eigen-Eigen1	−4.8856*	−3.1579*	−4.4210*	−2.6059*
Eigen-Eigen2	−1.8055	−1.9022	−1.7068	−1.2073
Eigen-Eigen3	−2.6978*	−2.0075*	−3.1627*	−1.3278
Eigen1-Eigen0	−1.8186	−0.2928	−1.1885	−0.4978
Eigen2-Eigen0	−1.0934	−1.0855	−1.2860	−1.0310
Eigen3-Eigen0	−0.9545	−1.2510	−1.1276	−1.1751
Eigen2-Eigen1	−4.9417*	−2.3846*	−4.1071*	−2.0129*
Eigen3-Eigen1	−4.5703*	−2.5499*	−4.1410*	−2.3572*
Eigen3-Eigen2	−0.1905	−0.7254	−0.1788	−0.8727

*Significant difference 5%.

because it minimizes the median and presents the smallest dispersion compared with Eigenvector and Eigenvector0. The five methods all present similar results for the 40% deletion rate.

Table 2 shows the Wilcoxon test results for the 10% and 20% percentage of missing values using $RMSPD_2(envmult)$. There are significant differences among Eigenvector and Eigenvector2, Eigenvector3, and Eigenvector1 for the 10% deletion rate. Differences were found between Eigenvector1 and Eigenvector0, Eigenvector2 and Eigenvector3, respectively. For 20%, Eigenvector1 was different from Eigenvector2 and Eigenvector3; besides, there is a difference between Eigenvector0 and Eigenvector2.

However, Table 3 shows the Wilcoxon test results of the standardized $RMSPD_3(envmult)$ and $RMSPD_3(genmult)$ values. In the 10% and 20% imputation percentages, there were significant differences between Eigenvector1 and Eigenvector, Eigenvector2 and Eigenvector3, respectively. Also, significant differences were detected between Eigenvector and the Eigenvector0 and Eigenvector3.

Finally, box plots were made for $RMSPD_2(envmult)$, $RMSPD_3(genmult)$, and $RMSPD_3(envmult)$, but are not presented here because they have similar behaviour to those in Figure 3, confirming that Eigenvector1 minimizes the median if it is compared with Eigenvector2 and Eigenvector3 and also has smaller dispersion than Eigenvector0. The method that always maximized all the statistics was Eigenvector, and for this reason it is the least recommended.

3.2. Brazilian Cotton Data. Figure 4 shows the $RMSPD(gen)$ distributions on the standardized scale for the "Farias" data set. The Eigenvector0 distribution is left asymmetric, and this asymmetry decreases as the missing values percentage increases. For the three percentages considered, the Eigenvector0 distribution is above one and very close to the other two, which means that this method had the biggest differences among the additive genotypic parameters of the real and completed (by imputation) data. With 10% imputation, the Eigenvector, Eigenvector2, and Eigenvector3 methods have very similar medians, but the smallest dispersion is achieved

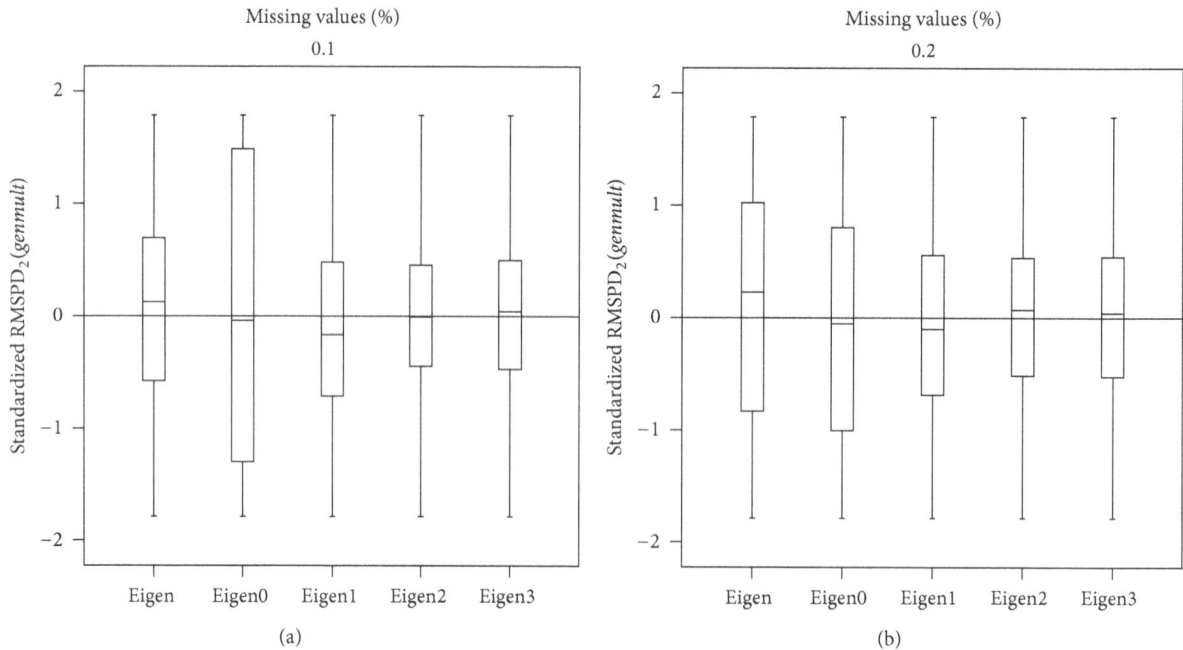

FIGURE 3: Box plot of the $RMSPD_2(genmult)$ distribution in Caliński data set.

TABLE 4: Friedman test for the standardized $RMSPD_i(\cdot)$—Farias data set.

Perc.	Statistic							
	$RMSPD_2(genmult)$		$RMSPD_2(envmult)$		$RMSPD_3(genmult)$		$RMSPD_3(envmult)$	
	Friedman	P value	Friedman	P value	Friedman	P value	Friedman	P value
10%	452.1168	0.0000	444.0952	0.0000	228.6352	0.0000	201.9368	0.0000
20%	313.0696	0.0000	295.0152	0.0000	193.6624	0.0000	173.3472	0.0000
40%	49.8712	0.0000	32.3296	0.0000	25.5240	0.0000	10.8736	0.0280

with Eigenvector2. Overall, when the missing values percentage increases Eigenvector achieves the best performance, because it minimizes RMSPD(gen). A similar behaviour is shown for RMSPD(env), as can be observed in Figure 5.

Table 4 shows the Friedman test statistics for $RMSPD_2(genmult)$, $RMSPD_2(envmult)$, $RMSPD_3(genmult)$, and $RMSPD_3(envmult)$. There is a significant difference among the imputation methods for all the percentages of missing values, so multiple pairwise comparisons were made with the Wilcoxon test.

Table 5 shows the Wilcoxon tests for $RMSPD_2(genmult)$ and $RMSPD_2(envmult)$. They indicate that with 10% imputation, the majority of the compared pairs have a significant difference, but, for example, Eigenvector1 is not significantly different from Eigenvector, Eigenvector2 or Eigenvector3. For the other two percentages, 20% and 40%, Eigenvector is not statistically different from Eigenvector2, and Eigenvector3 which have similar performances.

Table 6 shows the Wilcoxon test for $RMSPD_3(envmult)$. With 10% imputation, Eigenvector0 is different from all the others, while for $RMSPD_3(genmult)$ at the same percentage, Eigenvector1 was statistically different from Eigenvector2. With 20% and 40% of imputation, Eigenvector is not different

from Eigenvector2 or Eigenvector3, and likewise Eigenvector3 is not different from Eigenvector2.

In order to make a definitive conclusion, box plots were made for $RMSPD_3(genmult)$, $RMSPD_3(envmult)$, $RMSPD_2(genmult)$, and $RMSPD_2(envmult)$, but just one of them is presented because the distribution behaviour is similar for the others. From Figure 6, it can be concluded that the methods that minimize the median in all the percentages are Eigenvector, Eigenvector2, and Eigenvector3, and Tables 5 and 6 show that these methods are equivalent.

In summary, for the "Farias" data set, with the six standardized statistics, Eigenvector always showed good results and is therefore the recommended one.

3.3. Spanish Beans Data. Figure 7 shows the RMSPD(gen) distribution on the standardized scale for the "Flores" data set. Eigenvector has, in all the percentages, a left asymmetric distribution and maximizes the RMSPD(gen) median, therefore, it is the method that presents the biggest differences among the main genotypic parameters of the original and completed (by imputation) data. With 10% imputation, Eigenvector0 is the method which presents the best performance, while with 20% it is Eigenvector1 and

TABLE 5: Wilcoxon test for the standardized $RMSPD_2(\cdot)$—Farias data set.

Percentage comparison	$RMSPD_2(genmult)$			$RMSPD_2(envmult)$		
	10% Wilcoxon	20% Wilcoxon	40% Wilcoxon	10% Wilcoxon	20% Wilcoxon	40% Wilcoxon
Eigen-Eigen0	-12.7645^*	-11.8392^*	-5.5233^*	-12.0995^*	-11.3270^*	-4.4629^*
Eigen-Eigen1	-0.3505	-3.4137^*	-3.8021^*	-0.1890	-3.0716^*	-2.4969^*
Eigen-Eigen2	-2.6235^*	-0.7163	-1.4214	-2.4094^*	-0.4378	-0.3487
Eigen-Eigen3	-2.6720^*	-0.8664	-0.1908	-2.7633^*	-0.9311	-0.4897
Eigen1-Eigen0	-16.5991^*	-11.7349^*	-2.6190^*	-16.9885^*	-11.6590^*	-2.5653^*
Eigen2-Eigen0	-16.9878^*	-13.0576^*	-5.3009^*	-16.2317^*	-12.5528^*	-4.9312^*
Eigen3-Eigen0	-13.6133^*	-12.6292^*	-5.8550^*	-12.8970^*	-11.9226^*	-5.2543^*
Eigen2-Eigen1	-1.7703	-5.0465^*	-3.6028^*	-1.0721	-4.1600^*	-2.8340^*
Eigen3-Eigen1	-0.5797	-4.2466^*	-4.3441^*	-0.0083	-3.6872^*	-3.5903^*
Eigen3-Eigen2	-2.5865^*	-1.6592	-1.3257	-2.3910^*	-1.1199	-0.6422

*Significant difference 5%.

TABLE 6: Wilcoxon test for the standardized $RMSPD_3(\cdot)$—Farias data set.

Percentage comparison	$RMSPD_3(genmult)$			$RMSPD_3(envmult)$		
	10% Wilcoxon	20% Wilcoxon	40% Wilcoxon	10% Wilcoxon	20% Wilcoxon	40% Wilcoxon
Eigen-Eigen0	-9.2191^*	-9.2224^*	-4.1232^*	-8.2742^*	-8.9679^*	-2.1050^*
Eigen-Eigen1	-1.1084	-2.2832^*	-3.3175^*	-0.1224	-2.2990^*	-2.0120^*
Eigen-Eigen2	-0.6928	-0.1061	-0.8429	-1.1718	-0.2890	-0.3286
Eigen-Eigen3	-0.9784	-0.1836	-0.2097	-1.7433	-0.2468	-0.5434
Eigen1-Eigen0	-11.1032^*	-8.6574^*	-1.0162	-10.6797^*	-8.5424^*	-0.2532
Eigen2-Eigen0	-11.7189^*	-9.8996^*	-3.5702^*	-11.2791^*	-9.5638^*	-2.5008^*
Eigen3-Eigen0	-9.8820^*	-9.3163^*	-4.3048^*	-8.8932^*	-9.1492^*	-2.7670^*
Eigen2-Eigen1	-2.2248^*	-3.7149^*	-3.0406^*	-1.4067	-3.5119^*	-2.6124^*
Eigen3-Eigen1	-1.5319	-2.3342^*	-3.5506^*	-0.5309	-2.4132^*	-2.8674^*
Eigen3-Eigen2	-0.3787	-0.2394	-0.9666	-0.8871	-0.2512	-0.0848

*Significant difference 5%.

with 40% it is Eigenvector2, minimizing the median and taking the $RMSPD(gen)$ distribution to the bottom of the standardized scale. Figure 8 presents a similar result, but using $RMSPD(env)$. From the figure it can be said that with 20% imputation, Eigenvector0 and Eigenvector1 have similar medians, but Eigenvector1 is preferred because it has the smallest dispersion. With $RMSPD(env)$, Eigenvector0 has right asymmetric distributions and Eigenvector1, Eigenvector2, and Eigenvector3 have approximately symmetric distributions.

Table 7 shows the Friedman test for the statistics $RMSPD_2(genmult)$, $RMSPD_2(envmult)$, $RMSPD_3(genmult)$, and $RMSPD_3(envmult)$. It can be seen that significant differences exist among the methods only for 10% imputation. For this reason we restrict attention to this percentage.

Table 8 shows the 10 pairwise possible comparisons of imputation methods considering just 10% imputation and the statistics $RMSPD_2(genmult)$, $RMSPD_2(envmult)$, $RMSPD_3(genmult)$, and $RMSPD_3(envmult)$. Taken across the statistics all the methods are different except Eigenvector1 and Eigenvector0, but additionally, for $RMSPD_2(genmult)$ the pair Eigenvector1 and Eigenvector2 and for $RMSPD_3(genmult)$ the pair Eigenvector2 and Eigenvector0 are not significantly different.

Finally, to make a definitive conclusion about the four analyzed statistics in Tables 7 and 8, the box plot for $RMSPD_2(genmult)$ is presented in Figure 9. Plots were made of the other three statistics, but are not presented here because the behaviour is similar. According to the box plot, the best method is Eigenvector0 because it minimizes the median.

4. Discussion

We have presented five imputation methods and tested them through a simulation study based on three multienvironment trials and using six statistics derived from RMSPD. Overall, for big trials (i.e., 450 observations in the data matrix) Eigenvector should be used under convergence, while for small trials (i.e., 162 or 180 observations in the data matrix) two cycles of the process are enough in order to obtain good results without convergence.

We used experiments with different species, in different countries, and in different continents. Some of the results

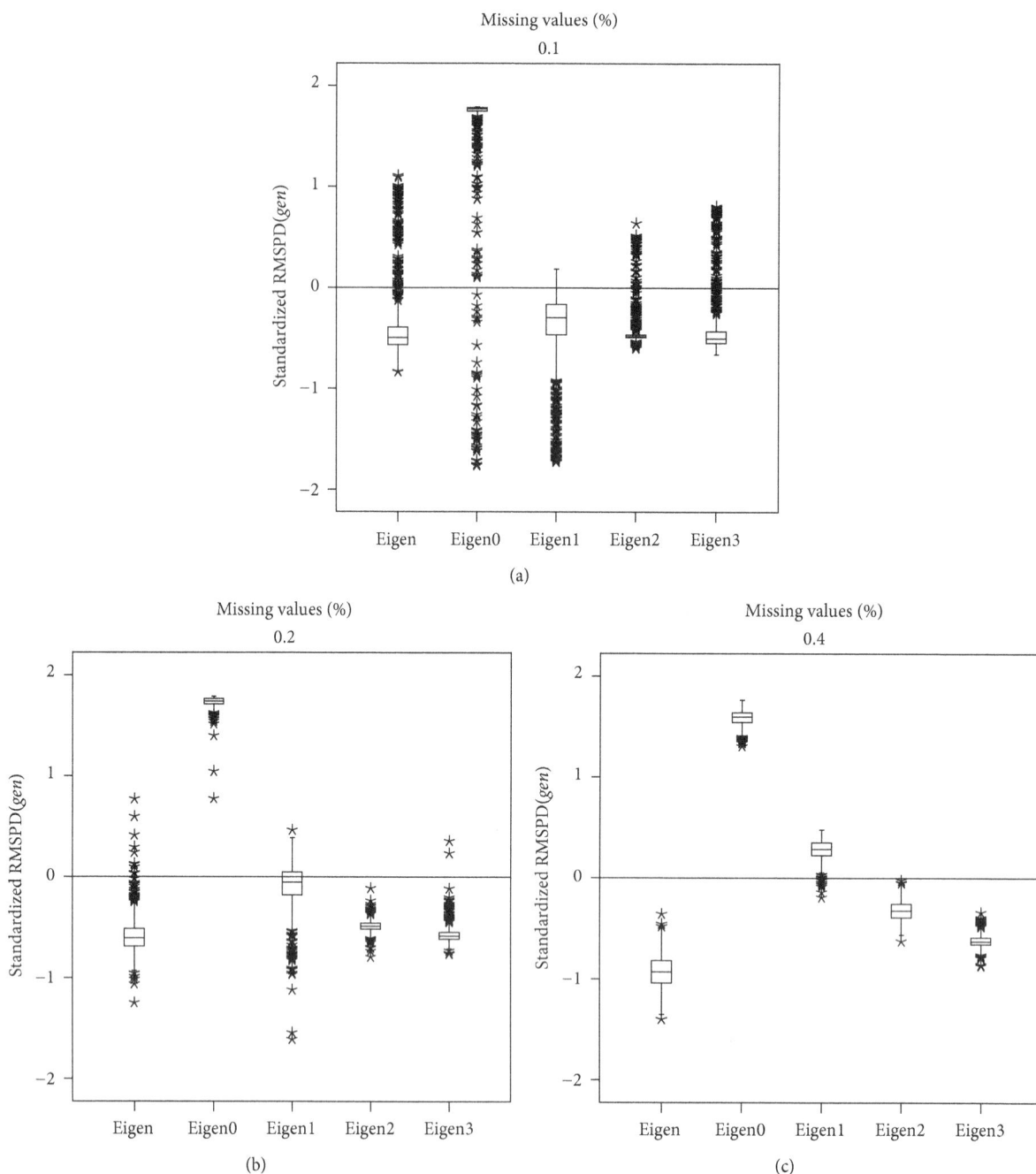

FIGURE 4: Box plot of the RMSPD(gen) distribution in Farias data set.

TABLE 7: Friedman test for the standardized $RMSPD_l(\cdot)$—Flores data set.

Perc.	RMSPD$_2$(genmult)		RMSPD$_2$(envmult)		RMSPD$_3$(genmult)		RMSPD$_3$(envmult)	
	Friedman	P value	Friedman	P value	Friedman	P value	Friedman	P value
10%	23.1512	0.0001	39.0136	0.0000	24.0736	0.0001	26.1888	0.0000
20%	5.0296	0.2843	2.2608	0.6879	1.8144	0.7698	1.0936	0.8953
40%	5.2256	0.2649	3.7480	0.4412	8.1944	0.0847	1.6856	0.7933

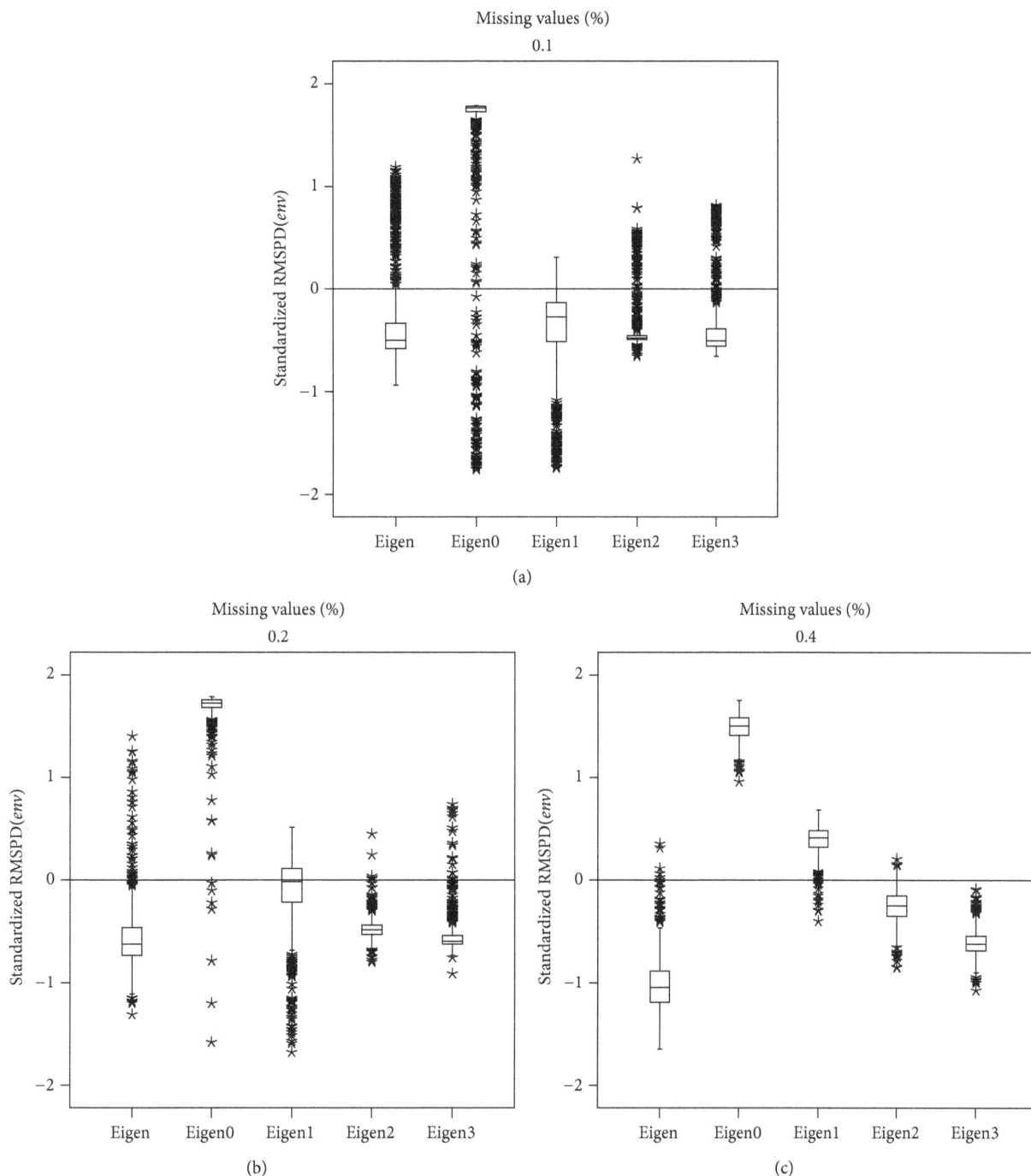

FIGURE 5: Box plot of the RMSPD(*env*) distribution in Farias data set.

were as expected, but one important outcome is that the iterative aspect of the proposed algorithms should be obligatory when missing values are imputed in $G \times E$ experiments.

So there is a natural question for the applied researcher: how to choose the appropriate Eigenvector imputation method for experiments with different size to those illustrated in this paper? The answer depends on the imputation objective, because the imputation can be used in several ways: to establish one or more genotype-environment combinations that for some reason were not observed, or to follow the imputation with some further statistical modeling. The choice criteria can be extensive, but for the first objective it would be natural to find the imputation errors associated with each Eigenvector method. To find these errors, we can employ cross-validation, using the methodology proposed by Piepho [18] and studied in more detail via simulations in real data by Arciniegas-Alarcón et al. [32]. This methodology is now briefly presented.

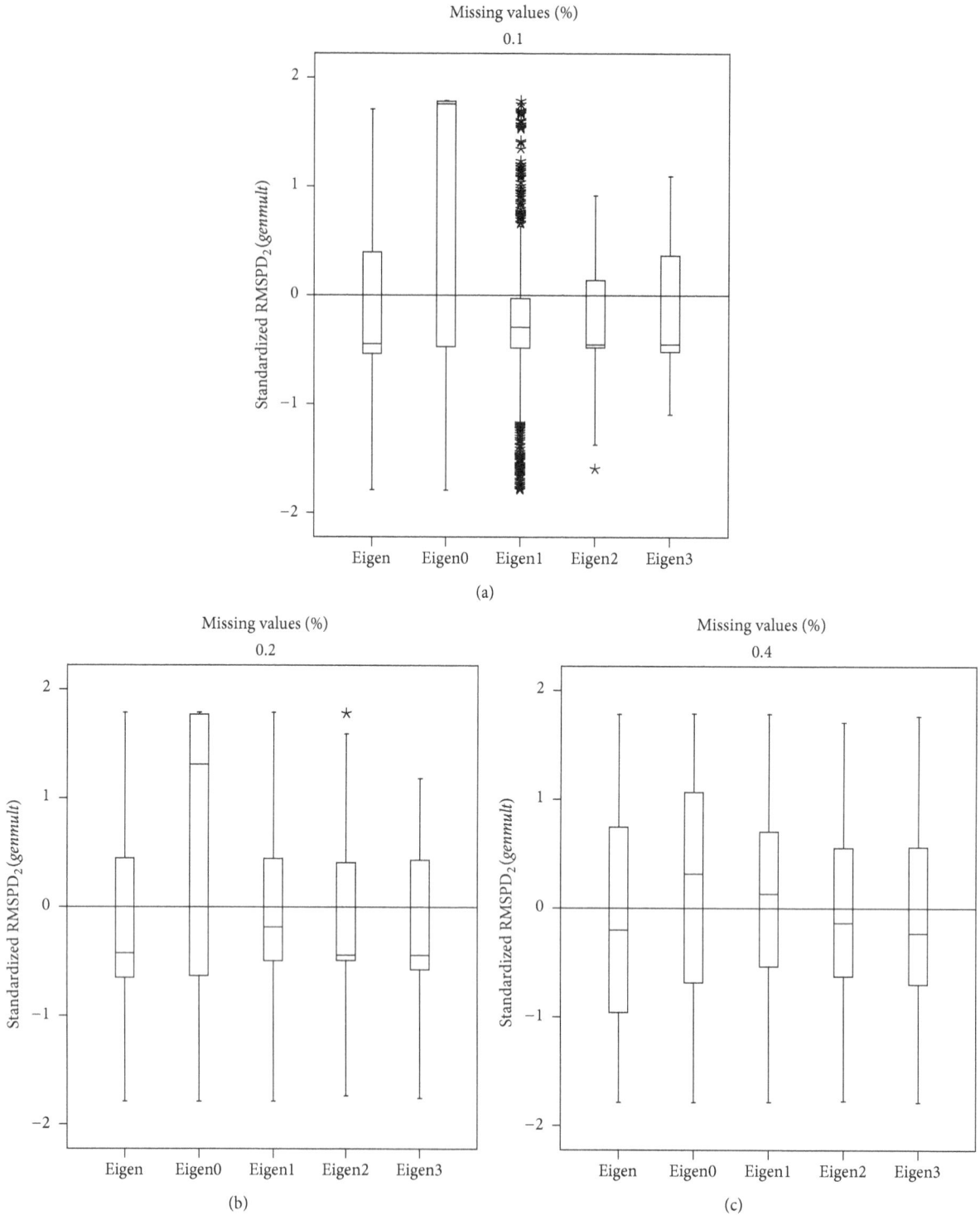

FIGURE 6: Box plot of the $RMSPD_2(genmult)$ distribution in Farias data set.

Suppose a $G \times E$ experiment is arranged in a table with missing values. From the table of observed values, delete one cell at a time, impute all the missing values, and record the difference between estimated and actual data for the cell under consideration. Do this for all observed cells, and take the average of the squared differences. Denote this quantity by D. D contains two components of variability: one due to predictive inaccuracy of the estimate, the other due to sampling error of the observed data. For this reason D may be corrected by subtracting an estimate of the error of a

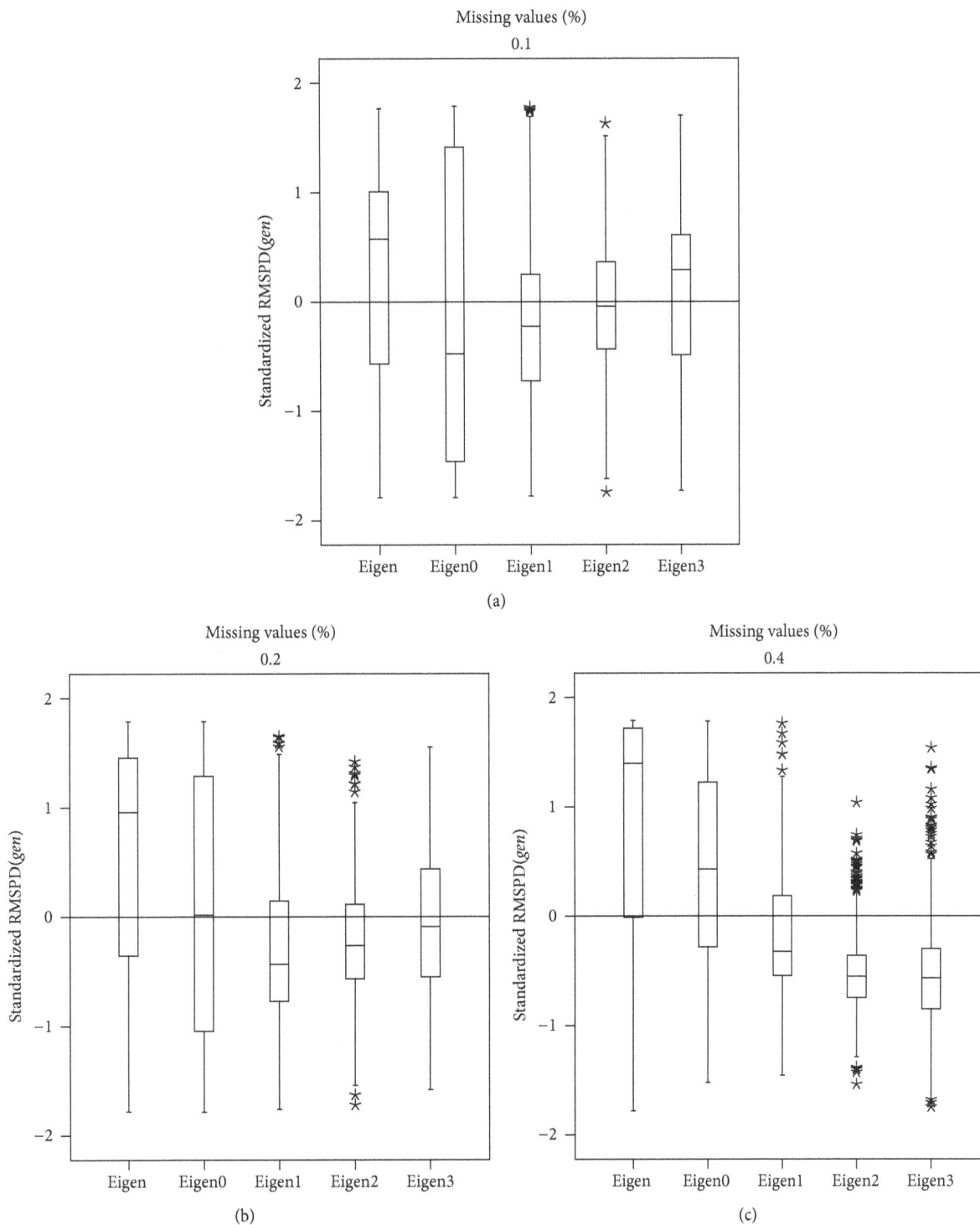

FIGURE 7: Box plot of the RMSPD(*gen*) distribution in Flores data set.

mean (s^2). The square root of ($D - s^2$) may be taken as the imputation error. The Eigenvector method with smallest imputation error is the method to choose.

On the other hand, if the objective after imputation is inference from the parameter estimates of a statistical model [53, 54], the criterion for choosing the best Eigenvector method can be the standard error of the statistic of interest.

The Eigenvector method that produces the smallest standard error will be the best. The modern treatment of missing values suggests multiple imputation as an alternative to find the standard error [55], but in the case of deterministic imputation a solution well known and tested with success can be applied. This is the proportional bootstrap method proposed by Bello [56], in which the proportion of present

Missing values (%)

0.1

(a)

Missing values (%)

0.2

(b)

Missing values (%)

0.4

(c)

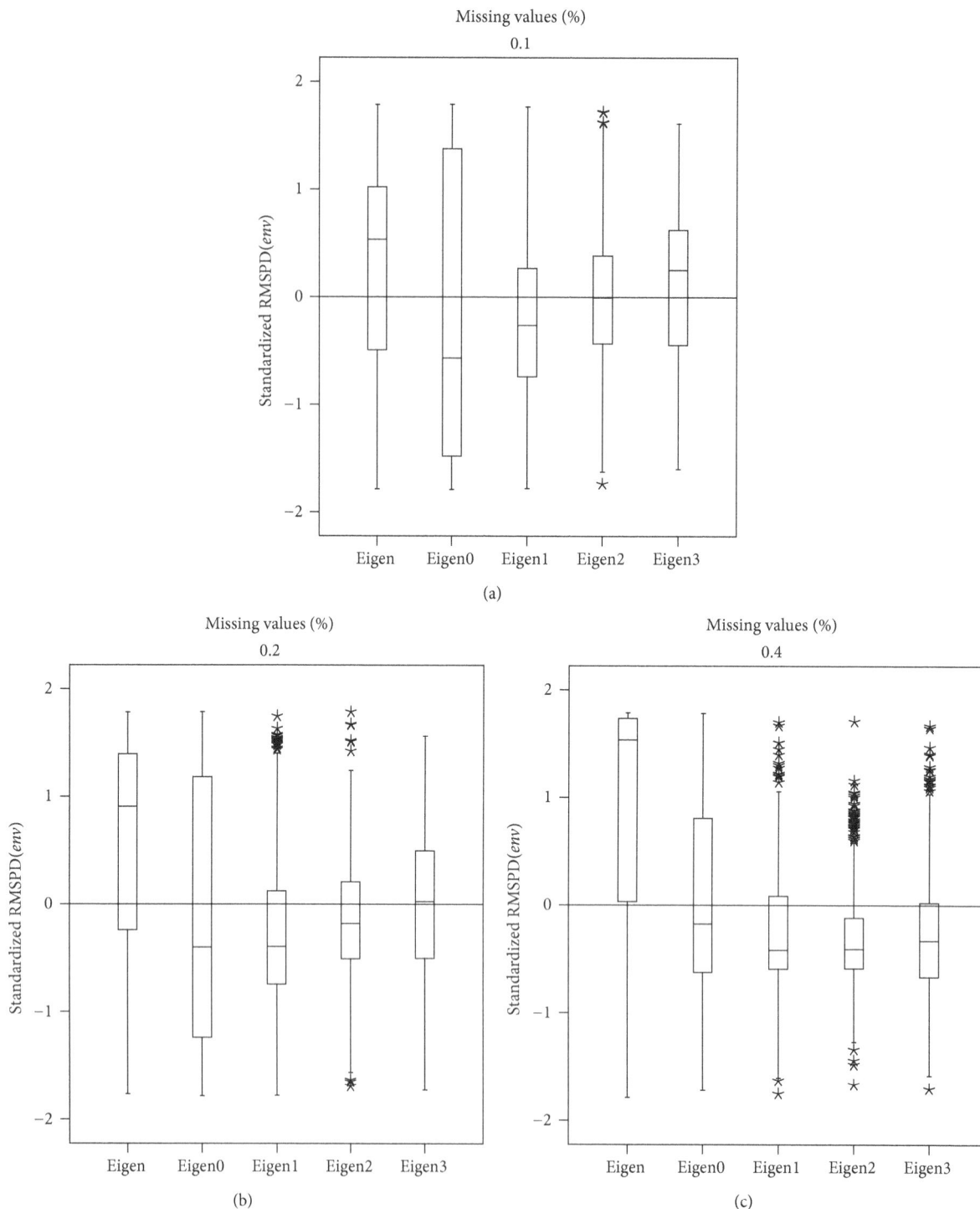

FIGURE 8: Box plot of the RMSPD(*env*) distribution in Flores data set.

and missing values that appear in each bootstrap sample is exactly equal to the proportion that appear in the original incomplete data.

Another aspect that can be of interest is the mechanism producing the missing data. Generally, in situations that involve the assessment of several genotypes in different environments, missing observations follow one of the definitions proposed by Little and Rubin [57], namely, missing completely at random (MCAR), missing at random (MAR), and missing not at random (MNAR). Values missing completely at random can occur, for example, when plants are damaged due to uncontrollable factors in the experiments,

TABLE 8: Wilcoxon test for the standardized $\text{RMSPD}_l(\cdot)$ (10% imputation)—"Flores" data set.

Comparison	Statistic			
	$\text{RMSPD}_2(genmult)$	$\text{RMSPD}_2(envmult)$	$\text{RMSPD}_3(genmult)$	$\text{RMSPD}_3(envmult)$
Eigen-Eigen0	-3.1132^*	-3.9729^*	-2.9800^*	-3.2533^*
Eigen-Eigen1	-2.7033^*	-3.4193^*	-2.6193^*	-2.6122^*
Eigen-Eigen2	-2.2662^*	-2.9110^*	-2.7950^*	-2.1989^*
Eigen-Eigen3	-2.2427^*	-3.0329^*	-2.5279^*	-2.8083^*
Eigen1-Eigen0	-1.3053	-1.9408	-0.6860	-0.9421
Eigen2-Eigen0	-2.4441^*	-3.2769^*	-1.6886	-2.1223^*
Eigen3-Eigen0	-3.2117^*	-3.8341^*	-2.3675^*	-2.7069^*
Eigen2-Eigen1	-1.8444	-2.8314^*	-2.0155^*	-2.5541^*
Eigen3-Eigen1	-2.3102^*	-2.9518^*	-2.3568^*	-2.3958^*
Eigen3-Eigen2	-2.2854^*	-2.4862^*	-2.4622^*	-2.0679^*

*Significant difference 5%.

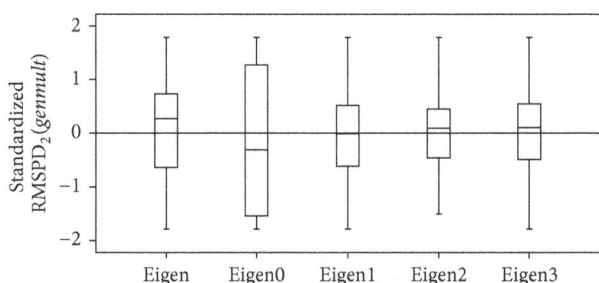

FIGURE 9: Box plot of the $\text{RMSPD}_2(genmult)$ distribution in Flores data set—with 10% imputation.

or by incorrect data measurement or transcription. In this case the cause of the missing value is not correlated with the variable that has it. However, in the genotypes test program in which the cultivars are chosen during each year, using only the observed data without considering the missing values, the missing mechanism is clearly random MAR [58]. The last type of missing, MNAR, can be seen usually when the same subset of genotypes can be missing in some environments of the same subregion, because the plant breeder in the location does not like these genotypes. So, a genotype missing in one environment possibly will be missing too in other environments. In these cases, the mechanism that produces missing values is naturally not at random. The present study has focused exclusively on the MCAR mechanism, and further research is needed to study the remaining mechanisms.

Finally, the proposed methods in this paper have easy computational implementation, but one of the main advantages is that they do not make any distributional or structural assumptions and do not have any restrictions regarding the pattern or mechanism of missing data in $G \times E$ experiments.

Acknowledgments

Sergio Arciniegas-Alarcón thanks the Coordenação de Aperfeiçoamento de Pessoal de Nível Superior, CAPES, Brazil, (PEC-PG program) for the financial support. Marisol García-Peña thanks the National Council of Technological and Scientific Development, CNPq, Brazil, and the Academy of Sciences for the Developing World, TWAS, Italy, (CNPq-TWAS program) for the financial support. Carlos Tadeu dos Santos Dias thanks the CNPq for financial support.

References

[1] H. G. Gauch Jr., "Statistical analysis of yield trials by AMMI and GGE," *Crop Science*, vol. 46, no. 4, pp. 1488–1500, 2006.

[2] F. A. van Eeuwijk, M. Malosetti, X. Yin, P. C. Struik, and P. Stam, "Statistical models for genotype by environment data: from conventional ANOVA models to eco-physiological QTL models," *Australian Journal of Agricultural Research*, vol. 56, no. 9, pp. 883–894, 2005.

[3] F. A. van Eeuwijk, M. Malosetti, and M. P. Boer, "Modelling the genetic basis of response curves underlying genotype environment interaction," in *Scale and Complexity in Plant Systems Research: Gene-Plant-Crop Relations*, J. H. J. Spiertz, P. C. Struik, and H. H. van Laar, Eds., Wageningen UR Frontier Series, pp. 115–126, Springer, New York, NY, USA, 2007.

[4] I. Romagosa, J. Voltas, M. Malosetti, and F. A. van Eeuwijk, "Interaccíon Genotipo por Ambiente," in *La Adaptación Ambiente y Los Estreses Abióticos en la Mejora Vegetal*, C. M. Avila, S. G. Atienza, M. T. Moreno, and J. I. Cubero, Eds., pp. 107–136, Instituto de Investigación y Formación Agraria y Pesquera; Consejería de Agricultura y Pesca, 2008.

[5] S. Arciniegas-Alarcón and C. T. S. Dias, "AMMI analysis with imputed data in genotype × environment interaction experiments in cotton," *Pesquisa Agropecuaria Brasileira*, vol. 44, no. 11, pp. 1391–1397, 2009.

[6] M. S. Kang, M. G. Balzarini, and J. L. L. Guerra, "Genotype-by-environment interaction," in *Genetic Analysis of Complex Traits Using SAS*, A. M. Saxton, Ed., pp. 69–96, SAS Institute nc, Cary, NC, USA, 2004.

[7] G. H. Freeman, "Analysis of interactions in incomplete two-ways tables," *Journal of Applied Statistics*, vol. 24, no. 1, pp. 47–55, 1975.

[8] H. G. Gauch Jr. and R. W. Zobel, "Imputing missing yield trial data," *Theoretical and Applied Genetics*, vol. 79, no. 6, pp. 753–761, 1990.

[9] A. J. R. Godfrey, G. R. Wood, S. Ganesalingam, M. A. Nichols, and C. G. Qiao, "Two-stage clustering in genotype-by-environment analyses with missing data," *Journal of Agricultural Science*, vol. 139, no. 1, pp. 67–77, 2002.

[10] A. J. R. Godfrey, *Dealing with Sparsity in Genotype × Environment Analysis [Dissertation]*, Massey University, 2004.

[11] B. M. K. Raju, "A study on AMMI model and its biplots," *Journal of the Indian Society of Agricultural Statistics*, vol. 55, pp. 297–322, 2002.

[12] J. Mandel, "The analysis of two-way tables with missing values," *Applied Statistics*, vol. 42, pp. 85–93, 1993.

[13] F. A. van Eeuwijk and P. M. Kroonenberg, "Multiplicative models for interaction in three-way ANOVA, with applications to plant breeding," *Biometrics*, vol. 54, no. 4, pp. 1315–1333, 1998.

[14] J. B. Denis, "Ajustements de modèles linéaires et bilinéaires sous contraintes linéaires avec données manquantes," *Revue de Statistique Appliquée*, vol. 39, pp. 5–24, 1991.

[15] T. Caliński, S. Czajka, J. B. Denis, and Z. Kaczmarek, "EM and ALS algorithms applied to estimation of missing data in series of variety trials," *Biuletyn Oceny Odmian*, vol. 24-25, pp. 7–31, 1992.

[16] J. B. Denis and C. P. Baril, "Sophisticated models with numerous missing values: the multiplicative interaction model as an example," *Biuletyn Oceny Odmian*, vol. 24-25, pp. 33–45, 1992.

[17] T. Caliński, S. Czajka, J. B. Denis, and Z. Kaczmarek, "Further study on estimating missing values in series of variety trials," *Biuletyn Oceny Odmian*, vol. 30, pp. 7–38, 1999.

[18] H. P. Piepho, "Methods for estimating missing genotype-location combinations in multilocation trials: an empirical comparison," *Informatik, Biometrie und Epidemiologie in Medizin und Biologie*, vol. 26, pp. 335–349, 1995.

[19] G. C. Bergamo, C. T. S. Dias, and W. J. Krzanowski, "Distribution-free multiple imputation in an interaction matrix through singular value decomposition," *Scientia Agricola*, vol. 65, no. 4, pp. 422–427, 2008.

[20] S. Arciniegas-Alarcón, *Data Imputation in Trials with Genotype by Environment Interaction: An Application on Cotton Data [Dissertation]*, University of São Paulo, 2008.

[21] S. Arciniegas-Alarcón and C. T. S. Dias, "Data imputation in trials with genotype by environment interaction: an application on cotton data," *Revista Brasileira de Biometria*, vol. 27, pp. 125–138, 2009.

[22] S. Arciniegas-Alarcón, M. García-Peña, C. T. S. Dias, and W. J. Krzanowski, "An alternative methodology for imputing missing data in trials with genotype-by-environment interaction," *Biometrical Letters*, vol. 47, pp. 1–14, 2010.

[23] B. M. K. Raju and V. K. Bhatia, "Bias in the estimates of sensitivity from incomplete GxE tables," *Journal of the Indian Society of Agricultural Statistics*, vol. 56, pp. 177–189, 2003.

[24] B. M. K. Raju, V. K. Bhatia, and V. V. Kumar, "Assessment of sensitivity with incomplete data," *Journal of the Indian Society of Agricultural Statistics*, vol. 60, pp. 118–125, 2006.

[25] B. M. K. Raju, V. K. Bhatia, and L. M. Bhar, "Assessing stability of crop varieties with incomplete data," *Journal of the Indian Society of Agricultural Statistics*, vol. 63, pp. 139–149, 2009.

[26] D. G. Pereira, J. T. Mexia, and P. C. Rodrigues, "Robustness of joint regression analysis," *Biometrical Letters*, vol. 44, pp. 105–128, 2007.

[27] P. C. Rodrigues, D. G. S. Pereira, and J. T. Mexia, "A comparison between joint regression analysis and the additive main and multiplicative interaction model: the robustness with increasing amounts of missing data," *Scientia Agricola*, vol. 68, no. 6, pp. 679–705, 2011.

[28] P. J. C. Rodrigues, *New Strategies to Detect and Understand Genotype-by-Environment Interactions and QTL-by-Environment Interactions [Dissertation]*, Universidade Nova de Lisboa, 2012.

[29] R. Bro, K. Kjeldahl, A. K. Smilde, and H. A. L. Kiers, "Cross-validation of component models: a critical look at current methods," *Analytical and Bioanalytical Chemistry*, vol. 390, no. 5, pp. 1241–1251, 2008.

[30] S. Wold, "Cross-validatory estimation of the number of components in factor and principal components models," *Technometrics*, vol. 20, pp. 397–405, 1978.

[31] H. T. Eastment and W. J. Krzanowski, "Cross-validatory choice of the number of components from a principal component analysis," *Technometrics*, vol. 24, no. 1, pp. 73–77, 1982.

[32] S. Arciniegas-Alarcón, M. García-Peña, and C. T. S. Dias, "Data imputation in trials with genotype × environment interaction," *Interciencia*, vol. 36, pp. 444–449, 2011.

[33] A. Smilde, R. Bro, and P. Geladi, *Multi-Way Analysis with Applications in the Chemical Sciences*, John Wiley and Sons, Chichester, UK, 2004.

[34] W. J. Krzanowski, "Missing value imputation in multivariate data using the singular value decomposition of a matrix," *Biometrical Letters*, vol. 25, pp. 31–39, 1988.

[35] I. J. Good, "Applications of the singular decomposition of a matrix," *Technometrics*, vol. 11, no. 4, pp. 823–831, 1969.

[36] T. Hastie, R. Tibshirani, G. Sherlock, M. Eisen, P. Brown, and D. Botstein, "Imputing missing data for gene expression arrays," Technical Report, Division of Biostatistics, Standford University, 1999.

[37] D. Hedderley and L. Wakeling, "A comparison of imputation techniques for internal preference mapping, using Monte Carlo simulation," *Food Quality and Preference*, vol. 6, no. 4, pp. 281–297, 1995.

[38] J. Josse, J. Pagès, and F. Husson, "Multiple imputation in principal component analysis," *Advances in Data Analysis and Classification*, vol. 5, no. 3, pp. 231–246, 2011.

[39] C. T. S. Dias and W. J. Krzanowski, "Model selection and cross validation in additive main effect and multiplicative interaction models," *Crop Science*, vol. 43, no. 3, pp. 865–873, 2003.

[40] A. L. Bello, "Choosing among imputation techniques for incomplete multivariate data: a simulation study," *Communications in Statistics*, vol. 22, pp. 853–877, 1993.

[41] T. Caliński, S. Czajka, Z. Kaczmarek, P. Krajewski, and W. Pilarczyk, "Analyzing the genotype-by-environment interactions under a randomization-derived mixed model," *Journal of Agricultural, Biological, and Environmental Statistics*, vol. 14, pp. 224–241, 2009.

[42] F. J. C. Farias, *Selection Index in Upland Cotton Cultivars [Dissertation]*, University of São Paulo, 2005.

[43] F. Flores, M. T. Moreno, and J. I. Cubero, "A comparison of univariate and multivariate methods to analyze $G \times E$ interaction," *Field Crops Research*, vol. 56, no. 3, pp. 271–286, 1998.

[44] A. L. Bello, "Imputation techniques in regression analysis: looking closely at their implementation," *Computational Statistics and Data Analysis*, vol. 22, pp. 853–877, 1995.

[45] R Development Core Team, *R: A Language and Environment For Statistical Computing. R Foundation For Statistical Computing*, Vienna, Austria, 2012, http://www.R-project.org/.

[46] H. G. Gauch, "Model selection and validation for yield trials with interaction," *Biometrics*, vol. 44, pp. 705–715, 1988.

[47] H. G. Gauch, *Statistical Analysis of Regional Yield Trials: AMMI Analysis of Factorial Designs*, Elsevier, Amsterdam, The Netherlands, 1992.

[48] K. R. Gabriel, "Le biplot-outil d'exploration de données multidimensionelles," *Journal de la Societe Francaise de Statistique*, vol. 143, pp. 5–55, 2002.

[49] C. T. S. Dias and W. J. Krzanowski, "Choosing components in the additive main effect and multiplicative interaction (AMMI) models," *Scientia Agricola*, vol. 63, no. 2, pp. 169–175, 2006.

[50] M. García-Peña and C. T. S. Dias, "Analysis of bivariate additive models with multiplicative interaction (AMMI)," *Revista Brasileira de Biometria*, vol. 27, pp. 586–602, 2009.

[51] K. Hongyu, *Empirical Distribution of Eigenvalues Associated with the Interaction Matrix of the AMMI Models by Non-Parametric Bootstrap Method [Dissertation]*, University of São Paulo, 2012.

[52] P. Sprent and N. C. Smeeton, *Applied Nonparametric Statistical Methods*, Chapman and Hall, London, UK, 2001.

[53] H. G. Gauch Jr., H.-P. Piepho, and P. Annicchiarico, "Statistical analysis of yield trials by AMMI and GGE: further considerations," *Crop Science*, vol. 48, no. 3, pp. 866–889, 2008.

[54] P. C. Rodrigues, S. Mejza, and J. T. Mexia, "Structuring genotype × environment interaction: an overview," *Bulletin of Plant Breeding and Acclimatization Institute*, vol. 250, pp. 225–236, 2009.

[55] J. L. Schafer and J. W. Graham, "Missing data: our view of the state of the art," *Psychological Methods*, vol. 7, no. 2, pp. 147–177, 2002.

[56] A. L. Bello, "A bootstrap method for using imputation techniques for data with missing values," *Biometrical Journal*, vol. 36, pp. 453–464, 1994.

[57] R. J. Little and D. B. Rubin, *Statistical Analysis With Missing Data*, John Wiley and Sons, New York, NY, USA, 2002.

[58] H.-P. Piepho and J. Möhring, "Selection in cultivar trials: is it ignorable?" *Crop Science*, vol. 46, no. 1, pp. 192–201, 2006.

20

Management of Palmer Amaranth (*Amaranthus palmeri*) in Glufosinate-Resistant Soybean (*Glycine max*) with Sequential Applications of Herbicides

Amy E. Hoffner, David L. Jordan, Aman Chandi, Alan C. York, E. James Dunphy, and Wesley J. Everman

Department of Crop Science, North Carolina State University, P.O. Box 7620, Raleigh, NC 27695-7620, USA

Correspondence should be addressed to David L. Jordan, david_jordan@ncsu.edu

Academic Editors: D. Chikoye, O. Ferrarese-Filho, and C. Ramsey

Palmer amaranth (*Amaranthus palmeri* S. Wats.) is one of the most difficult weeds to control in soybean (*Glycine max* (L.) Merr.) in North Carolina. Research was conducted during 2010 and 2011 to determine if Palmer amaranth control and soybean yield were affected by soybean plant population and combinations of preemergence (PRE) herbicides followed by a single application of glufosinate postemergence (POST) versus multiple applications of glufosinate POST. Palmer amaranth was controlled more, and soybean yield was greater when soybean was established at 483,000 plants ha^{-1} in 3 of 4 experiments compared with soybean at 178,000 plants ha^{-1} irrespective of herbicide treatments. In separate experiments, application of PRE herbicides followed by POST application of glufosinate or multiple POST applications of glufosinate provided variable Palmer amaranth control, although combinations of PRE and POST herbicides controlled Palmer amaranth the most and provided the greatest soybean yield. In 1 of 3 experiments, sequential applications of glufosinate were more effective than a single application. Yield was higher in 2 of 3 experiments when glufosinate was applied irrespective of timing of application when compared with the nontreated control. In the experiment where glufosinate was applied at various POST timings, multiple applications of the herbicide provided the best control and the greatest yield compared with single applications.

1. Introduction

Resistance of Palmer amaranth to acetolactate synthase (ALS-) inhibiting herbicides and glyphosate is well documented in North Carolina [1–3]. Controlling herbicide-resistant Palmer amaranth can be challenging because alternatives to glyphosate and ALS-inhibiting herbicides require timely application [4, 5]. Row pattern and plant population can influence weed management in soybean [6, 7]. Herbicides that applied PRE or POST other than glyphosate are important in managing glyphosate-resistant Palmer amaranth in soybean [8–10]. Under weed-free conditions, soybean at populations below those established in many fields in North Carolina yield as well as soybean at higher populations considered "standard" (exceeding 300,000 plants ha^{-1}) (E. J. Dunphy, personal communication). Establishing soybean at lower populations can reduce production costs if yields are maintained. Results comparing the economic value of high soybean populations versus increasing the number of POST herbicides are mixed [11–16]. Additional research is needed to determine if Palmer amaranth control is more effective when soybean is established at higher populations in glufosinate-resistant soybean.

Glufosinate-resistant soybean allows producers to apply glufosinate to control problematic weeds like Palmer amaranth [17]. Sequential programs, either as PRE followed by POST applications or multiple POST applications of herbicides, can be effective in controlling weeds, especially glyphosate-resistant Palmer amaranth [18]. Timely applications of glufosinate in glufosinate-resistant soybean can be effective in controlling glyphosate-resistant and glyphosate-susceptible Palmer amaranth [17]. Reed et al. [19] reported

Management of Palmer Amaranth (Amaranthus palmeri) in Glufosinate-Resistant Soybean (Glycine max) with Sequential Applications of Herbicides

173

that multiple applications of glufosinate in a total POST herbicide program controlled Palmer amaranth effectively. However, most recommendations by Cooperative Extension Service representatives include sequential applications of PRE herbicides followed by timely POST herbicides, especially in fields where glyphosate-resistant Palmer amaranth is suspected.

Given that Palmer amaranth has become more prevalent in North Carolina, especially herbicide-resistant biotypes, research is needed to better define the role of seeding rate on weed management programs for soybean. Determining if sequential applications of PRE and POST herbicides are as effective as multiple POST applications of glufosinate alone will be important, especially in terms of response to soybean population. Research is also needed to better define the time between multiple applications of glufosinate. Therefore, research was conducted in North Carolina to compare Palmer amaranth control and soybean yield with combinations of PRE and POST herbicides applied to glufosinate-resistant soybean at two soybean populations and to determine the most effective timing of multiple POST applications of glufosinate.

2. Materials and Methods

2.1. Interactions of Plant Population and Herbicide Program. The experiment was conducted in North Carolina in two separate fields during each year (2010 and 2011) at the Upper Coastal Plain Research Station located near Rocky Mount with natural and relatively high populations of Palmer amaranth. Populations of weed species other than Palmer amaranth were low and inconsistent across fields compared with density and distribution of Palmer amaranth. One field consisted of a Lynchburg fine sandy loam soil (fine-loamy, siliceous, semiactive, thermic, and Aeric Paleaquults) while the second field was a Nahunta loam soil (fine-silty, siliceous, thermic, and Aeric Paleaquults). Corn (*Zea mays* L.) and cotton (*Gossypium hirsutum* L.) preceded soybean on these respective soil series. Soybean cultivar LL 595 N (Southern States Cooperative, Inc., Farmville, NC 27828, USA) was planted between May 24 and June 7 after disking and field cultivation in rows spaced 20 cm apart in plots 12 rows wide by 9 m.

Treatments consisted of two levels of soybean population (178,000 plants ha^{-1} versus 483,000 plants ha^{-1}) and four levels of herbicide program. Herbicide treatments included glufosinate POST applied 2 weeks after planting (WAP), glufosinate applied sequentially 2 and 4 WAP, *S*-metolachlor plus fomesafen applied PRE, and *S*-metolachlor plus fomesafen applied PRE followed by glufosinate POST 4 WAP. A nontreated control also was included. *S*-metolachlor plus fomesafen (Prefix herbicide, Syngenta Crop Protection, Greensboro, NC 27709, USA) was applied at 1200 + 270 g ae ha^{-1} while glufosinate-ammonium (Ignite 280 herbicide, Bayer CropScience, Research Triangle Park, NC 27709) was applied at 670 g ae ha^{-1}. Herbicides were applied using a CO_2-pressurized backpack sprayer calibrated to deliver 145 L ha^{-1} using 8002 regular flat-fan

nozzles (Teejet Corporation, Wheaton, IL 60187, USA) at 275 kPa.

Palmer amaranth was 2 to 6 cm in height when glufosinate was applied POST. Visible estimates of percent Palmer amaranth control were recorded 6 WAP using a scale of 0 to 100, where 0 is no control and 100 is complete control. Soybean was machine harvested, and final yield was adjusted to 15% moisture.

The experimental design was a randomized complete block with treatments replicated four times. Data for Palmer amaranth control and soybean yield were subjected to ANOVA using the Proc GLM procedure in SAS (SAS Institute, GLM Procedure, Cary, NC 27513, USA) for a four (site/year combinations) by a four (herbicide program) factorial arrangement of treatments. The nontreated control was not included in this analysis because soybean was not harvestable when herbicide was not included. Means of significant main effects and interactions were separated using Fisher's Protected LSD test at $P \leq 0.05$.

2.2. Efficacy of Sequential Herbicide Programs in a Single Plant Population. Two experiments were conducted with either sequential applications of PRE followed by POST herbicides or multiple applications of glufosinate only. The experiment with sequential applications of PRE and POST herbicides was conducted during 2010 and 2011 at the Upper Coastal Plain Research Station near Rocky Mount in the fields discussed previously (Nahunta soil series during 2010 and Lynchburg soil series during 2011). The experiment with multiple applications of glufosinate was conducted during 2010 on a Norfolk loam sand soil (fine-loamy, kaolinitic, thermic, and Typic Kandiuldults) and in two separate fields during 2011 on both the Nahunta and Lynchburg soil series described previously. The glufosinate-resistant soybean cultivar LL 595N (Southern States Cooperative, Inc., Farmville, NC 27828, USA) was planted after disking and field cultivation in rows spaced 20 cm apart with a final plant population of 480,000 plants ha^{-1}. Plot size was the same as described previously.

In the experiment with sequential application of PRE and POST herbicides, treatments consisted of seven levels of PRE herbicides, including no PRE, flumioxazin (Valor SX, Valent USA Corporation, Walnut Creek, CA 94596, USA) alone at 70 g ai ha^{-1} or with chlorimuron (Valor XLT herbicide, E. I. DuPont de Nemours and Co., Wilmington, DE 19898, USA) at 21 g ai ha^{-1}, the sodium salt of fomesafen (Reflex herbicide, Syngenta Crop Protection, Greensboro, NC 27709, USA) at 280 g ae ha^{-1}, *S*-metolachlor (Dual II Magnum, Syngenta Crop Protection, Greensboro, NC 27709, USA) at 1500 g ai ha^{-1}, *S*-metolachlor plus fomesafen (Prefix herbicide, Syngenta Crop Protection, Greensboro, NC 27709, USA) at 1200 + 270 g ha^{-1}, and sulfentrazone plus cloransulam-methyl (Authority First herbicide, FMC Corporation, Philadelphia, PA 19103, USA) at 280 + 36 g ai ha^{-1}, respectively. Each of these PRE herbicides was followed by no POST herbicide or a single application of glufosinate ammonium (Ignite 280 herbicide, Bayer CropScience, Research Triangle Park, NC 27709, USA) at 670 g ha^{-1} 2 WAP or sequential applications of glufosinate at $400 \, g \, ha^{-1}$ at 2 and 4 WAP. In the experiment with only POST applications of glufosinate, treatments consisted of glufosinate applied 2 WAP at 400 and $670 \, g \, ha^{-1}$ (referred to as POST 1) followed by no additional applications of glufosinate or followed by a second application of glufosinate at $400 \, g \, ha^{-1}$ at 3 WAP (referred to as POST 2) or 4 WAP (referred to as POST 3). Treatments also included three sequential applications of glufosinate at $400 \, g \, ha^{-1}$.

Palmer amaranth size ranged from 5 to 10 cm in height when glufosinate was applied in the experiment with PRE and POST herbicides. In the experiment with multiple applications of glufosinate only, Palmer amaranth height was 5 to 10 cm at POST 1, 5 to 20 cm at POST 2, and 5 to 40 cm at POST 3. The range in size of Palmer amaranth reflected differences in emergence of weeds caused by PRE herbicides prior to the POST application in experiment. In both experiments, glufosinate was applied using the procedures described previously.

Percent Palmer amaranth control was estimated visually and was recorded 10 WAP in both experiments as described previously. Soybean yield was determined in each plot and adjusted to a final moisture of 15%.

The experimental design was a randomized complete block with four replications. Data for Palmer amaranth control and soybean yield were subjected to ANOVA using the Proc GLM procedure in SAS (SAS Institute, Cary, NC 27513, USA) considering the factorial arrangement of treatments. Means of significant main effects and interactions were separated using Fisher's Protected LSD test at $P \leq 0.05$.

3. Results and Discussion

3.1. Interactions of Plant Population and Herbicide Program. Interactions of experiment (site/year combination) by soybean population and experiment by herbicide program were significant for both Palmer amaranth control ($P = 0.0525$ and ≤ 0.0001, resp.) and soybean yield ($P \leq 0.0001$ and 0.0344, resp.). The interaction of experiment by soybean population by herbicide program was not significant for Palmer amaranth control ($P = 0.5630$) and soybean yield ($P = 0.8912$). Additionally, the interaction of soybean population by herbicide program was not significant for these parameters ($P = 0.4373$ and 0.2719, resp.).

Palmer amaranth was controlled more effectively when soybean was established at the higher seeding rate by 11 to 13 percentage points at 3 of 4 locations (Table 1). At a fourth location control was numerically higher by 6 percentage points at the higher plant population. Soybean yield was higher in 3 of 4 experiments when established at the higher population (Table 1). However, there was some discrepancy in Palmer amaranth control and soybean yield among experiments when comparing soybean population. The relatively low soybean yields on the Nahunta soil series during 2010 due to dry weather most likely minimized the possibility of separating yields based on Palmer amaranth control. The higher soybean yield during 2010 on the Lynchburg soil series or during 2011 on the Nahunta soil

series most likely reflected differences in Palmer amaranth control. On the Lynchburg soil series during 2011 there was a slight but nonstatistical difference in Palmer amaranth control but a clear increase in yield at the higher plant population. Other factors could have contributed to higher yields at a higher soybean population that are not related to weed control.

Considerable variation in Palmer amaranth control was noted among herbicide treatments across experiments (Table 2). Control by multiple applications of glufosinate exceeded that of a single application of glufosinate in 3 of 4 experiments. Sequential applications of glufosinate only and S-metolachlor plus fomesafen PRE controlled Palmer amaranth similarly in 3 of 4 experiments. Applying glufosinate following S-metolachlor plus fomesafen was more effective than S-metolachlor plus fomesafen alone, sequential applications of glufosinate, and a single application of glufosinate in 1, 3, and 4 experiments, respectively.

Soybean yield during 2010 on the Nahunta soil series was low and did not differ among herbicide programs, most likely because yield was relatively low compared to the Lynchburg soil series during 2010 and both soil series during 2011 (Table 2). On the Lynchburg series during 2010, soybean yield was higher following sequential applications of glufosinate compared with a single application; yield with S-metolachlor plus fomesafen alone or followed by glufosinate was intermediate. On the Nahunta soil series during 2011, soybean yield with the more intensive program of S-metolachlor plus fomesafen followed by glufosinate yield exceeded yield following the other herbicide programs. On the Lynchburg soil series, yield with this program also was higher than a single application of glufosinate.

Although some variation in Palmer amaranth control and soybean yield was noted when comparing herbicide programs, lack of a soybean population by herbicide program interaction for these parameters suggests that higher soybean populations will increase Palmer amaranth control and soybean yield irrespective of herbicide program.

3.2. Efficacy of Sequential Herbicide Programs in a Single Plant Population. The interaction of year by PRE by POST herbicide was significant for Palmer amaranth control ($P \leq 0.0001$) and for soybean yield ($P \leq 0.0001$). Therefore, data were analyzed by year. The interaction of PRE by POST was not significant for Palmer amaranth control ($P = 0.1466$) but was significant for soybean yield ($P = 0.0268$) during 2010. However, the interaction of PRE by POST herbicides was significant for both Palmer amaranth control ($P = 0.0011$) and soybean yield ($P = 0.0049$).

Palmer amaranth control by flumioxazin alone or with chlorimuron was 93 to 98% while fomesafen alone or with S-metolachlor provided 77 to 80% control (Table 3). Sulfentrazone plus cloransulam methyl controlled Palmer amaranth 66%, while S-metolachlor provided only 56% control. Whitaker et al. [20] found similar results for Palmer amaranth control with flumioxazin and fomesafen PRE.

Palmer amaranth control was similar when glufosinate was applied either once or sequentially (Table 3). Lack of a difference in single and sequential applications most likely

Management of Palmer Amaranth (Amaranthus palmeri) in Glufosinate-Resistant Soybean (Glycine max) with Sequential Applications of Herbicides

175

TABLE 1: Interaction of experiment and soybean population for Palmer amaranth control and soybean yield[a].

Year	Soil series	Palmer amaranth control		Soybean yield	
		Soybean plant population[b]		Soybean plant population[b]	
		Low	High	Low	High
		%		kg ha^{-1}	
2010	Nahunta	57	69*	220	430
2010	Lynchburg	80	93*	720	1190*
2011	Nahunta	82	93*	2640	3710*
2011	Lynchburg	83	89	3810	4500*

[a] Indicates a significant difference at $P \leq 0.05$ within an experiment comparing low and high plant populations for Palmer amaranth control and soybean yield. Data are pooled over herbicide programs.
[b] The low soybean population was approximately 178,000 plants ha^{-1}. The high soybean plant population was approximately 483,000 plants ha^{-1}.

TABLE 2: Interaction of experiment and herbicide program for Palmer amaranth control and soybean yield[a].

Herbicide program	Palmer amaranth control				Soybean yield			
	Soil series				Soil series			
	2010		2011		2010		2011	
	Nahunta	Lynchburg	Nahunta	Lynchburg	Nahunta	Lynchburg	Nahunta	Lynchburg
	%				kg ha^{-1}			
Glufosinate	46 c	74 c	81 b	71 c	230 a	770 b	2640 c	3840 b
Glufosinate followed by glufosinate	62 b	100 a	74 b	85 b	360 a	1130 a	3170 b	4120 ab
S-metolachlor plus fomesafen	70 b	93 ab	95 a	89 ab	360 a	1110 ab	3020 bc	4220 ab
S-metolachlor plus fomesafen followed by glufosinate	100 a	87 b	100 a	99 a	370 a	950 ab	3870 a	4420 a

[a] Means within a year and field followed by the same letter are not different according to Fisher's Protected LSD test at $P \leq 0.05$. Data are pooled over soybean plant population.

reflects limited weed emergence after the first POST application and rapid closure of the soybean canopy.

All PRE herbicides increased soybean yield during 2010 compared with the no-PRE herbicide control (Table 3). Yield of soybean receiving S-metolachlor plus fomesafen exceeded that of soybean treated with flumioxazin alone, sulfentrazone plus cloransulam methyl, or flumioxazin plus chlorimuron. Surprisingly, yield following a single application of glufosinate was lower than soybean yield without a POST application of glufosinate or soybean receiving multiple applications of glufosinate.

Several differences in Palmer amaranth control among PRE herbicides were observed during 2011, especially in absence of glufosinate (Table 4). Flumioxazin, flumioxazin plus chlorimuron, S-metolachlor plus fomesafen, and fomesafen alone provided the best control. Sulfentrazone plus cloransulam-methyl and S-metolachlor alone provided intermediate control between the most effective PRE herbicides and the nontreated control. When glufosinate was applied only once, no difference in control among PRE herbicides was observed.

There were no significant differences in soybean yield among PRE herbicide treatments when glufosinate was applied, regardless of the number of glufosinate applications (Table 5). In the absence of glufosinate, soybean yield following fomesafen, S-metolachlor, and sulfentrazone plus cloransulam methyl was intermediate between the nontreated control and flumioxazin alone or with chlorimuron or S-metolachlor plus fomesafen.

In the second experiment with total POST applications of glufosinate applied sequentially 2, 3, and 4 WAP, the interaction of glufosinate rate with sequence of glufosinate application was not significant for Palmer amaranth control ($P = 0.1064$) or soybean yield ($P = 0.7318$). Although glufosinate rate as a main effect or the interaction with year or sequence of glufosinate application was not significant ($P \leq 0.05$), the interaction of experiment by sequence of glufosinate application was significant for Palmer amaranth control ($P = 0.0077$) and soybean yield ($P = 0.0100$). In 2010, multiple applications of glufosinate controlled Palmer amaranth 97 to 99% than a single application 82% (Table 6). In systems with two applications, timing of the second application had no effect on Palmer amaranth control. Also, Palmer amaranth control was similar with two and three applications. Soybean yield was not affected by the number of glufosinate applications. However, glufosinate POST increased yield over nontreated soybean. During 2011, on the Nahunta soil series, Palmer amaranth control was similar with all glufosinate applications and was at least 97%. In the second field on a Lynchburg soil series, control was greater where glufosinate was applied sequentially compared with a single application.

TABLE 3: Palmer amaranth control and soybean yield as influenced by main effects of PRE and POST herbicide treatments during 2010 on the Nahunta soil series[a].

Herbicides	Herbicide rate $g\,ha^{-1}$	Palmer amaranth control %	Soybean yield $kg\,ha^{-1}$
PRE herbicides			
None		29 e	1150 c
Flumioxazin	700	98 a	1460 b
Sulfentrazone plus cloransulam methyl	28 + 36	66 cd	1450 b
S-metolachlor plus fomesafen	1200 + 270	77 c	1680 a
Flumioxazin plus chlorimuron	70 + 21	93 ab	1420 b
Fomesafen	280	80 bc	1570 ab
S-metolachlor	1500	56 d	1500 ab
POST herbicides			
None		64 b	1500 a
Glufosinate	670	73 a	1340 b
Glufosinate then glufosinate	400 then 400	76 a	1530 a

[a] Means within a treatment factor (PRE or POST herbicides) for each parameter followed by the same letter are not different according to Fisher's Protected LSD test at $P \leq 0.05$. Data for PRE herbicides are pooled over levels of POST herbicides. Data for POST herbicides are pooled over levels of PRE herbicides.

TABLE 4: Palmer amaranth control as influenced by the interaction of PRE and POST herbicide treatments during 2011 on the Lynchburg soil series[a].

PRE herbicides	PRE herbicide rate $g\,ha^{-1}$	Palmer amaranth control Glufosinate rate ($kg\,ha^{-1}$)		
		0	0.67	0.40 then 0.40
		%		
None		0 e	88 a–d	81 bcd
Flumioxazin	700	94 abc	100 a	100 a
Sulfentrazone plus cloransulam methyl	280 + 36	80 cd	100 a	93 a–d
S-metolachlor plus fomesafen	1200 + 270	90 a–d	96 abc	97 abc
Flumioxazin plus chlorimuron	70 + 21	100 a	100 a	96 abc
Fomesafen	280	91 a–d	100 a	98 ab
S-metolachlor	1500	76 d	91 a–d	91 a–d

[a] Means followed by the same letter are not different according to Fisher's Protected LSD test at $P \leq 0.05$.

TABLE 5: Soybean yield as influenced by the interaction of PRE and POST herbicide treatments during 2011 on the Lynchburg soil series[a].

PRE herbicides	PRE herbicide rate $g\,ha^{-1}$	Soybean yield Glufosinate rate ($kg\,ha^{-1}$)		
		0	0.67	0.40 then 0.40
		$kg\,ha^{-1}$		
None		1340 e	3800 abc	3850 abc
Flumioxazin	700	3940 ab	3500 a–d	3770 a–d
Sulfentrazone plus cloransulam methyl	28 + 36	2880 d	3850 abc	3740 a–d
S-metolachlor plus fomesafen	1200 + 270	3400 a–d	3550 a–d	4090 a
Flumioxazin plus chlorimuron	70 + 21	3720 a–d	4000 ab	3870 abc
Fomesafen	280	3130 bcd	3890 abc	4240 a
S-metolachlor	1500	3020 cd	3350 a–d	3530 a–d

[a] Means followed by the same letter are not different according to Fisher's Protected LSD test at $P \leq 0.05$.

Management of Palmer Amaranth (Amaranthus palmeri) in Glufosinate-Resistant Soybean (Glycine max) with Sequential Applications of Herbicides

177

TABLE 6: Palmer amaranth control and soybean yield as influenced by the glufosinate rate and number of applications[a].

| Glufosinate application timing (WAP) | | | Palmer amaranth control or soybean yield | | |
| | | | | 2011 | |
2	3	4	2010[b]	Nahunta soil series	Lynchburg soil series
			Palmer amaranth control		
Yes	No	No	82 b	98 a	89 b
Yes	No	Yes	97 a	100 a	100 a
Yes	Yes	No	99 a	98 a	100 a
Yes	Yes	Yes	98 a	100 a	100 a
			Soybean yield (kg ha^{-1})		
Yes	No	No	1040 a	2950 b	3030 a
Yes	No	Yes	1020 a	3350 ab	2990 a
Yes	Yes	No	970 a	3360 a	2940 a
Yes	Yes	Yes	1240 a	3030 ab	2530 b
No	No	No	600 b	2780 b	870 c

[a] Means within a year or year and soil series combination for each parameter followed by the same letter are not significantly different according to Fisher's Protected LSD test at $P \leq 0.05$. Data are pooled over glufosinate rates. Data for the nontreated control were not included in the statistical analyses to allow consideration of factorial treatment structure.
[b] Norfolk soil series.

During 2011, on the Nahunta soil series, soybean receiving a single application of glufosinate and nontreated soybean yielded similarly. Yield was increased when a second application of glufosinate was made 3 WAP (Table 6). On the Lynchburg series during 2011, soybean yield exceeded that of the nontreated control when glufosinate was applied irrespective of sequence of application. However, soybean yield was lower with three applications of glufosinate compared with two applications only. Given Palmer amaranth control was high with the program containing three applications, lower yield most likely reflected soybean injury. The total amount of glufosinate applied in this treatment exceeds the manufacturer's recommendations [21].

4. Conclusions

The most consistent Palmer amaranth control was obtained with multiple applications of herbicides either as PRE followed by POST or POST applications of glufosinate. Additionally, timing of glufosinate was not critical when two applications were made within the first 4 weeks after planting. Concern for overreliance on protoporphyrinogen oxidase (PPO-) inhibiting herbicides and subsequent development of PPO-resistant biotypes in fields with glyphosate-resistant Palmer amaranth has been expressed [9]. Results from this research indicate that while treatments including PPO herbicides controlled Palmer amaranth more effectively when applied alone for the entire season, the PRE herbicide treatment with a non-PPO mode of action, in this case S-metolachlor alone, performed as well in a comprehensive program including one or two applications of glufosinate. While multiple applications of glufosinate are currently effective in controlling Palmer amaranth as described here and from other research [17–19], the long-term sequential

application of PRE and POST herbicides with a greater diversity of modes of action is the best approach when considering herbicide resistance management.

Conflict of Interests

None of the authors have any conflict of interests in terms of the products mentioned in the paper.

Acknowledgments

The North Carolina Soybean Growers Association and Bayer CropScience provided partial funding for this research. Appreciation is expressed to Jamie Hinton, Rick Seagroves, Dewayne Johnson, Peter Eure, and staff at the Upper Coastal Plain Research Station for assistance with this research.

References

[1] I. Heap, "The international survey of herbicide resistant weeds," 2012, http://www.weedscience.org/In.asp.

[2] J. R. Whitaker, Distribution, biology, and management of glyphosate-resistant palmer amaranth in North Carolina [PhD dissertation], North Carolina State University, 2009.

[3] A. E. Hoffner, D. L. Jordan, and A. C. York, "Geographical distribution of herbicide resistance in Palmer amaranth (Amaranthus palmeri) populations across North Carolina," in Proceedings of the 7th International IPM Symposium: IPM on the World Stage, pp. 27–29, Memphis, Tenn, USA, 2012.

[4] A. S. Culpepper, A. C. York, A. W. MacRae, and J. Kichler, "Glyphosate-resistant Palmer amaranth response to weed management programs in Roundup Ready and Liberty Link Cotton," in Proceedings of the Beltwide Cotton Conferences, National Cotton Council, Nashville, Tenn, USA, January 2008.

[5] L. M. Sosnoskie, J. M. Kichler, R. D. Wallace, and A. S. Culpepper, "Multiple resistance in Palmer amaranth to glyphosate and pyrithiobac confirmed in Georgia," *Weed Science*, vol. 59, no. 3, pp. 321–325, 2011.

[6] D. B. Harder, C. L. Sprague, and K. A. Renner, "Effect of soybean row width and population on weeds, crop yield, and economic return," *Weed Technology*, vol. 21, no. 3, pp. 744–752, 2007.

[7] G. T. Place, S. C. Reberg-Horton, J. E. Dunphy, and A. N. Smith, "Seeding rate effects on weed control and yield for organic soybean production," *Weed Technology*, vol. 23, no. 4, pp. 497–502, 2009.

[8] T. C. Mueller, P. D. Mitchell, B. G. Young, and A. S. Culpepper, "Proactive versus reactive management of glyphosate-resistant or -tolerant weeds," *Weed Technology*, vol. 19, no. 4, pp. 924–933, 2005.

[9] M. D. K. Owen, "Evolved glyphosate-resistant weeds and weed shifts: weed species in glyphosate-resistant crops," *Pest Management Science*, vol. 64, no. 4, pp. 377–387, 2008.

[10] D. R. Shaw, M. D. Owen, P. M. Dixon et al., "Benchmark study on glyphosate-resistant cropping systems in the United States—part 1: introduction to 2006–2008," *Pest Management Science*, vol. 25, pp. 183–191, 2011.

[11] D. A. Guillermo, P. Pedersen, and R. G. Hartzler, "Soybean seeding rate effects on weed management," *Weed Technology*, vol. 23, no. 1, pp. 17–22, 2009.

[12] O. W. Howe and L. R. Oliver, "Influence of soybean (*Glycine max*) row spacing on pitted morningglory (*Ipomoea lacunose*) interference," *Weed Science*, vol. 35, pp. 185–193, 1987.

[13] G. R. W. Nice, N. W. Buehring, and D. R. Shaw, "Sicklepod (*Senna obtusifolia*) response to shading, soybean (*Glycine max*) row spacing, and population in three management systems," *Weed Technology*, vol. 15, no. 1, pp. 155–162, 2001.

[14] J. K. Norsworthy and J. R. Frederick, "Reduced seeding rate for glyphosate-resistant, drilled soybean on the southeastern Coastal Plain," *Agronomy Journal*, vol. 94, no. 6, pp. 1282–1288, 2002.

[15] J. K. Norsworthy and L. R. Oliver, "Effect of seeding rate of drilled glyphosate resistant soybean (*Glycine max*) on seed yield and gross profit margin," *Weed Technology*, vol. 15, pp. 284–292, 2001.

[16] R. J. Kratochvil, J. T. Pearce, and M. R. Harrison Jr., "Row spacing and seeding rate effects on glyphosate-resistant soybean for mid-Atlantic production systems," *Agronomy Journal*, vol. 96, no. 4, pp. 1029–1038, 2004.

[17] E. Coetzer, K. Al-Khatib, and D. E. Peterson, "Glufosinate efficacy on Amaranthus species in glufosinate-resistant soybean (*Glycine max*)," *Weed Technology*, vol. 16, pp. 326–331, 2002.

[18] A. S. Culpepper, A. C. York, and L. E. Steckel, "Glyphosate-resistant Palmer amaranth management in cotton," in *Proceedings of the Southern Weed Science Society*, vol. 64, p. 226, 2011.

[19] J. D. Reed, P. A. Dotray, and J. W. Keeling, "Palmer amaranth (*Amaranthus palmeri* S. Wats.) management in Glytol + LibertyLink cotton," in *Proceedings of the Southern Weed Science Society*, vol. 64, p. 260, 2011.

[20] J. R. Whitaker, A. C. York, D. L. Jordan, and A. S. Culpepper, "Palmer amaranth (*Amaranthus palmeri*) control in soybean with glyphosate and conventional herbicide systems," *Weed Technology*, vol. 24, no. 4, pp. 403–410, 2010.

[21] Ignite 280 herbicide label. Bayer CropScience, Research Triangle Park, NC, USA, 2012.

Growth and Physiological Responses of Maize and Sorghum Genotypes to Salt Stress

Genhua Niu,[1] Wenwei Xu,[2] Denise Rodriguez,[1] and Youping Sun[1]

[1] *Texas AgriLife Research, The Texas A&M University System, 1380 A&M Circle, El Paso, TX 79927, USA*
[2] *Texas AgriLife Research, The Texas A&M University System, 1102 East FM 1294, Lubbock, TX 79403, USA*

Correspondence should be addressed to Genhua Niu, gniu@ag.tamu.edu

Academic Editors: W. P. Williams and L. Zeng

The growth and physiological responses of four maize inbred lines (CUBA1, B73, B5C2, and BR1) and four sorghum hybrids (SS304, NK7829, Sordan 79, and KS585) to salinity were determined. Fifteen days after sowing, seedlings were irrigated with nutrient solution (control) at electrical conductivity (EC) of $1.5\,dS\,m^{-1}$ or saline solution at EC of $8.0\,dS\,m^{-1}$ (salt treatment) for 40 days. Dry weight of shoots in maize was reduced by 58%, 65%, 62%, and 69% in CUBA1, B73, B5C2, and BR1, respectively, while that of sorghum was reduced by 51%, 56%, 56%, and 76% in SS304, NK7829, Sordan79, and KS585, respectively, in the salt treatment compared to their respective control. Salinity stress reduced all or some of the gas exchange parameters, leaf transpiration (E), stomatal conductance (g_s), and net photosynthetic rate (P_n) in the late part of the experiment for both crops. Salinity treatment greatly increased Na^+ uptake in all maize genotypes but did not affect the Na^+ uptake in sorghum, regardless of genotype. In maize, CUBA1 was slightly more resistant to salt stress, while BR1 was more sensitive to salt stress. In sorghum, Sordan79 was the most tolerant genotype, and KS585 was the least tolerant genotype.

1. Introduction

Soil salinity is one of the major environmental stresses that adversely affect plant growth and development. More than 800 million hectares of land throughout the world are salt affected, either by salinity (397 million ha) or the associated condition of sodicity (434 million ha) [1]. Effects of salinity on crop productivity are more severe in arid and semiarid regions where limited rainfall, high evapotranspiration, high temperature, poor water quality, and poor soil management practices exacerbate salinity effect [2]. As world population increases rapidly, the demand for maize and sorghum to meet the food and nonfood requirement necessitates crop production in marginal lands. Marginal land refers to land with low inherent productivity, that has been abandoned or degraded, or is of low quality for agricultural uses [3]. Most marginal lands are located in arid and semiarid regions where soil salinity is often too high for optimal production for most common economic crops and groundwater with high salinity is the primary water source. Therefore, identifying salt-tolerant crops and

improving salt tolerance for salt-affected lands are critically important.

Salinity affects plants through osmotic stress and ion imbalance and toxicity [4]. Osmotic effects are due to salt-induced decrease in the soil water potential. High salts inside the plant take time to accumulate before they affect plant function. Plants have developed a wide range of mechanisms to sustain productivity under salt stress environment. These mechanisms are osmotic adjustment, Na^+ and/or Cl^- exclusion, and tissue tolerance of high concentrations of Na^+ and/or Cl^- [4]. Research on salt tolerance of various crops has indicated that salt tolerance depends largely on genera and species and even on cultivars within certain species.

Maize (*Zea mays* L.) was considered moderately salt sensitive [5–7], while sorghum (*Sorghum bicolor* (L.) Moench)) was characterized as moderately tolerant to salinity [8, 9]. Selection and breeding have always been conducted to achieve high yield and better quality of crops under stressful conditions. Maize is a highly cross-pollinated crop and has become highly polymorphic through the course of natural and domesticated evolution and thus contains enormous variability in which salinity tolerance may exist [5]. Maize is not only a food product; more importantly, maize-derived products have been used in various aspects in our daily life. Sorghum is a major grain and forage, crop and both maize and sorghum are considered potential bioenergy crops in recent years. Large variations in salt tolerance among genotypes have been reported for sorghum [10–12]. With this economic importance and variability in salt tolerance among genotypes, a high-throughput method to screen salt tolerance and the development of maize and sorghum varieties for salt tolerance for salt-affected areas is urgently needed.

The purpose of this study was to assess the salt tolerance of four maize inbred lines and four sorghum hybrids. Growth, gas exchange rates, leaf chlorophyll fluorescence, relative chlorophyll content, and tissue ion accumulation of the selected maize and sorghum genotypes were investigated when irrigated with saline or nonsaline solutions. Physiological response of crops to salinity is valuable information for breeding programs.

2. Materials and Methods

2.1. Experimental Design and Treatments. Seeds of four maize inbred lines (CUBA1, B73, B5C2, and BR1) and four sorghum hybrids (SS304, NK7829, Sordan79, and KS585) were sown in 2.6 L containers, 4 seeds per container, filled with commercial potting mix (Sunshine Mix number 4, SunGro Hort., Bellevue, WA). B73 was a temperate line developed by the Iowa State University, while the other three lines were developed by Wenwei Xu using the temperate and tropical crosses. Four sorghum hybrids were provided by Sorghum Partners, Inc. [13]. KS585 and NK7829 are grain type hybrids. SS304 and Sordan 79 are forage sorghum hybrids. Sordan 79 is a sorghum x sudangrass hybrid and good for alkali soils due to its salt tolerance. Seedlings were thinned to one per container 10 days after sowing. Two weeks after sowing, treatments were initiated by irrigating seedlings with nutrient solution or saline solution, 1 L per container. The nutrient solution with electrical conductivity (EC) of $1.5\,dS\,m^{-1}$ was prepared by adding $0.5\,g\,L^{-1}$ of 20N-8.6P-16.7 K (Peters 20-20-20; Scotts) to tap water. The major ions in the tap water were Na^+, Ca^{2+}, Mg^{2+}, Cl^-, and SO_4^{2-} at 184, 52.0, 7.5, 223.6, and $105.6\,mg\,L^{-1}$, respectively. Saline solutions at EC of $3.0\,dS\,m^{-1}$ (first irrigation) or $8.0\,dS\,m^{-1}$ (second irrigation and after) were prepared by adding calculated amounts of sodium chloride (NaCl), magnesium sulfate ($MgSO_4 \cdot 7H_2O$), and calcium chloride ($CaCl_2$) at 87:8:5 (weight ratio) to the nutrient solution. The experiment followed a split-plot design with salinity as the main plot and genotype as subplot. Greenhouse environmental conditions were maintained at air temperature at $33.6\pm1.1^\circ C$ during the day and $20.4\pm1.5^\circ C$ at night, relative air humidity at $20.4 \pm 3.3\%$, and daily light integral (photosynthetically active radiation) at $21.4 \pm 2.3\,mol\,m^{-2}\,d^{-1}$.

2.2. Measurement. Upon termination of the experiment (40 days after the initiation of treatment), shoots were severed at the substrate surface and were separated into leaves and stems for maize or separated into stalks and tillers for sorghum. The number of tillers was recorded for sorghum. Dry weights of separated tissue were determined after oven dried at $70^\circ C$ to constant weight. In order to monitor salt accumulation in the root zone, leachate was collected periodically and the EC of the leachate was measured using an EC meter (Model B-173, Horiba, Ltd., Japan). Solution was diluted properly whenever the leachate EC exceeded $20\,dS\,m^{-1}$ by adding deionized water to obtain the actual EC accurately because the maximum range of the EC meter is $20\,dS\,m^{-1}$. To reduce the salt accumulation, plants were flushed with tap water to lower the salinity in the root zone.

2.2.1. Gas Exchange Rates. Leaf net photosynthesis (P_n), transpiration (E), and stomatal conductance (g_s) were measured on four plants per genotype per treatment on 15, 30, and 35 days after the initiation of treatment by placing the recently matured leaf in the cuvette of a portable gas exchange measurement system (CIRAS-2, PP Systems, Amesbury, MA). The environmental conditions in the cuvette were controlled at leaf temperature = $25^\circ C$, photosynthetic photon flux (PPF) = $1000\,\mu mol\,m^{-2}\,s^{-1}$, and CO_2 concentration = $400\,\mu mol\,mol^{-1}$. Data were recorded when the environmental conditions and gas exchange parameters in the cuvette became stable. These measurements were taken on sunny days between 1000 HR and 1400 HR and plants were well watered to avoid water stress.

2.2.2. Chlorophyll Fluorescence. In order to examine the influence of progressively increased salt stress on leaf photosynthetic apparatus among the genotypes, leaf chlorophyll fluorescence values, minimal fluorescence F_o, maximum fluorescence F_m, variable fluorescence F_v, and the maximal photochemical efficiency of photosynthesis system II, F_v/F_m ($F_v = F_m - F_o$), were measured on three days during the experiment on young, fully expanded leaves using a Plant Efficiency Analyzer (Hansatech Instruments Ltd., Kings Lynn, UK). Before the measurement, leaves were dark-adapted for 10 min by using the light-exclusion clips.

2.2.3. Relative Chlorophyll Content. Leaf greenness (or relative chlorophyll content) was measured using a hand-held chlorophyll meter (measured as the optical density, SPAD reading, Minolta Camera Co., Osaka, Japan) at the end of the experiment for all plants (10 plants per treatment) in each treatment [14]. SPAD readings of three leaves per plant selected from the middle sections of the plant were measured. All plants were well watered when this measurement was taken.

2.2.4. Mineral Analysis. Four samples per tissue per treatment were collected for mineral analysis of Na^+, Ca^{2+}, Mg^{2+}, and Cl^- at the end of the experiment. For maize genotypes, leaves and stems were separately sampled while for sorghum, stalks and tillers were separately sampled. Dried tissue was ground with a stainless Wiley mill (Thomas Scientific, Swedesboro, NJ), and ground samples were sent

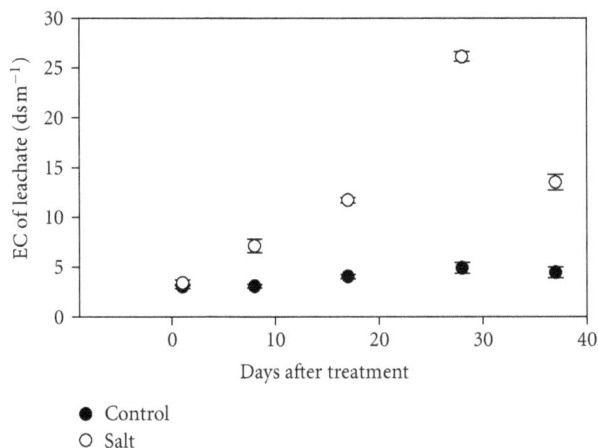

FIGURE 1: Leachate EC pooled from four maize genotypes (CUBA1, B73, B5C2, and BR1) and four sorghum genotypes (SS304, NK7829, Sordan79, and KS585) measured during the treatment period when irrigated with nutrient solution at EC of 1.5 dS m^{-1} or saline solution at EC of 8.0 dS m^{-1} for 40 days (replication of 4).

to an analytical lab for mineral analysis (SWAT laboratory at New Mexico State University, Las Cruces, NM). The Na$^+$, Ca^{2+}, and Mg^{2+} concentrations were determined by EPA method 200.7 [15] and analyzed on an ICAP Trace Analyzer (Thermo Jarrell Ash, Franklin, MA). Chloride was determined by EPA method 300.0 [15] and analyzed using an Ion Chromatograph (Dionex, Sunnyville, CA).

2.3. Statistical Analysis. All analyses were carried out separately for maize and sorghum due to obvious differences in growth. Analysis of variance was carried out to determine the effects of salt and genotype for each crop. When genotype effect was significant, means were separated by Student-Newman-Keuls (SNK) multiple comparisons at $P = 0.05$. When salt effect was significant, t-test was carried out to determine the significance. All statistical analyses were performed using SAS (version 9.1.3; SAS Institute, Cary, NC).

3. Results and Discussion

3.1. Growth, Foliar Salt Damage, and Substrate Salinity. Leachate salinity in the salt treatment increased with time due to salt accumulation in the peat-based substrate, while that of control did not change substantially (Figure 1). Four weeks after the initiation of treatment, leachate salinity in the salt treatment increased to 26 dS m^{-1}. The substrate was flushed with tap water to leach out salts and prevent excessive salt accumulation. At the end of the experiment, the EC was decreased to 13.5 dS m^{-1}. Salt accumulation depends on the substrate property, salinity of the irrigation water, leaching fraction, and frequency of the irrigation. As indicated in this study, irrigating with saline water led to salt accumulation in the root zone. In reality, the salinity of irrigation water would not be as high as 8.0 dS m^{-1} as used in this study. The reason for choosing this EC was to distinguish the salt

tolerance among the genotypes in a relatively short term, in this case, 40 days.

Among the four genotypes of maize, BR1, an inbred line with 50% tropical germplasm, had the most obvious leaf salt damage with leaf rolling and yellowing in some young leaves (data not shown). B73 and B5C2 had leaf rolling. CUBA1 did not exhibit any visible salt damage. Therefore, in terms of foliar salt damage, BR1 was the least tolerant, while CUBA1 was the least sensitive to salt stress. CUBA1 is an inbred line developed with a cross between a temperate line and the tropical Cuba flint, and was selected for heat and drought tolerance. Among the four genotypes of sorghum, KS585 had the most severe leaf edge burn and leaf yellowing, followed by NK7929; SS304 had minor leaf edge burn, while Sordan79 looked healthy without any salt damage. It was obvious that Sordan 79 was the most tolerant and KS585 was the least tolerant among the sorghum and maize genotypes. The high salt tolerance of Sordan 79 in the greenhouse agreed with extensive field testing under various soil conditions. Sordan 79 is a sorghum-and-sudangrass hybrid for forage production and well adapted to alkali soils [13].

For maize, dry weights of leaves and stems were reduced by elevated salinity in all genotypes compared to those of the control (Figure 2). Total dry weight of shoots was reduced by 58%, 65%, 62%, and 69% in CUBA1, B73, B5C2, and BR1, respectively, in the salt treatment compared to their respective control. Therefore, in term of growth, BR1 was less tolerant to salt stress among the four genotypes, while CUBA1 was relatively more tolerant to salt stress. The relative salt tolerance based on growth was in agreement with that in terms of foliar salt damage.

For sorghum, salinity treatment did not affect the number of tillers (not presented) but affected the dry weight of tillers except for NK7829 where no tiller was observed in the control (Figure 3). The reduction of dry weight of stalk (shoots excluding tillers) due to elevated salinity was highest in KS585 (79%) and lowest in Sordan79 and SS304 (38% and 39%). Total dry weight of shoots was reduced by 51%, 56%, 56%, and 76% in SS304, NK7829, Sordan79, and KS585, respectively, in the salt treatment compared to their respective control. Therefore, combined with the visual salt damage ratings, Sordan79 was the most tolerant, followed by SS304, while KS585 was the least tolerant among the eight genotypes (both maize and sorghum). Although total shoot dry weight reduction was smaller in SS304 compared to that of Sordan79, Sordan79 was still considered to be the most tolerant because SS304 did exhibit some leaf edge burn.

3.2. Gas Exchange Rates. For maize, gas exchange rates E, g_s, and P_n of all genotypes on Day 15 were not affected by salt stress (Figure 4). However, on Day 30 and Day 35, all gas exchange rates were reduced significantly by salt stress. The reduction percentages caused by elevated salt stress were approximately 60% in E, 80% in g_s, and 45% in P_n, indicating that effect of salt stress on P_n was the least, while that on g_s was the greatest. No differences in E, g_s, and P_n among genotypes were found on all measurement days for the same treatment. For the control plants, E was higher

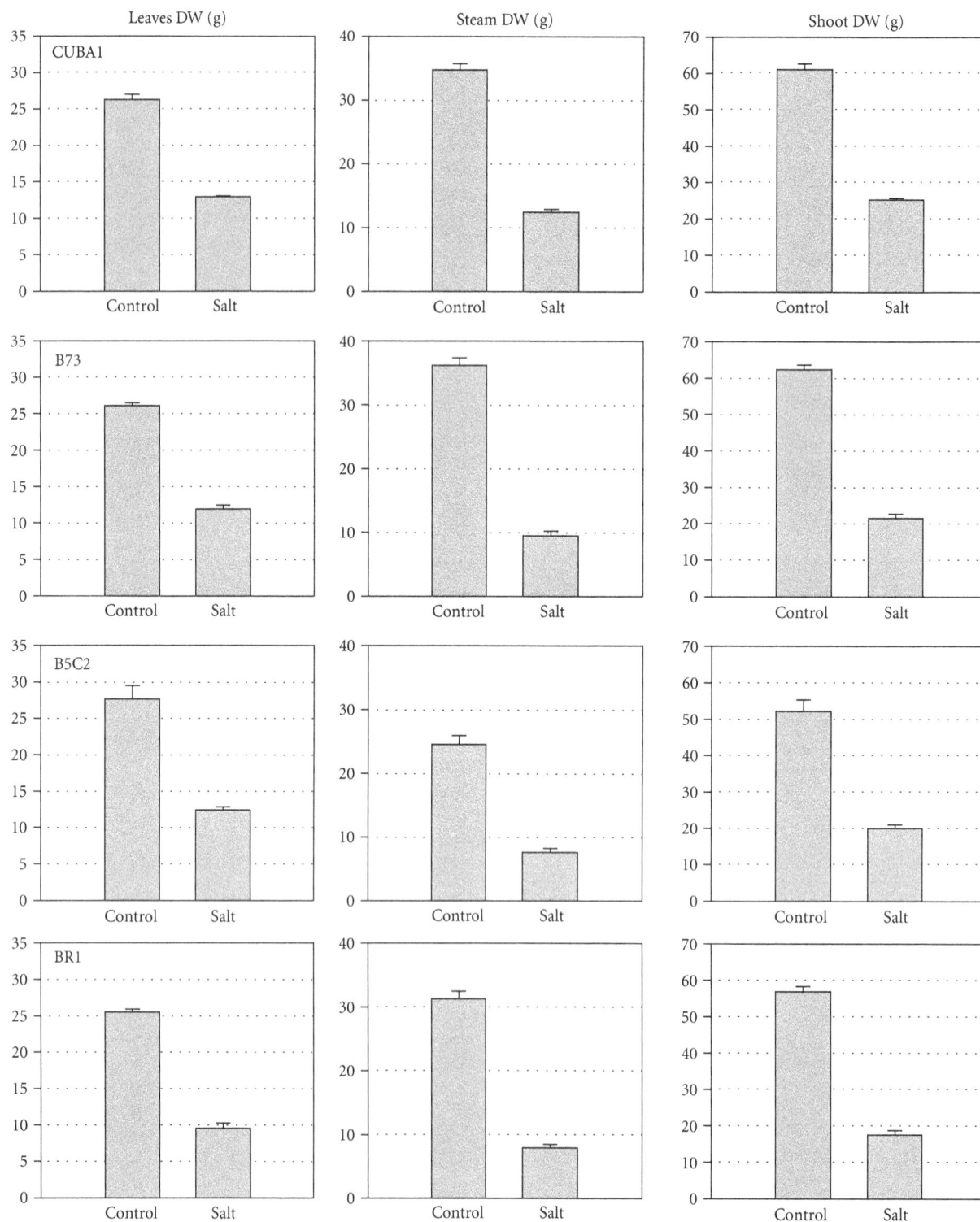

FIGURE 2: Dry weight of leaves, stems, and shoot of four maize genotypes (CUBA1, B73, B5C2, and BR1) when irrigated with nutrient solution at EC of 1.5 dS m^{-1} or saline solution at EC of 8.0 dS m^{-1} for 40 days. Vertical bars represent standard errors (replication of 10).

on Day 30 and Day 35 for CUBA1 and B5C2, although not statistically significant, those of B73 and BR1 were also numerically higher on Day 30 and Day 35 compared to Day 15. For BR1, E and P_n did not change significantly over time, although numerically they did decrease compared to those on Day 15. Generally, salt stress reduced gas exchange rates, while no substantial differences were found among genotypes.

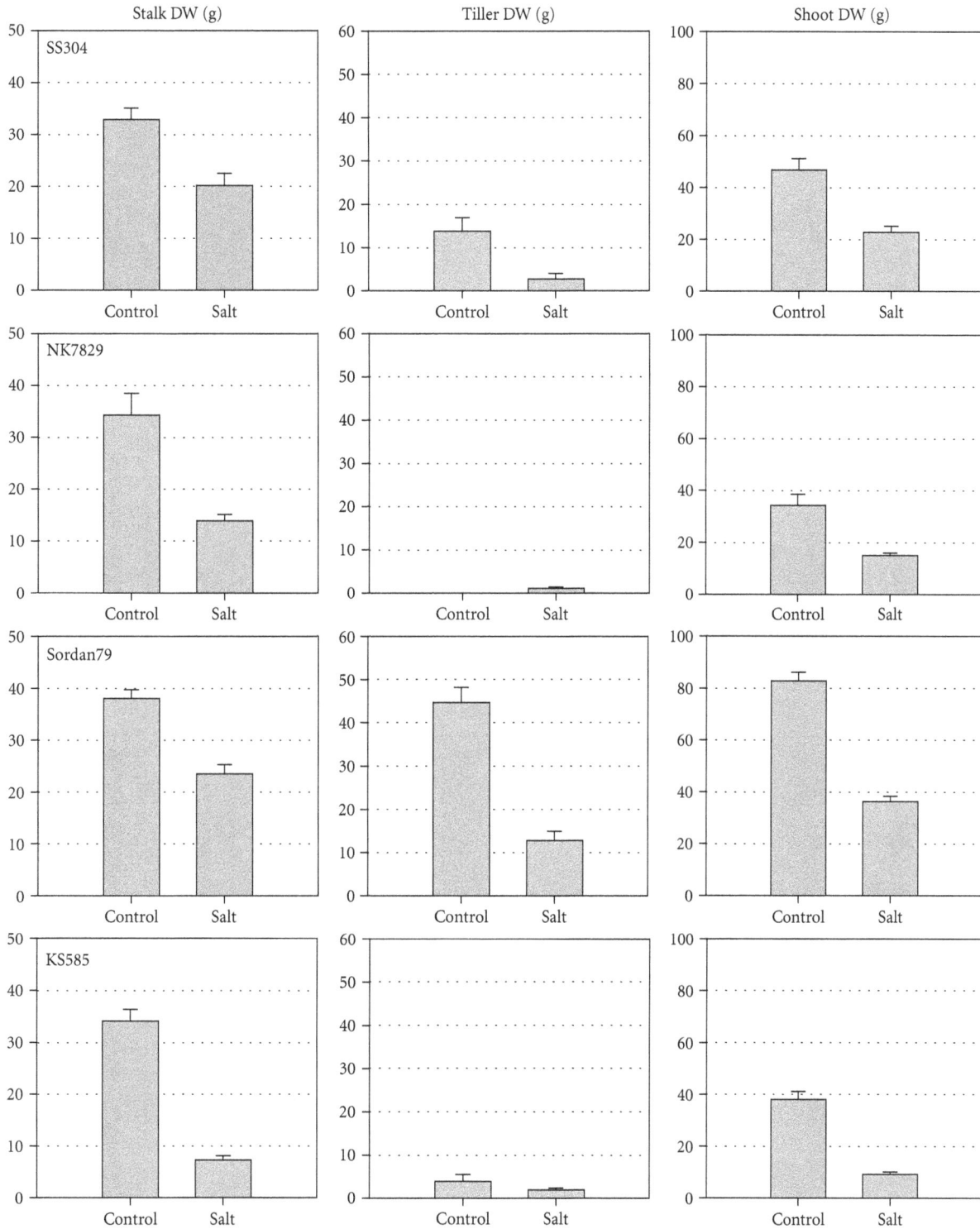

FIGURE 3: Dry weight of stalks, tillers, and shoots of four sorghum genotypes (SS304, NK7829, Sordan79, and KS585) when irrigated with nutrient solution at EC of $1.5\,dS\,m^{-1}$ or saline solution at EC of $8.0\,dS\,m^{-1}$ for 40 days. Vertical bars represent standard errors (replication of 10).

For sorghum, on Day 15, salt treatment did not affect E, g_s, and P_n of the plants, except for P_n of NK 7829 (Figure 5). On Day 30, salt treatment did not affect E, g_s, and P_n, except for g_s and P_n of KS585. The g_s and P_n of KS585 were reduced by 50% and 16% in salt treatment compared to control. On Day 35, salt treatment significantly reduced E, g_s, and P_n of NK7829 and KS585, E and P_n of Sordan79, and P_n of SS304.

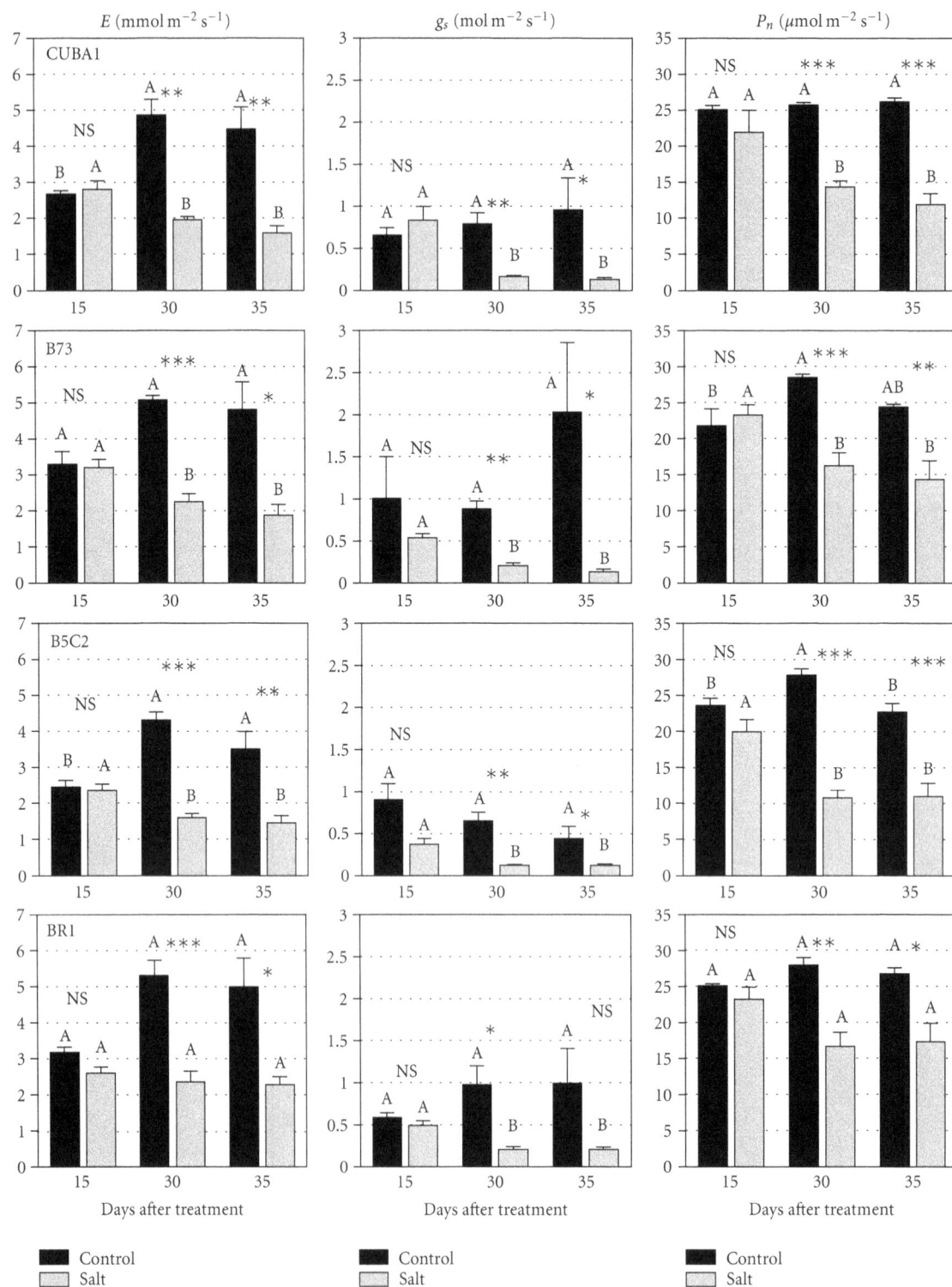

FIGURE 4: Leaf gas exchange rates, transpiration (E), stomatal conductance (g_s), and net photosynthetic rate (P_n) of four maize genotypes (CUBA1, B73, B5C2, and BR1) when irrigated with nutrient solution at EC of 1.5 dS m^{-1} or saline solution at EC of 8.0 dS m^{-1} for 40 days. Means with the same letters on different days are not significantly different tested by Student-Newman-Keuls (SNK) multiple comparisons at $P = 0.05$. ***, **, *, and NS are significant at $P = 0.0001$, 0.01, 0.05, or nonsignificant between the two treatments by t-test. Vertical bars represent standard errors (replication of 4).

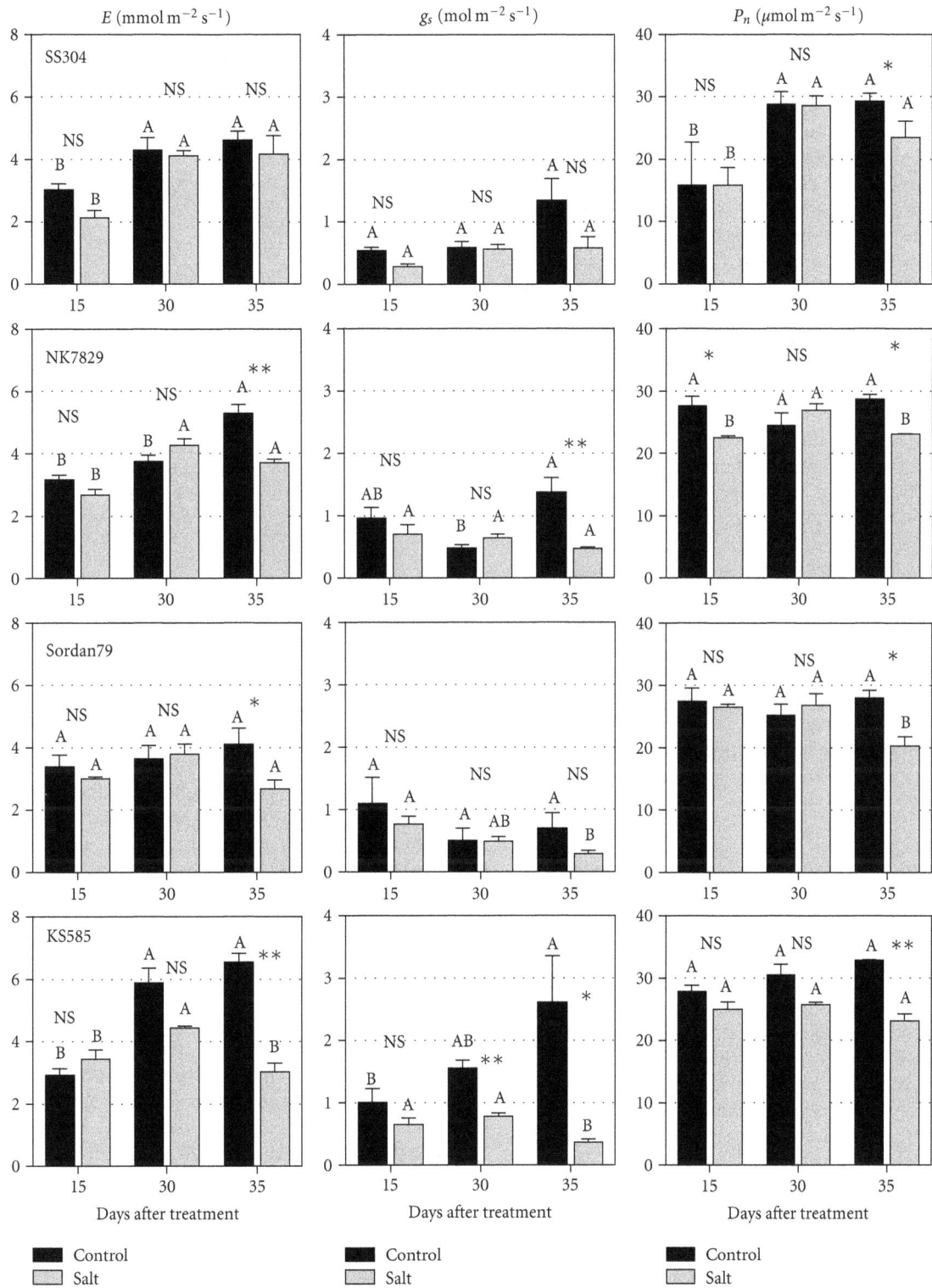

FIGURE 5: Leaf gas exchange rates, transpiration (E), stomatal conductance (g_s), and net photosynthetic rate (P_n) of four sorghum genotypes (SS304, NK7829, Sordan79, and KS585) when irrigated with nutrient solution at EC of 1.5 dS m^{-1} or saline solution at EC of 8.0 dS m^{-1} for 40 days. Means with the same letters on different days are not significantly different tested by Student-Newman-Keuls (SNK) multiple comparisons at $P = 0.05$. ***, **, *, and NS are significant at $P = 0.0001$, 0.01, 0.05, or nonsignificant between the two treatments by t-test. Vertical bars represent standard errors (replication of 4).

As soil salinity increases, leaf gas exchange rates decrease for many crops. At low or moderate soil salinity, decreased growth is primarily associated with a reduction in photosynthetic area rather than a reduction in net photosynthetic rate per unit leaf area [16]. At high salinity, however, leaf photosynthesis can be reduced by lowered stomatal conductance or by nonstomatal factors that may be caused by toxic ions, as indicated by many researchers [17, 18]. It must be pointed out that when a portable gas exchange instrument such as CIRAS-2 (used in the current study) or LI-6400 (LI-COR, Inc., Lincoln, NE) is used for gas exchange measurement, the potential rates of a selected single leaf, in most cases, a fully expanded healthy leaf, at the specified cuvette environmental conditions are measured. The negative effect of salinity on actual gas exchange rates of a whole plant is probably greater than that of a single healthy leaf because older leaves are more affected by salinity than newly developed leaves. Also, not all the leaves have the same potentials as the one measured. In the current study, a PPF of $1000 \, \mu mol \, m^{-2} \, s^{-1}$ was chosen because during that time period inside the greenhouse, the maximum instant PPF was between 800 to nearly $1000 \, \mu mol \, m^{-2} \, s^{-1}$. Gas exchange rates of several cultivars of ornamental peppers (Niu, unpublished data) and roses [19] grown under nonsaline and moderate saline conditions were statistically the same when measured on a single young leaf with the same instrument as used in this study, while shoot growth was reduced significantly by the elevated salinity as seen in this study. These results may indicate that under low-to-moderate salinity, gas exchange rates of a healthy leaf are often not affected. Therefore, single leaf gas exchange rates measured under specified optimal conditions are less effective indicators to assess salt tolerance of the crop compared to visual salt damage and growth.

3.3. Chlorophyll Fluorescence and Relative Chlorophyll Content.
For maize genotypes, effect of salt on chlorophyll fluorescence parameters F_o, F_m, F_v, and F_v/F_m was not consistent over time (Table 1). For example, 21 days after the treatment, salinity significantly reduced F_m, F_v, and F_v/F_m of B5C2, while for all other genotypes, no effect was observed. On Day 31, all genotypes were affected by the salt stress on one or more of the parameters. On Day 37, only F_o of CUBA1, B73, and B5C2 was affected (increased) by salt stress, while all other parameters were not. The longer the treatment is, the more stressed the plants should be. However, chlorophyll fluorescence parameters did not indicate any sign of progressive salinity stress. This may be because every time the fully expanded new leaf was measured, instead of the same leaf on different days.

For sorghum genotypes, similar to that found in maize genotypes, the effect of salt on chlorophyll fluorescence parameters F_o, F_m, F_v, and F_v/F_m was not consistent over time and even had few significances among these parameters (data not shown). These results may indicate that the salt stress on both crops may not be severe enough to cause consistent and significant damage on PSII.

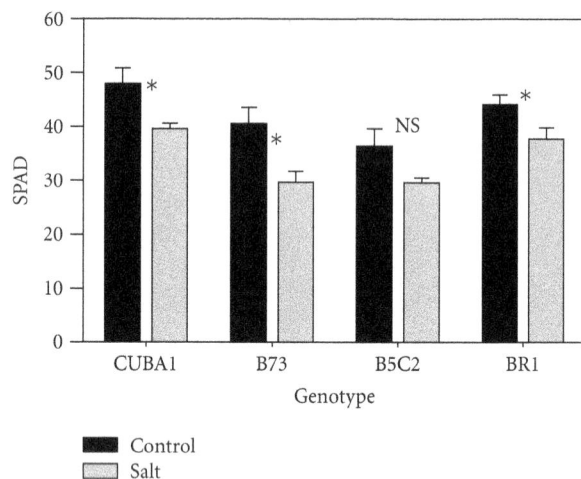

FIGURE 6: Relative chlorophyll content measured as SPAD index of four maize genotypes (CUBA1, B73, B5C2, and BR1) when irrigated with nutrient solution at EC of $1.5 \, dS \, m^{-1}$ or saline solution at EC of $8.0 \, dS \, m^{-1}$ for 40 days. * and NS are significant at $P = 0.05$ or nonsignificant between the two treatments by t-test. Vertical bars represent standard errors (replication of 10).

Under the combined salinity-alkalinity stress, F_v/F_m of maize seedlings decreased only at high salinity-alkalinity, which is NaCl of $100 \, mmol \, L^{-1}$ and $NaHCO_3$ of $100 \, mmol \, L^{-1}$ [20]. Another study also reported that a decline in F_v/F_m of maize was minimal when plants were exposed to salinity levels lower than $10 \, dS \, m^{-1}$, while a significant difference in F_v/F_m occurred at the higher salinity levels [21]. Similar effect of salinity on F_v/F_m in sorghum was reported [22]. These studies may indicate that F_v/F_m was an appropriate tool to screen tolerance to salt stress for maize and sorghum at low salinity level; however, it may be useful at high salinity levels.

For maize genotypes, relative chlorophyll content measured as SPAD value at the end of the experiment was reduced by salt stress in CUBA1, B73, and BR1, while that of B5C2 was numerically reduced but not significant statistically ($P = 0.0677$, Figure 6). For sorghum, no differences were found in SPAD values among control and salt treatment, regardless of genotype (data not shown). Reduction of leaf chlorophyll content at high salinity stress was reported in maize [20, 23–25] and other crops such as wheat [26], radish [27], and basil [28]. Again, the salinity stress was not severe enough to cause more significant differences in SPAD values for both crops between the treatments.

3.4. Ion Accumulation.
For maize genotypes, significant effects of salt treatment and genotype on tissue mineral contents were observed, especially on Na^+ and Cl^- (Table 2). Na^+ concentrations were generally higher in stems than in leaves for all genotypes, except for B5C2 and BR1 in the control. Na^+ concentrations in leaves and stems in the salt treatment were 20 to 200 times that of control. The increase in Na^+ concentrations in leaves and stems was even

TABLE 1: Summary of t-test results on the effect of salinity on chlorophyll fluorescence parameters (initial fluorescence F_o, maximum fluorescence F_m, variable fluorescence F_v, and ratio of F_v/F_m) of four maize genotypes (CUBA1, B73, B5C2, and BR1) when irrigated with nutrient solution at EC of $1.5\,dS\,m^{-1}$ or saline solution at EC of $8.0\,dS\,m^{-1}$ for 40 days (replication of 6).

Genotype	F_o	F_m	F_v	F_v/F_m
		Day 22		
CUBA1	NS	NS	NS	NS
B73	NS	NS	NS	NS
B5C2	NS	0.0177	0.015	0.0231
BR1	NS	NS	NS	NS
		Day 31		
CUBA1	0.0029	0.0029	0.0085	NS
B73	<0.0001	NS	NS	0.0061
B5C2	0.0013	NS	0.0474	0.0007
BR1	NS	NS	NS	0.037
		Day 37		
CUBA1	0.0232	NS	NS	NS
B73	0.0112	NS	NS	NS
B5C2	0.0468	NS	NS	NS
BR1	NS	NS	NS	NS

NS: non-significant.

TABLE 2: Ion concentrations of leaves and stems of four maize genotypes (CUBA1, B73, B5C2, and BR1) irrigated with nutrient or saline solutions for 40 days (replication of 4).

Genotype	Tissue	Treatment	Na^+	Cl^-	Ca^{2+}	Mg^{2+}
				($mg\,g^{-1}$)		
CUBA1	Leaves	Control	0.44	11.41	4.25	3.03
		Salt	10.05	19.01	4.28	2.18
	Stems	Control	1.15	17.57	3.15	3.65
		Salt	31.93	47.23	1.88	2.08
B73	Leaves	Control	0.43	12.02	4.73	4.88
		Salt	8.83	26.87	4.95	2.95
	Stems	Control	0.73	19.05	2.45	4.53
		Salt	34.98	54.83	1.88	1.95
B5C2	Leaves	Control	0.38	12.70	3.78	5.13
		Salt	22.18	37.50	3.95	3.43
	Stems	Control	0.23	19.52	4.28	8.13
		Salt	41.58	73.06	2.73	3.18
BR1	Leaves	Control	0.09	8.47	3.95	2.95
		Salt	7.35	18.43	3.95	2.55
	Stems	Control	0.14	14.99	3.65	4.68
		Salt	32.45	50.77	2.78	2.38
ANOVA Summary						
Genotype			0.0013	<0.0001	NS	<0.0001
Tissue			<0.0001	<0.0001	<0.0001	NS
Genotype × tissue			NS	NS	0.0188	0.0278
Treatment			<0.0001	<0.0001	NS	<0.0001
Genotype × treatment			0.001	<0.0001	NS	0.0104
Tissue × treatment			<0.0001	<0.0001	0.0377	0.0013
Genotype × tissue × treatment			NS	NS	NS	NS

NS: non-significant.

TABLE 3: Ion concentrations of stalks and tillers of four sorghum genotypes (SS304, NK7829, Sordan79, and KS585) irrigated with nutrient or saline solutions for 40 days (replication of 4).

Genotype	Tissue	Treatment	Na$^+$	Cl$^-$	Ca^{2+}	Mg^{2+}
				(mg g^{-1})		
SS304	Stalk	Control	0.88	24.61	3.78	6.50
		Salt	0.40	19.87	2.50	4.20
	Tiller	Control	0.33	8.83	2.90	3.98
		Salt	0.83	17.37	2.47	3.80
NJ7829	Stalk	Control	0.28	18.48	3.68	6.13
		Salt	0.58	18.26	2.38	4.00
	Tiller	Control	—	—	—	—
		Salt	2.29	24.30	4.35	4.75
Sordan79	Stalk	Control	1.28	13.02	4.93	7.23
		Salt	2.33	25.87	2.98	4.88
	Tiller	Control	1.65	14.53	4.18	6.05
		Salt	1.53	16.81	2.33	3.03
KS585	Stalk	Control	0.28	21.55	4.03	5.98
		Salt	0.75	22.42	3.63	4.18
	Tiller	Control	0.85	13.62	3.60	4.30
		Salt	1.23	18.91	4.08	3.65
ANOVA summary						
Genotype			0.0070	NS	NS	NS
Tissue			NS	0.0022	NS	0.0008
Genotype × tissue			NS	0.0170	NS	NS
Treatment			NS	0.0075	0.0018	<0.0001
Genotype × treatment			NS	NS	NS	NS
Tissue × treatment			NS	NS	NS	NS
Genotype × tissue × treatment			NS	0.0026	NS	NS

NS: non-significant.

higher in BR1 because BR1 in the control had very low concentrations of Na$^+$ in leaves and stems. Same to Na$^+$, Cl$^-$ concentrations were higher in stems than in leaves for all genotypes. Cl$^-$ concentrations in leaves and stems in the salt treatment were 1.7 to 3.7 times that of control. No significant differences in Cl$^-$ concentrations among genotypes were found. As for Ca^{2+}, genotype and treatment did not affect Ca^{2+} concentration. B5C2 had higher Mg^{2+} concentrations compared to those of other genotypes in both control and salt treatment. Salt treatment reduced Mg^{2+} concentration by 30% to 50%, depending on genotype and tissue.

For sorghum genotypes, Na$^+$ concentration was not affected by salinity, while all other mineral (Cl$^-$, Ca^{2+}, and Mg^{2+}) concentrations were affected by salinity (Table 3). Salinity increased uptake of Cl$^-$ but decreased Ca^{2+} and Mg^{2+} uptake in the stalk and tiller, although the differences in these mineral concentrations between the control and salt treatment were small. Genotype affected Na$^+$ concentrations, but not Cl$^-$, Ca^{2+}, and Mg^{2+} concentrations. The Na$^+$ concentrations in Sordan79 tissue were very low compared to maize genotypes; however, they were higher compared to those of SS304 and KS585. There was no difference in Na$^+$ between Sordan79 and NK7829, or between NK7829 and

those of SS304 and KS585. Compared to maize genotypes, both Na$^+$ and Cl$^-$ concentrations in sorghum genotypes were very low, while there were no substantial differences in Ca^{2+} and Mg^{2+} between the two crops. Sorghum had high ability of Na$^+$ exclusion from shoots, while maize genotypes had extremely high uptake of Na$^+$ in shoots.

Plant adaptations to salinity are of three distinct types: osmotic stress tolerance, Na$^+$ and/or Cl$^-$ exclusion, and the tissue tolerance of high concentrations of Na$^+$ and/or Cl$^-$ [4]. Some species tolerate salt stress by avoiding uptake of certain ions or by tolerating high ion concentrations in the tissue. In maize, all genotypes had high Na$^+$ concentrations in stems and leaves, ranged from 7.35 mg g^{-1} to 22.18 mg g^{-1} in the leaves and from 31.93 mg g^{-1} to 41.58 mg g^{-1} in stems. These concentrations are in the high range for most glycophyte. Similar high Na$^+$ concentrations in maize genotypes were reported [5, 18, 21, 29]. However, at similar NaCl salinity (100 mM), Turan et al. [25] reported a lower shoot Na$^+$ concentration of 4.46 mg g^{-1} for a maize cv: RX 947, while its shoot Cl$^-$ concentration was 44.16 mg g^{-1}, which was not substantially different from those in this study. These differences could be due to genotype, experimental duration, growth stage, and fertility.

Sorghum genotypes had extremely low Na^+ concentrations with little extra uptake of Na^+ in the salt-treated plants compared to those in the control. Low Na^+ concentrations in sorghum leaves and stems at similar salinity were reported by other researchers [18]. Other cereal crops such as wheat had lower tissue Na^+ concentrations than those of maize cultivars [29]. Apparently, maize genotypes coped with salt stress by tolerating high Na^+ and Cl^- concentrations, while sorghum genotypes had high ability of excluding Na^+ from shoots.

4. Conclusion

Responses of maize and sorghum to salinity differed among genotypes. Based on growth and visual salt damage, in maize, CUBA1 was relatively tolerant to salinity, followed by B73 and B5C2; BR1 was the least tolerant, although the differences among the four genotypes were small. In sorghum, Sordan79 was the most tolerant, followed by SS304; KS585 was the least tolerant among the four sorghum genotypes and was less tolerant than BR1 in terms of its visual salt damage and great shoot growth reduction. Maize genotypes accumulated Na^+ excessively in shoots, while sorghum had high ability to exclude Na^+ uptake from shoots. Both visual foliar salt damage of the seedlings and growth parameters are reliable criteria for assessing salt tolerance among genotypes for both crops, while physiological responses to salinity are useful information for breeding programs and help understand the mechanisms of salt tolerance of the crops.

References

[1] F.A.O., *Global Network on Integrated Soil Management for Sustainable Use of Salt-Affected Soils*, FAO Land and Plant Nutrient Management Service, 2005.

[2] A. D. Azevedo Neto, J. T. Prisco, J. Enéas-Filho, C. E. B. D. Abreu, and E. Gomes-Filho, "Effect of salt stress on antioxidative enzymes and lipid peroxidation in leaves and roots of salt-tolerant and salt-sensitive maize genotypes," *Environmental and Experimental Botany*, vol. 56, no. 1, pp. 87–94, 2006.

[3] X. Cai, X. Zhang, and D. Wang, "Land availability for biofuel production," *Environmental Science and Technology*, vol. 45, no. 1, pp. 334–339, 2011.

[4] R. Munns and M. Tester, "Mechanisms of salinity tolerance," *Annual Review of Plant Biology*, vol. 59, pp. 651–681, 2008.

[5] E. B. Carpici, N. Celik, G. Bayram, and B. B. Asik, "The effects of salt stress on the growth, biochemical parameter and mineral element content of some maize (*Zea mays* L.) cultivars," *African Journal of Biotechnology*, vol. 9, no. 41, pp. 6937–6942, 2010.

[6] E. V. Maas and G. J. Hoffman, "Crop salt tolerance—current assessment," *Journal of the Irrigation and Drainage Division*, vol. 103, no. 2, pp. 115–134, 1977.

[7] S. A. E. Ouda, S. G. Mohamed, and F. A. Khalil, "Modeling the effect of different stress conditions on maize productivity using yield-stress model," *International Journal of Natural and Engineering Sciences*, vol. 2, no. 1, pp. 57–62, 2008.

[8] E. Igartua, M. P. Gracia, and J. M. Lasa, "Field responses of grain sorghum to a salinity gradient," *Field Crops Research*, vol. 42, no. 1, pp. 15–25, 1995.

[9] E. V. Maas, "Crop tolerance to saline sprinkling water," *Plant and Soil*, vol. 89, no. 1–3, pp. 273–284, 1985.

[10] F. M. Azhar and T. McNeilly, "Variability for salt tolerance in *Sorghum bicolor* (L) Moench under hydroponic conditions," *Journal of Agronomy and Crop Science*, vol. 159, no. 4, pp. 269–277, 1987.

[11] F. M. Azhar and T. McNeilly, "The genetic basis of variation for salt tolerance in *Sorghum bicolor* (L) Moench seedlings," *Plant Breeding*, vol. 101, no. 2, pp. 114–121, 1988.

[12] L. Krishnamurthy, R. Serraj, C. T. Hash, A. J. Dakheel, and B. V. S. Reddy, "Screening sorghum genotypes for salinity tolerant biomass production," *Euphytica*, vol. 156, no. 1-2, pp. 15–24, 2007.

[13] S. Partners, *Product Profiles—KS585, NK7829, SS304, and Sordan79*, Sorghum Partners, INC, New Deal, Tex, USA, 2009.

[14] Q. Wang, J. Chen, R. H. Stamps, and Y. Li, "Correlation of visual quality grading and SPAD reading of green-leaved foliage plants," *Journal of Plant Nutrition*, vol. 28, no. 7, pp. 1215–1225, 2005.

[15] U.S. Environmental Protection Agency, Methods of Chemical Analysis of Water and Wastes (EPA-600/4-79-020), Cincinnati, Ohio, USA, 1983.

[16] R. Munns, "Physiological processes limiting plant growth in saline soils: some dogmas and hypotheses," *Plant, Cell & Environment*, vol. 16, no. 1, pp. 15–24, 1993.

[17] R. Munns, "Comparative physiology of salt and water stress," *Plant, Cell and Environment*, vol. 25, no. 2, pp. 239–250, 2002.

[18] G. W. Netondo, J. C. Onyango, and E. Beck, "Sorghum and salinity: I. Response of growth, water relations, and ion accumulation to NaCl salinity," *Crop Science*, vol. 44, no. 3, pp. 797–805, 2004.

[19] G. Niu and D. S. Rodriguez, "Responses of growth and ion uptake of four rose rootstocks to chloride- or sulfate-dominated salinity," *Journal of the American Society for Horticultural Science*, vol. 133, no. 5, pp. 663–669, 2008.

[20] C. N. Deng, G. X. Zhang, X. L. Pan, and K. Y. Zhao, "Chlorophyll fluorescence and gas exchange responses of maize seedlings to saline-alkaline stress," *Bulgarian Journal of Agricultural Science*, vol. 16, no. 1, pp. 49–58, 2010.

[21] M. Akram, M. Y. Ashraf, M. Jamil, R. M. Iqbal, M. Nafees, and M. A. Khan, "Nitrogen application improves gas exchange characteristics and chlorophyll fluorescence in maize hybrids under salinity conditions," *Russian Journal of Plant Physiology*, vol. 58, no. 3, pp. 394–401, 2011.

[22] G. W. Netondo, J. C. Onyango, and E. Beck, "Sorghum and salinity: II. Gas exchange and chlorophyll fluorescence of sorghum under salt stress," *Crop Science*, vol. 44, no. 3, pp. 806–811, 2004.

[23] Y. Demir and I. Kocacaliskan, "Effects of proline on maize embryos cultured in salt stress," *Fresenius Environmental Bulletin*, vol. 17, no. 5, pp. 536–542, 2008.

[24] A. M. A. Magnaye, P. J. A. Santos, and P. C. S. Cruz, "Responses of yellow maize (*Zea mays* L.) inbreds to salinity," *Asia Life Sciences*, vol. 20, no. 2, pp. 521–533, 2011.

[25] M. A. Turan, A. H. A. Elkarim, N. Taban, and S. Taban, "Effect of salt stress on growth, stomatal resistance, proline and chlorophyll concentrations on maize plant," *African Journal of Agricultural Research*, vol. 4, no. 9, pp. 893–897, 2009.

[26] R. Chaabane, H. Bchini, H. Ouji et al., "Behaviour of Tunisian durum wheat (*Triticum turgidum* L.) varieties under saline stress," *Pakistan Journal of Nutrition*, vol. 10, no. 6, pp. 539–542, 2011.

[27] M. Jamil, S. U. Rehman, J. L. Kui, M. K. Jeong, H. S. Kim, and S. R. Eui, "Salinity reduced growth PS2 photochemistry and chlorophyll content in radish," *Scientia Agricola*, vol. 64, no. 2, pp. 111–118, 2007.

[28] M. Heidari, "Effects of salinity stress on growth, chlorophyll content and osmotic components of two basil (*Ocimum basilicum* L.) genotypes," *African Journal of Biotechnology*, vol. 11, no. 2, pp. 379–384, 2012.

[29] M. Goudarzi and H. Pakniyat, "Comparison between salt tolerance of various cultivars of wheat and maize," *Journal of Applied Sciences*, vol. 8, no. 12, pp. 2300–2305, 2008.

Permissions

The contributors of this book come from diverse backgrounds, making this book a truly international effort. This book will bring forth new frontiers with its revolutionizing research information and detailed analysis of the nascent developments around the world.

We would like to thank all the contributing authors for lending their expertise to make the book truly unique. They have played a crucial role in the development of this book. Without their invaluable contributions this book wouldn't have been possible. They have made vital efforts to compile up to date information on the varied aspects of this subject to make this book a valuable addition to the collection of many professionals and students.

This book was conceptualized with the vision of imparting up-to-date information and advanced data in this field. To ensure the same, a matchless editorial board was set up. Every individual on the board went through rigorous rounds of assessment to prove their worth. After which they invested a large part of their time researching and compiling the most relevant data for our readers. Conferences and sessions were held from time to time between the editorial board and the contributing authors to present the data in the most comprehensible form. The editorial team has worked tirelessly to provide valuable and valid information to help people across the globe.

Every chapter published in this book has been scrutinized by our experts. Their significance has been extensively debated. The topics covered herein carry significant findings which will fuel the growth of the discipline. They may even be implemented as practical applications or may be referred to as a beginning point for another development. Chapters in this book were first published by Hindawi Publishing Corporation; hereby published with permission under the Creative Commons Attribution License or equivalent.

The editorial board has been involved in producing this book since its inception. They have spent rigorous hours researching and exploring the diverse topics which have resulted in the successful publishing of this book. They have passed on their knowledge of decades through this book. To expedite this challenging task, the publisher supported the team at every step. A small team of assistant editors was also appointed to further simplify the editing procedure and attain best results for the readers.

Our editorial team has been hand-picked from every corner of the world. Their multi-ethnicity adds dynamic inputs to the discussions which result in innovative outcomes. These outcomes are then further discussed with the researchers and contributors who give their valuable feedback and opinion regarding the same. The feedback is then collaborated with the researches and they are edited in a comprehensive manner to aid the understanding of the subject.

Apart from the editorial board, the designing team has also invested a significant amount of their time in understanding the subject and creating the most relevant covers. They scrutinized every image to scout for the most suitable representation of the subject and create an appropriate cover for the book.

The publishing team has been involved in this book since its early stages. They were actively engaged in every process, be it collecting the data, connecting with the contributors or procuring relevant information. The team has been an ardent support to the editorial, designing and production team. Their endless efforts to recruit the best for this project, has resulted in the accomplishment of this book. They are a veteran in the field of academics and their pool of knowledge is as vast as their experience in printing. Their expertise and guidance has proved useful at every step. Their uncompromising quality standards have made this book an exceptional effort. Their encouragement from time to time has been an inspiration for everyone.

The publisher and the editorial board hope that this book will prove to be a valuable piece of knowledge for researchers, students, practitioners and scholars across the globe.

List of Contributors

Imen Ben Ammar and Bouthaina Al Mohandes Dridi
High Institute of Agronomy of Chott-Mariem, University of Sousse, 4042 Chott Mariem, Tunisia

Fethia Harzallah-Skhiri
High Institute of Biotechnology of Monastir, University of Monastir, Tunisia

Antônio Heriberto de Castro Teixeira
Embrapa, 13070-115 Campinas, SP, Brazil

Jorge Tonietto and Giuliano Elias Pereira
Embrapa, 95700-000 Bento Gonc,alves, RS, Brazil

Fernando Braz Tangerino Hernandez
Sao Paulo State University, 15385-000 Ilha Solteira, SP, Brazil

Abel Chemura
School of Agricultural Sciences and Technology, Chinhoyi University of Technology, Private Bag 7724, Chinhoyi, Zimbabwe

Caleb Mahoya and Pardon Chidoko
DR&SS Coffee Research Institute, Chipinge, Zimbabwe

Dumisani Kutywayo
DR&SS Head Office, Agricultural Research Centre, Harare, Zimbabwe

Nkeki Kamai, Nuhu Adamu Gworgwor and Joshua Wasinaninda Wabekwa
Department of Crop Production, University of Maiduguri, Maiduguri 600001, Nigeria

Iqbal Hussain, Shamim Akhtar, Muhammad Arslan Ashraf and Rizwan Rasheed
Department of Botany, Government College University, Faisalabad 38000, Pakistan

Ejaz Hussain Siddiqi
Department of Botany, University of Gujarat, Gujarat, Pakistan

Muhammad Ibrahim
Department of Applied Chemistry, Government College University, Faisalabad 38000, Pakistan

Dibyendu Talukdar
Department of Botany, R.P.M. College, University of Calcutta, Uttarpara, Hooghly, West Bengal 712 258, India

Asmo Saarinen
Berner Oy, Etelaranta 4B, 00130 Helsinki, Finland

Tuomas Uusitalo
Department of Agricultural Science, P.O. Box 27, University of Helsinki, 00014 Helsinki, Finland
Raisioagro Oy, P.O. Box 101, 21201 Raisio, Finland

Pirjo S. A. Mäkelä
Department of Agricultural Science, P.O. Box 27, University of Helsinki, 00014 Helsinki, Finland

Ferzana Islam and Shoji Ohga
Department of Agro-Environmental Sciences, Faculty of Agriculture, Kyushu University, Fukuoka 811-2415, Japan

Tulole Lugendo Bucheyeki
University of KwaZulu-Natal, Private Bag X01, Scottsville 3209, South Africa
Tumbi Agricultural Research and Development Institute, P.O. Box 306, Tabora, Tanzania

Tuaeli Emil Mmbaga
Selian Agricultural Research and Development Institute, P.O. Box 6024, Arusha, Tanzania

Suraji Senanayake and KKDS Ranaweera
Department of Food Science & Technology, University of Sri Jayewardenepura, Sri Lanka

Anil Gunaratne
Faculty of Agricultural Sciences, Sabaragamuwa University of Sri Lanka, Belihuloya, Sri Lanka

Arthur Bamunuarachchi
"ON−SITE" Consultancy, Training & Trade Systems, 128/22, Poorwarama Road, Kirulapone, 5 Colombo, Sri Lanka

Mashezha Ian and Manyangarirwa Walter
Africa University, P.O. Box 1320, Mutare, Zimbabwe

Svotwa Ezekia and Rukuni Dzingai
Tobacco Research Board, Kutsaga Farm, P.O. Box 1909, Harare, Zimbabwe

Liu Xianzhao
College of Architecture and Urban Planning, Hunan University of Science and Technology, Xiangtan 411201, China
College of Geography and Planning, Ludong University, Yantai 264025, China
Department of Geography, Linyi University, Linyi 264000, China

Wang Chunzhi
College of Geography and Planning, Ludong University, Yantai 264025, China

Su Qing
College of Life Science, Hunan University of Science and Technology, Xiangtan 411201, China

Jorge J. Casal
IFEVA, Facultad de Agronomıa, Universidad de Buenos Aires and CONICET, Avenida San Martın 4453, 1417 Buenos Aires, Argentina
Fundacion Instituto Leloir, Instituto de Investigaciones Bioquımicas de Buenos Aires, CONICET, 1405 Buenos Aires, Argentina

Shuxian Li
United States Department of Agriculture-Agricultural Research Service, Crop Genetics Research Unit, Stoneville, MS 38776, USA

Pengyin Chen
University of Arkansas, Fayetteville, AR 72701, USA

Ezekia Svotwa
Department of Crop Science, University of Zimbabwe, Harare, Zimbabwe

J. Anxious Masuka
Department of Geography and Environmental Studies, University of Zimbabwe, Harare, Zimbabwe

Barbara Maasdorp and Amon Murwira
Tobacco Research Board/Kutsaga Research Station, Harare, Zimbabwe

Ren Yinzhe
College of Chemistry and Materials, Shanxi Normal University, Linfen 041004, China

Zhang Shaoying
College of Engineering, Shanxi Normal University, Linfen 041004, China

Abdollah Khadivi-Khub
Department of Horticultural Sciences, Faculty of Agriculture and Natural Resources, Arak University, Arak 38156-8-8349, Iran

Ezekia Svotwa, Anxious J. Masuka and Munyaradzi Shamudzarira
Tobacco Research Board, Kutsaga Research Station, Harare, Zimbabwe

Barbara Maasdorp
Department of Crop Science, University of Zimbabwe, Zimbabwe

Amon Murwira
Department of Geography and Environmental Studies, University of Zimbabwe, Zimbabwe

Sergio Arciniegas-Alarcón, Marisol García-Peña and Carlos Tadeu dos Santos Dias
Departamento de Ciencias Exatas, Universidade de Sao Paulo/ESALQ, Cx.P.09, CEP. 13418-900, Piracicaba, SP, Brazil

Wojtek Janusz Krzanowski
College of Engineering, Mathematics and Physical Sciences Harrison Building, University of Exeter, North Park Road, Exeter, EX4 4QF, UK

Amy E. Hoffner, David L. Jordan, Aman Chandi, Alan C. York, E. James Dunphy and Wesley J. Everman
Department of Crop Science, North Carolina State University, P.O. Box 7620, Raleigh, NC 27695-7620, USA

Genhua Niu, Denise Rodriguez and Youping Sun
Texas Agri Life Research, The Texas A&M University System, 1380 A&M Circle, El Paso, TX 79927, USA

Wenwei Xu
Texas Agri Life Research, The Texas A&M University System, 1102 East FM 1294, Lubbock, TX 79403, USA

www.ingramcontent.com/pod-product-compliance
Lightning Source LLC
Chambersburg PA
CBHW050451200326
41458CB00014B/5146